Technikzukünfte, Wissenschaft und Gesellschaft / Futures of Technology, Science and Society

Edited by
A. Grunwald, Karlsruhe, Germany
R. Heil, Karlsruhe, Germany
C. Coenen, Karlsruhe, Germany

Diese interdisziplinäre Buchreihe ist Technikzukünften in ihren wissenschaftlichen und gesellschaftlichen Kontexten gewidmet. Der Plural „Zukünfte" ist dabei Programm. Denn erstens wird ein breites Spektrum wissenschaftlich-technischer Entwicklungen beleuchtet, und zweitens sind Debatten zu Technowissenschaften wie u.a. den Bio-, Informations-, Nano- und Neurotechnologien oder der Robotik durch eine Vielzahl von Perspektiven und Interessen bestimmt. Diese Zukünfte beeinflussen einerseits den Verlauf des Fortschritts, seine Ergebnisse und Folgen, z.B. durch Ausgestaltung der wissenschaftlichen Agenda. Andererseits sind wissenschaftlich-technische Neuerungen Anlass, neue Zukünfte mit anderen gesellschaftlichen Implikationen auszudenken. Diese Wechselseitigkeit reflektierend, befasst sich die Reihe vorrangig mit der sozialen und kulturellen Prägung von Naturwissenschaft und Technik, der verantwortlichen Gestaltung ihrer Ergebnisse in der Gesellschaft sowie mit den Auswirkungen auf unsere Bilder vom Menschen.

This interdisciplinary series of books is devoted to technology futures in their scientific and societal contexts. The use of the plural "futures" is by no means accidental: firstly, light is to be shed on a broad spectrum of developments in science and technology; secondly, debates on technoscientific fields such as biotechnology, information technology, nanotechnology, neurotechnology and robotics are influenced by a multitude of viewpoints and interests. On the one hand, these futures have an impact on the way advances are made, as well as on their results and consequences, for example by shaping the scientific agenda. On the other hand, scientific and technological innovations offer an opportunity to conceive of new futures with different implications for society. Reflecting this reciprocity, the series concentrates primarily on the way in which science and technology are influenced social and culturally, on how their results can be shaped in a responsible manner in society, and on the way they affect our images of humankind.

Prof. Dr. Armin Grunwald, Physiker, Mathematiker und Philosoph, lehrt Technikphilosophie und Technikethik am Karlsruher Institut für Technologie (KIT), ist Leiter des Instituts für Technikfolgenabschätzung und Systemanalyse (ITAS) in Karlsruhe und Leiter des Büros für Technikfolgen-Abschätzung beim Deutschen Bundestag (TAB) in Berlin. / Professor Armin Grunwald, physicist, mathematician and philosopher, teaches the philosophy and ethics of technology at the Karlsruhe Institute of Technology (KIT), and is the director of the Institute for Technology Assessment and Systems Analysis (ITAS) in Karlsruhe and of the Office of Technology Assessment at the German Bundestag (TAB) in Berlin.

Reinhard Heil, Philosoph, ist wissenschaftlicher Mitarbeiter am KIT-ITAS. / Reinhard Heil, philosopher, is a researcher at KIT-ITAS.

Christopher Coenen, Politikwissenschaftler, ist wissenschaftlicher Mitarbeiter am KIT-ITAS und Herausgeber der Zeitschrift 'NanoEthics: Studies of New and Emerging Technologies'. / Christopher Coenen, political scientist, is a researcher at KIT-ITAS and the editor-in-chief of the journal 'NanoEthics: Studies of New and Emerging Technologies'.

J. Benjamin Hurlbut
Hava Tirosh-Samuelson (Eds.)

Perfecting Human Futures

Transhuman Visions and Technological Imaginations

Springer VS

Editors
J. Benjamin Hurlbut
Arizona State University
Tempe, USA

Hava Tirosh-Samuelson
Arizona State University
Tempe, USA

Technikzukünfte, Wissenschaft und Gesellschaft / Futures of Technology, Science and Society
ISBN 978-3-658-11043-7 ISBN 978-3-658-11044-4 (eBook)
DOI 10.1007/978-3-658-11044-4

Library of Congress Control Number: 2015960799

Springer VS

Lektorat: Frank Schindler, Monika Mülhausen

Printed on acid-free paper

Springer VS is a brand of Springer Fachmedien Wiesbaden
Springer Fachmedien Wiesbaden GmbH is part of Springer Science+Business Media
(www.springer.com)

Acknowledgments

This volume is the result of a grant "The Transhumanist Imagination: Innovation, Secularization and Eschatology." Funded by the program in Religion and Innovation in Human Affairs (RIHA) of the Historical Society with support from the John Templeton Foundation, this grant supported researchers who conducted case studies; an international research symposium in Karlsruhe, Germany; seminars at Arizona State University (ASU) with faculty members and invited experts; public lectures; publications in academic journals; and public outreach to disseminate the findings of the project, including a website and podcast of the conference. We are very grateful to Dr. Donald A. Yerxa, the program leader of RIHA, for his support of the project and for providing the opportunity for thinking through the challenges that technology poses to the future of humanity.

Most of the essays in the volume were first delivered at a conference at Karlsruhe castle, hosted by Karlsruhe Institute of Technology's Institute for Technology Assessment and Systems Analysis (KIT-ITAS) in Karlsruhe, Germany. We are truly grateful to Christopher Coenen, the co-organizer of the conference, and his colleagues for hosting the conference, assembling an international cadre of scholars, and publicizing the event in Germany. Christopher Coenen was also critical in linking this project to Springer, and we are glad that this book is part of the new book series that this publisher is devoting to futures of science, technology and society. We are grateful to John Evans, Nasser Zakariya, Brice Laurent, and Margarita Boenig-Liptsin for their contributions to the case studies and for the unique and valuable intellectual perspectives that each of them brought to the project.

The intellectual labor that carried the program for two years was done at Arizona State University, in the context of faculty seminars supported by the Center for the Study of Religion and Conflict (CSRC) at ASU. We would like to thank Linell Cady, director of the center, for her interest and support of this project; Carolyn Forbes, assistant director of the center, for all her hard work in the logistics of converting our visions for the project into reality; and Laurie Perko and Maureen Olmsted for providing essential administrative assistance throughout the two-year-long project. The seminars at ASU could not have been successful without the active involvement and intellectual commitment of faculty members at ASU, some of whom have been studying the challenges of trans/posthumanism for over a decade.

Edited volumes take a lot of organizational work and editorial expertise. We are indebted to Mary Lou Bertucci, our wonderful copyeditor, whose utmost professionalism enabled us to submit the manuscript to Springer. Without her dedicated work, this volume could not have seen the light of day. Finally, we thank Ina Gravitz, who indexed the manuscript, a task that exceeds our expertise.

Contents

Introduction: Technology, Utopianism and Eschatology

Hava Tirosh-Samuelson and J. Benjamin Hurlbut

Humans are tool-making animals: making and using tools are expressions of being human. Yet technology is a creation of the modern age. Although there has never been a pretechnological human society, the figure of "technology" in the sense of scientific technics is a modern invention, emerging in the late 19th century alongside the fundamental transformations of social life affected by the rise of industrial manufacturing and mass production of consumer products (Hughes 2005). The advent of technology has been fundamental to the transformations of the last century, altering the material conditions of modern societies, but also reshaping the imaginations of power and progress that permeate contemporary social life.

Technology is not reducible to machines and material objects. With the technological developments of the last century, "technology" has become progressively more vague, abstract, and slippery, encompassing not only human-made devices but complex systems of machines, processes, and techniques that are simultaneously social and material. With the social and environmental consequences of industrialization, two world wars, and the development of military technologies capable of annihilating the human race, the promise of progress and the imagination of human flourishing as an inevitable consequence of advancing technological capacity were inflected with a newfound ambivalence (Beck 1992). In the second half of the 20th century, life itself came to be a locus of new forms of technological intervention. The advent of cybernetics and artificial intelligence in the mid-20th century perturbed the material and conceptual demarcations between machine and living beings. The development of molecular biology and biotechnology opened the way to new forms of mastery and control over life and, with it, imaginations of the power to direct evolution to desirable ends.

In this context, technology (both as material reality and concept) acquired a new cultural significance. It demanded theorizing in relation to other aspects of culture. A sense of urgency emerged around the task of making sense of technology as a constitutive feature of modernity. For Jacques Ellul (1964), "technology" denoted "the totality of methods rationally arrived at and aiming at absolute efficiency (for a given stage of development) in every field of human activity" (p. 24). Ellul's expansive definition embraced dimensions of social life and moral order that he saw as inflected with—and configured by—the essential logic of "technics." For Herbert Marcuse (1991), an oppressive "technological rationality" remade society in its image, crushing the human impulse to transcendence into repressive social conformity. Thorstein Veblen (1906) lamented the subjection of "spiritual and intellectual life" to the logic of "machine technology," whereby the "ideal of human life" risked reduction to "the finikin skeptic in the laboratory [or] the animated slide rule" (p. 609). Other philosophers of technology in the mid-20th century, perhaps most notably Martin Heidegger, Ortega y Gasset, and Lewis Mumford, attempted various appraisals of the essence of technology and its place in modern life (Scharff and Dusek, 2005). Though these lines of thought were divergent in many respects, taken together they highlight the central position that "technology" acquired as a locus of moral imagination and a preoccupation of cultural critique. They also highlight the centrality of the figure of technology in imaginations of authenticity and moral perfectibility, even where conceived as contradicting such achievements, for instance, in worrying over technocracy as the threatening other to a perfected deliberative politics (Habermas, 1971).

Toward the end of the 20th century, grand theories of technology and the cultural anxieties that they refracted gave way to new engagements that interrogated the embeddedness of the technological in configurations of social life (Winner, 1986). This turn to contextualization, reflected most clearly in developments in history and sociology of technology, offered a far more empirical and granular picture of technological orders (Bijker, Hughes, & Pinch, 2012; Hughes, 2005; Latour, 1992). To the threat of technological determinism or domination, these accounts offered a rejoinder that revealed unacknowledged agency where technology had been seen as imposing structure and rejected the notion that technology drives history (Smith & Marx, 1994). Technology in these accounts is more plural and more partial than the term in the singular suggests, and the separation of humans and their machine creations is a (false) myth of modernity (Latour, 1993).

The rediscovery of humans in the dehumanized realm of "technological rationality" simultaneously offered emancipation from the anxieties of the Veblens and Marcuses of the world. In the picture advanced by actor-network theorist Bruno Latour, from the suffocating constraints of technological domination emerged a world picture suffused with agency, one that rejected a static imagination of technology for a ceaseless whirring of the making and unmaking of hybrid configurations of social and material, human and nonhuman, man and machine.

The overriding metaphysical tenet of this world picture is particularism: sociotechnical configurations are situated, contingent, and transitory. They are held together by the forms of life that suffuse them and by the coordinated behaviors of the actors (human and machine alike) that give them regularity. Technology is technologies, and any given technology can only be understood by interrogating the web of social relations in which it is embedded and which give it form. The macroscopic totality of "technological rationality" gives way to a microscopic attention to myriad particularities, and the discoveries of the latter become grounds to reject the narratives of the former as a false consciousness that ignores the indeterminacy—and openness—of technological orders.

While the question of how this theoretical reorientation stands in relation to wider cultural sensibilities is a difficult one, we need not address it here in order to note the following: there is a consonance between the above theoretical sensibilities and the notion of technological innovation as an entrepreneurial space in which sociotechnical orders can be intervened in and forms of life can be experimented with. The metaphysics of an innovation imaginary in which we can remake ourselves and our worlds through new hybridities with novel technologies is grounded in a similar particularism. For example, the iPhone is an infinitely plastic platform, customizable by the individual to integrate into life by exercising choice over an extensive menu of apps. The technology becomes a platform of new expressions of agency and emergent sociotechnical "imbroglios" (Latour, 1993).

Yet the turn to the particular occludes the stakes of the question concerning technology. It risks neglecting the role that technological orders, whether already instantiated or imagined, play in imaginaries of progress and perfectibility. An emphasis on agency and the plasticity of technological regimes privileges the contingencies of behavior over the ordering effects of belief. As "technology" becomes unthinkable, displaced by always already-situated technologies, the ways in which societies imagine themselves in the image of technological orders drop out of view. The penchant for abstraction notwithstanding, critical theories of technology of the mid-20th century recognized that the modern project of technology was centrally a project of meaning whereby visions of perfectibility found expression in aspirations to power and control. The "flight from ambivalence" that, according to Zygmunt Bauman (1993), informed the project of modernity was at once an imagination of order achieved through technique and an aspiration to the secular perfection that it offered.

Recovering this dimension while retaining the important theoretical advances of more granular, constructivist analyses requires recognizing that technology is always already embedded within and an expression of the social and the moral (Jasanoff, 2015). Technologies encode and express human values, norms, and ideals (Nye,

2006). They embody images of human life as it is and aspirations toward human life as it ought to be. What is true about particular technologies also applies to the idea of technology itself. As products of human imagination, technological orders are neither determined nor inevitable but are extensions of human agency configured in turn by collective imaginations. This applies equally to existing technological regimes and to imagined technological futures. With imagined technological futures, we tell a certain story about our aspirations and anxieties. Such stories reveal not what the future will inevitably be, but our postures in the present that invite particular futures: our aspirations, our fears, and our shared moral imaginations.

Human beings, of course, are not only tool-making animals but are also reflexive animals, deeply aware of their vulnerability and finitude. Unable to escape the fact of death, humans dream of immortality and have articulated immensely complex worldviews, religions, and philosophies that portray eternal life and promise a means to attain it. Because human eternity is not achievable in the spatial-temporal order, humans have perpetually imagined better futures for themselves. Whether these ideal scenarios are projected backward into a mythological Golden Age or forward to the remote end of time, or eschaton (Kumar, 1991, pp. 1–19) the utopian impulse expresses the "principle of hope,"—as Ernst Bloch famously called it (Bloch, 1973[1959])—offering humanity a means to cope with a less-than-ideal existence. In agreement with Bloch, Paul Tillich (1971) insightfully observed that utopia "expresses man's essence, the inner aim of his existence" (cited in Sargent, 2010, p. 98) because human beings seek to refuse what is negative in human existence. Utopian narratives, too, are neither necessary nor inevitable: they reveal the commitments, values, and norms of those who construct them.

This volume explores the nexus of technological and utopian imaginations. Its central objects of interrogation are visions of the human future wherein powers of technological control emancipate human beings from the constraints of their mortal bodies. The desire to overcome human finitude and the aspiration to control the world through technological means are as ancient as humanity itself. Yet our contemporary moment has brought a strange marriage of these impulses. Our remarkable technological achievements of the last century invite us to imagine a future in which we have the technological capacity to radically transcend the limitations of our bodies and our environments. In embracing technology as the medium for reimagining what the human is and, thus, what aspirations and forms of life are appropriate to it, these visions of human perfectibility privilege "innovation" as the agent of progress. The aim of this volume is to interrogate the stakes of such visions by attending to technological projects that promise to transcend the constraints of human body and being.

Imagined Technological Futures

Today, the human future is thrown into question by our technological capacities. The convergence of new technologies (for example, biotechnology, robotics, informatics, and nanotechnology) in projects of controlling life has radically reconfigured our sense of the human condition, both through technological capacities already at our disposal and through emergent imaginations of what human futures are possible, desirable, and good. We have begun to achieve unprecedented capacities to manipulate not only our external environment but the internal environments of our bodies as well. In light of these emerging and anticipated capacities, questions about human progress, redemption, or demise are increasingly asked in relation to imagined technological futures. For instance, the Anthropocene has been invoked to mark an epochal transition when the technological powers of *homo faber* are felt in every ecosystem on the planet. Importantly, the Anthropocene marks a shift not only in human technological capacity but also in moral imagination whereby trajectories of technological change, and the forms of life they engender, become the central figures in political controversy over what must be done in the here and now. Put simply, imagined technological futures inform imperatives in the present.

Opposite this dystopian imagination of a violated earth stands the utopian imagination of a posthuman future in which the human body and the world it inhabits become loci of profound control and human beings enjoy extraordinary new powers of self-determination. In the light of this imagined technological future, questions of what the human is and what it ought to be take new form. Fifty years ago, Julian Huxley observed that evolutionary biology had simultaneously dismantled the philosophical notion of human essence, secularized the remaining vestiges of theological transcendence, and given humanity custodial responsibility for—and the potential for technological mastery of— its own evolutionary future. This epochal shift demanded a response to the question "what are people for?" (Huxley, 1963, p. 17). In the half-century since Huxley asked this question, the answers have come to be ever more inflected with imaginations of technological control. Questions about intrinsic human dignity have been formulated against imaginations of biotechnological self-transformation (Bennett, 2015). Accounts of justice have been revised to reflect the (anticipated) fact of technological choice in distributing natural talents (Buchanan, Brock, & Daniels, 2000). And social, political, and moral subjectivities have been remade in the image of biotechnological futures (Rajan, 2006; Jasanoff, 2005; Hurlbut, 2015a).

Indeed, technological visions of human life have come to occupy a powerful position in contemporary constructions of history and social change, yet in ways that sometimes occlude and sometimes overprivilege the technological. Charles Taylor

(2003) observes that one of the central transitions marked by secular modernity was a profaning of time. Dissociated from theological origins and destinations, time became a homogenous container of contingent events. For Taylor, the imaginary of the market is exemplary of secular time, a space of human activity and imagination wherein the dynamics of movement and change are precisely what gives it durability and permanence. When Francis Fukuyama famously declared the end of history, he was asserting that the triumph of capitalism over communism foreclosed imaginations of historical progress, thereby in effect rendering the secularizing process of profaning time complete. The last powerful teleological vision of history had given way to "the end point of mankind's ideological evolution and the universalization of Western liberal democracy as the final form of human government" (Fukuyama, 1989, p. 4).

Yet only a few short years later, Fukuyama declared the "recommencement of history" under the guise of "our posthuman future" (Fukuyama, 2002). In remarkable contrast to *The End of History and the Last Man*, Fukuyama treated technology as an agent of historical change, once again giving direction (for good or ill) to human history. In his revised picture, questions about the good life became questions of what sorts of technological futures are desirable. Visions of progress (or dystopia) were articulated in the idiom of imagined technological futures.

This reflex to attribute agency to technology and to imagine the good in terms of technological futures is a defining feature of contemporary knowledge societies: technology is characterized as progressing through its own, autonomous momentum. Imagining technological futures has become a particularly consequential locus for both religious and secular engagements with notions of the human person. The theological reflections on the human person are increasingly undertaken in response to—and in terms of—the possibilities of human biotechnology (Smith, 2010), even as efforts to constitute appropriately secular public deliberation render the moral questions subsidiary to scientific and technological accounts (Hurlbut, 2015b).

Transitional and Transhumanist Visions

These stakes are particularly evident in visions of human enhancement. For some, the ultimate aim of human enhancement is to achieve the power to imbue human bodies with superior physical and mental traits, the ability to live a longer and happier life and, ultimately, to postpone death indefinitely. Indeed, for some the defeat of death is the ultimate justification for the project of human enhancement (De Grey, 2007). On these views, if until now the human species was transformed by evolution, which was a slow, uncontrolled, and unpredictable process, now evolution will be fast, controlled, and directed by humans. This imagined technological future is at once a vision of radical human agency and control engendered through newfound technological powers. Notions of "[e]nhancement revolution" (Buchanan, Brock, & Daniels, 2000; Buchanan, 2011), "radical evolution" (Garreau, 2004), "designer evolution" (Young, 2006), or "conscious evolution" (Chu, 2014) anticipate a world in which chance is transcended by control, determining the trajectory of future generations. Human self-transformation will be achieved, so we are told, by means of "genetic engineering, life-extending biosciences, intelligence intensifiers, smarter interfaces to swifter computers, neural compute integration, worldwide data networks, virtual reality, intelligent agents, swift electronic communication, artificial intelligence, neuroscience, neural networks, artificial life, off-planet migration and molecular nanotechnology" (More & Vita-More, 2003). Thus, the enhanced human, or H+, will become *transhuman*.

This tale of the future is at once utopian imagination and technological prediction. It trades in the myth of immortality on the one hand and in putatively realistic visions of what is technologically achievable on the other. It imbues notions of technological progress with eschatological significance. At the same time, it offers an ethical vision in which technological innovation is *the* central human achievement and thereby becomes the medium for achieving authenticity, emancipation, and transcendence. Transhumanism anticipates the application of science and technology to the amelioration of the human condition through genetic engineering, robotics, informatics, and nanotechnology. According to transhumanists who vigorously advocate technological self-transformation, augmented transhumans are not going simply to enjoy longer and happier lives devoid of pain and suffering but will make the technology that will render biological humans obsolete. The transhuman age will be characterized by new forms of human–machine interface in which "chemical, computational, genetic, bio-mechanical, or nanotechnological" technologies will "augment and alter, mind, sensing, and body" (Clark, 2013 [2007], p. 125). At some point in this anticipated human–machine merger, the superintelligent machines will become autonomous, decision-making entities that will make their own independent decisions. Futurists who predict this sequence of events (for example, Clark, 1997, 2003; Minsky, 1986, 2006; Moravec, 1988, 1999; Kurzweil, 1999, 2005; Chu, 2014) claim that it will come about after an irreversible turning point—the Singularity— the result of exponential, accelerated process of technological progress.

Coined by Vernor Vinge (1993), the term "technological singularity" refers to a variety of processes, including accelerated change, self-improving technology, intelligence explosion, emergence of superintelligence, shifts to new forms of organization, or increased complexity and interconnectedness (Sandberg, 2013, p. 377). As James Barrat observes, those who are convinced of the coming of the Singularity (known as Singularitarians) "tend to be twenty- and thirty-somethings, male, and childless.... A lot are autodidacts probably in part because no undergraduate program offers a major in computer science, ethics, bioengineering, neuroscience, psychology, and philosophy, in short, Singularity Studies" (2013, p. 136). Barrat goes on to note that "the Singularity is often called Rapture of the Geeks" and that as a movement it "has the hallmarks of an apocalyptic religion, including rituals of purifications, eschewing frail human bodies, anticipating eternal life, and an uncontested (somewhat) charismatic leader" (ibid., p. 137), namely, Ray Kurzweil. But regardless of the precise meaning of "Singularity" or its religious aspects, Singularity is imagined as an inevitable and irrevocable shift from the biological to the mechanical, a shift that will inaugurate a *posthuman* phase. In these futuristic narratives, the Singularity is not merely a hypothesis that can be subject to philosophic critique and scientific analysis (Eden, Moor, Soraker, & Steinhart, 2012), but rather a fact that accounts for how the future *must and will* develop. On the basis of his own calculation of the Law of Accelerating Returns (LOAR), which links evolution to innovation, competition, and market dynamics, Kurzweil has predicted that the Singularity will take place in 2045.

In this narrative of the anticipated future, the transition from the transhuman to the posthuman is imagined as a gradual and voluntary but ultimately inevitable process. It is a radically individualist vision in which freedom is reimagined as the agency to radically transform—and thereby transcend— the body. This understanding of freedom does not measure the present in terms of improvement over the past, but rather as an incremental progress toward a future in which transcendence is achieved by rendering the body utterly subordinate to the individual, creative will. At first, humans will upload their minds (presumably the most salient aspect of their personalities) into supercomputers who will serve the material needs of humanity. With the transfer of human minds into computer systems "different post-biological embodiment options" will emerge. Some will choose to live in robotic bodies and synthetic biological bodies, similar to our current bodies, but with vastly enhanced capabilities," but others "will choose to live as pure software without a permanent physical body" (Prisco, 2013, p. 235). Eventually, the machines "will tire of caring for humanity and will decide to spread throughout the universe in the interest of discovering all the secrets of the cosmos" (Geraci, 2008, p. 151). As Hans Moravec imagines it, machines will convert the entire universe into an "extended thinking entity" (1988, p. 116). As the "Age of Robots" will be supplanted by the "Age of Mind," machines will create space for a "subtler world" (Moravec, 1999, 163) in which computations alone remain. In the Virtual Kingdom, the "Mind Fire" will render earthly life meaningless, ultimately swallowed by cyberspace (ibid., p. 167). This is the ultimate *telos* of the transformation of the human to the posthuman. Imagining the future in this way, the enhanced transhuman will not only usher the way for the Virtual Kingdom in which humans will be irrevocably transformed but will also bring about the planned demise of embodied biological humans. The posthuman will be disembodied, nonlocalized, nonindividuated, Artificial General Intelligence (AGI) (Goertzel, 2006, 2007).

Trans/posthumanists eagerly anticipate this future because they consider human biological existence to be a burden and a curse. Thus, taking charge of the evolutionary process will enable humans to liberate themselves from their own biological limitations, freeing them to experience the bliss of immortality in the here and now, albeit digitally. Cyberspace will be the arena for immortality, since there the imagined selves, or avatars, will continue to live "forever" in a disembodied form creating a truly alternative digital reality (Bell & Kennedy, 2000). In the trans/posthumanist narrative of self-liberation, mechanical creativity itself is the salvation of humanity, since it will destroy that which is most problematic about the biologically evolved human body. While human reason itself has created artificial intelligence, the very product of human rationality— the robots— will do what biological life cannot do, eventually making biological humans obsolete. The anticipated future is, thus, a narrative about the human self-liberation from the shortcomings of carbon-based humanity, a story that will culminate in posthuman cyber existence. As a movement, transhumanism is relatively small and membership in the World Transhumanist Organization is only about 5,000 people worldwide and the intellectual coherence of the movement is still subject to ongoing discussion (More & Vita-More, 2013). However, trans/posthumanist ideas are increasingly widespread in contemporary culture, permeating movies, novels, science fiction, horror genre, performance art, and video games (Geraci, 2010, 2012).

At stake in the trans/posthumanist narrative are ancient philosophical questions about the nature and groundings of the human. For transhumanists, the will to technological mastery is the defining feature of human being. The human self, then, is reimagined as infinitely malleable, integrated into and ontologically indistinguishable from its own technological productions. This celebration of contingency over essential continuity has consonances with certain cultural forms that are defined not by computer scientists and artificial intelligence engineers who celebrate rationality but by literary critics and critical theorists who reject the traditional ideal of rationality as an instrument of social hierarchy, inequality, and injustice. Inspired by various critical movements including feminism, postcolonialism, environmentalism, queer theory, race studies, and poststructuralism, the so-called critical humanism "has demolished the myth of the unified, coherent, autonomous, self-identical human subject. It has posited the subject, and biology, as a construct of discourses, of enmeshed and co-evolved species and technologies," thereby preparing "the ground for the new form of the human, the posthuman" (Nayar, 2014, p. 29).

Technoscientific posthumanism and critical (or cultural) posthumanism hold different philosophical presuppositions (Tirosh-Samuelson, 2014a) but invite the same result: the end (whether conceptual or biological) of humanity and a transition to a posthumanism that embraces the remaking of the human as a creative, agential project. In literary critical discourse, the posthuman future is creatively and artistically imagined through narratives and performance art that erase the boundaries between humans and machines, transform or eliminate the human body, disengage reproduction and procreation from biology, reconfigure the relationship with nonhuman species, and welcome encounters with aliens, zombies, and monsters (Badmington, 2000, 2004; Braidotti, 2013; Gomel, 2014). Although there is relatively limited interaction between these two communities that anticipate the posthuman future, together these discourses reinforce the notion that the posthuman future is both desirable and inevitable: our "exit to the posthuman future" is already well underway, and there is little we can do to halt it (Kroker, 2014). Indeed, these are ideas with consequences: for instance, one need only look to projects of the techno-military-industrial complex, especially in the United States. The Defense Advanced Research Project Agency (DARPA) is deeply invested in the project of human enhancement. The making of "better humans," then, is not an idle utopian fantasy about the ideal future but an expensive program that consumes resources in the name of human betterment, certain forms of which inhabit powerful public institutions.

Both boosters and detractors tend to see transhumanism as a radical project, different in kind from the more circumscribed visions of technological progress that inform mainstream approaches to innovation. It has been characterized as a techno-utopian project, an ideology of extreme progress, and a heretical cult. Francis Fukuyama has labeled transhumanism "the most dangerous idea" of our time (Fukuyama, 2004). There is much in transhumanist discourse to recommend these sorts of characterizations. From predictions of an imminent technological eschaton to theologically inflected ideas of human perfectibility achieved through technological means, there is ample warrant to see transhumanism as culturally other. Philosopher of technology Langdon Winner observes that in transhumanism, "one finds level of self-indulgence and megalomania that are off the charts. The greatest puzzle about this fin de siècle fad is how tawdry notions could have attracted such a large audience at all" (Winner, 2005, p. 410).

Yet this is a puzzle worth puzzling over. What is notable about transhumanism is not merely that it is radically other but that it simultaneously participates in widely held imaginations of technological innovation as a source of progress and a force of social and historical change. The construction of transhumanism as an extreme ideology and a dangerous departure from mainstream views distracts from those features of the transhumanist imagination that are commonplace features of contemporary cultural postures toward technology. Transhumanist visions may indulge in grandiose production, but they do so by drawing upon tropes about technological change that are eminently familiar. The same applies to the notions of progress and human benefit that are offered to justify mainstream innovation agendas. As transhumanists are sometimes quick to point out, apparent differences in kind are often merely differences in degree. They note that their ideas are consonant with mainstream sensibilities about the purpose and value of technological innovation as a response to human frailty and insecurity: the transhumanist imagination merely extends this logic beyond the limited horizon of near-term technological change. For instance, responding to Fukuyama's characterization of transhumanism as among the world's most dangerous ideas, Nick Bostrom, a prominent transhumanist philosopher and then-director of the World Transhumanist Association, wrote:

> Transhumanists advocate increased funding for the research and development of medical and technological means that might extend human life and improve memory, concentration and other human capacities. The agenda is a natural extension of the traditional aims of medicine and technology and offers a great humanitarian opportunity to genuinely improve the human condition (Solomon & Bostrom, 2005, p. 4).

This statement reflects a characteristic logic for locating transhumanist visions in relation to mainstream views about the imperative of technological innovation. Framed in this way, transhumanist visions simply extend the horizon of imagination to more distant and ambitious technological futures: they are evolutionary extensions of mainstream sensibilities rather than radically disjunctive.

Surely Bostrom's is a rhetorical move to defuse Fukuyama's critique, and clearly much of what goes under the heading of transhumanism does indeed imagine radical departure from the state of human life in the present. But the important point to observe is that, in either case, the same underlying imaginary is at work: the promise of progress is seen as technological, and the human condition is progressively ameliorated through advances in technological capacity (Hurlbut, 2015b).

This imaginary underwrites notions of what ought to be done to improve human life but also of what human life is in essence. For instance, Ray Kurzweil asserts, "[T]he essence of being human is not our limitations—although we do have many—it's our ability to reach beyond our limitations" (Kurzweil, 2005, p.

311). This is a defining feature of the transhumanist imagination: the notion that human quintessence lies in the ability to transcend limitations through technological means. In this sense, technology is human essence, such that the project of reweaving the fabric of the human body through technologies of enhancement is an expression, rather than a negation, of human essence. This appraisal is both ethical and ontological—it gives an account of what human beings are and how they ought, therefore, to comport themselves.

For the past decade, the engagement with trans/posthumanist visions of the future has been undertaken by various academic institutions, involving several academic disciplines, and supported by public and private funders. Under the leadership of Hava Tirosh-Samuelson, Arizona State University has been at the forefront of these systematic, critical, and interdisciplinary reflections, generating workshops, public lectures, and publications (for example, Hansell & Grassy, 2011; Tirosh-Samuelson & Mossman, 2012; Transhumanism, 2012; Mercer & Trothen, 2014, p. xi). Building on these activities, a more recent project, "The Transhumanist Imagination: Innovation, Secularization and Eschatology," was awarded to Hava Tirosh-Samuelson and J. Benjamin Hurlbut by the Religion and Innovation in Human Affairs (RIHA) initiative of the Historical Society funded by the John Templeton Foundation. Funded by this grant, the volume consists of papers delivered in the international conference at Karlsruhe Institute of Technology, Germany, on July 8–9, 2013, along with essays solicited for the volume. The conference and the granted project examined the connection between visions of the future of humanity and technological innovation. By focusing on the beliefs, narratives, practices, and projects of techno-enthusiasts who anticipate the "transhuman" phase of the evolution of humanity, this project has interrogated the ways in which concepts such as progress, agency, and perfection are articulated in relation to imagined technological futures.

This volume builds upon this prior scholarship on trans/posthumanism by interrogating visions of human perfectibility in contemporary cultures of innovation and imaginaries of progress. It enriches critical reflections on trans/posthumanism in the following ways. First, the volume engages transhumanism by situating it within a broader context of ideas, institutions, and practices of innovation. Second, it develops robust empirical analysis that advances our understanding of transhumanism as a movement, as a philosophical frame, and as a vision of the human future that trades in more widespread imaginations of technology as an agent of transformation. As such, the volume illuminates contemporary cultural sensibilities that privilege putatively secular and morally neutral constructions of technology in shared moral imaginations of progress and human good. And, third, the volume interrogates the ways in which transhumanism reflects secularized religious motifs, thereby offering deeper understanding of the contemporary postsecular moment. Taken together, the volume contributes a deeper understanding of how religion (secularized or not) figures in projects of technological innovation and informs imaginations of the future in contemporary society.

The essays locate imaginations of human perfectibility in visions of the role of technological change in human progress, teasing out the utopian and eschatological dimensions of trans/posthumanism. Transhumanism is utopian in that it expresses "the desire for a better way of being" and is critical of the way we now live, "suggesting what needs to be done to improve things" (Sargent, 2010, p. 5). As a "forward dream" that expresses "anticipatory consciousness," as Ernst Bloch (2000 [1923]) called the utopian spirit (Levitas, 2011, pp. 97–122), the transhumanist project not only wants things to improve but also gives that want a concrete direction. And since the locus of the posthuman is the future, it is simultaneously a "non-existent place" (that is, *outopia)* and an aspirational better place (that is, *eutopia*). Transhumanism envisions a transitional society in which individuals will prosper and flourish because human physical and mental abilities will be augmented. Social ills such as poverty, sickness, pain, and suffering will be eliminated, and even death will be perpetually postponed. Because trans/posthumanism is not only about the power of technology to enhance the human body but also a vision about the destiny of humanity, it is important to appreciate its eschatological import. What makes trans/posthumanism eschatological is the claim that the ideal future is imminent and that it will come about as a result of radical, irreversible break from the present (Tirosh-Samuelson, 2014b). The Singularity is that event that will radically transform reality and the Virtual Kingdom is a mechanical eschaton in which super-intelligent machines constitute "last things."

This volume also situates trans/posthumanism in the context of wider sociotechnical imaginaries wherein technological innovation is attributed a progressive and even redemptive role in contemporary societies. In so doing, the volume looks at trans/posthumanism in light of ideas of innovation-driven social justice, in notions of governance, in ideas of technology that have figured in late modern societies' notions of secularism and progress, and in characterizations of religious postures toward technological innovation that underwrite distinctions between "religion" and the "secular" in contemporary cultural vernaculars. In examining these areas, the volume sharpens a set of questions and an emergent research program around the intertwined cultural threads of secularism, religion, and technology. These threads extend well beyond trans/posthumanism itself, but it offers a rich case for laying bare relationships that tend to be systematically overlooked in contemporary social theory and in contemporary culture more broadly. Trans/posthumanism embraces technology as the medium for reimagining what the human is and which aspirations and forms of life are appropriate to it. Thus, transhumanism exploits a collective imagination that privileges "innovation" as the agent of progress, while at once revealing the ways in which this imagination displaces and appropriates other imaginations of progress, including religious ones.

Part I: Technological Imaginations

The essays in Part I of the volume frame our attempts to think about the interplay between humans and technology in the future. The opening essay by Armin Grunwald, the director of the institute that hosted the conference (the Institute for Technology Assessment and Systems Analysis within Karlsruhe Institute of Technology), considers the current state of the conversation that preoccupies scholars in several fields: technology assessment (TA); science, technology, and society (STS) research; ethical, legal, and social implications research (ELSI); and applied ethics. Grunwald argues that "[w]hile futuristic visions as in the field of human enhancement often appear somewhat fictitious in content, such visions can and will have real impact on scientific and public discussions." Given the importance and power of futuristic visions, Grunwald calls for "increasing reflexivity and transparency in these debates," but he notes that the field of technological assessment falls short of the task. To offer sound orientation today for technological scenarios of the future, "new tools for structuring, interpretation, criticism, rationalization, and assessment of visionary future communication are needed to overcome the described aporia." These aporias arise because predictions and projections of the future always reflect not what future reality will be but only our present pictures of the future. What is needed is a new hermeneutics of the future that is "based on philosophical and social science methods such as discourse analysis" and that will "prepare the groundwork for anticipatory governance informed by applied ethics and technology assessment." The proposed hermeneutical approach will not only shed new light on the on-going debate concerning human enhancement but will also link that debate to other dimensions of contemporary culture, reminding us that "futures are . . . not something separate from the present, but a specific part of each present."

There are many ways to characterize our present moment (for example, "late-capitalism," "postmodernism," or "globalization"). But one label in particular—"post-secularism"—has emerged as a new analytical category to understand the unique feature of the early 21st century (Gorski et al., 2012; Habermas et al., 2010; Nynäs, Lasssander & Utrianen, 2012). Postsecularism signals the acknowledgment that the secularization theory that predicated modernity on secularization has failed to account for the resilience or religious traditions and the revival of religious presence in the public sphere. In this volume, Elaine Graham, who has written extensively on postsecularism (for example, Graham, 2002, 2004, 2007, 2013) links the debate about postsecularism to the debate about posthumanism and offers a framework that links technoscience studies and critical theory. For Graham, the "posthuman reflects the emergence over the past half-century of a technoscientific culture in which, thanks to cybernetics, artificial intelligence, virtual reality, plastic surgery, gene therapies, and assisted reproduction, biological humans are everywhere surrounded—and transformed—into mixtures of machine and organism." According to Graham, these new technologies compel us to reflect "what the ethical, political, and cultural implications are likely to be." Considering negative and positive attitudes toward these developments, Graham presents the posthuman as a kind of "thought experiment," namely, "an opportunity to think anew about the relationship between humans and our environments, artifacts and tools in a digital and biotechnological age." For Graham, the posthuman is "an inevitable condition" and interrogating it is necessary to understand "what it means to be human." Building on her previous studies, the essay concludes that critical posthumanism can fulfill a positive and necessary cultural task by transcending binary dichotomies between "religion" and "secular," the "human," "nonhuman," and "more than human" so as to celebrate life "in all its fullness and diversity."

Whereas technological posthumanism gestures toward a technological future, it also reflects longstanding relationships between technological aspirations and imaginations of human agency and perfectibility. Shifting the focus to empirically grounded investigations of culturally situated technological projects, Sheila Jasanoff demonstrates that technological visions are always and already imbued with morally inflected imaginations of human life as it is and as it ought to be. Jasanoff examines a series of high-modern technological projects that were each undertaken in the name of a national collective. These projects reflect sociotechnical imaginaries: "collectively held, institutionally stabilized, and publicly performed visions of desirable futures, animated by shared understandings of forms of social life and social order, attainable through and supportive of advances in science and technology" (as quoted in her essay). The framework of sociotechnical imaginaries illuminates the complex interconnections of the technical, the social, and the normative. It furthermore reveals the cultural specificity of the political meaning of technological undertakings, even where technologies themselves are very similar. Jasanoff offers a comparative analysis of nuclear power in Germany, South Korea, and the United States as an example of a "constitutional technology, implicated in realizing the kinds of futures that nations imagine for themselves." This kind of comparative analysis elicits subtle cross-cultural differences that tend to be occluded by notions that technological systems shape, but are not shaped by, the social and political worlds they inhabit. It thereby reveals the subtle but powerful normative commitments that inhabit technological projects and the central role of imagined technological futures in articulating visions of human flourishing. Indeed, comparison is a "technology of humility." By attending to technological futures as sociotechnical imaginaries—as the imagined worlds through which societies give material form to their particular desires, aspirations, and visions of the good— we come to be attuned to the "the enmeshing of the technological with the social and the moral" that underwrites our technological projects. For "the technologies with which we hope to better ourselves are our own creations, and, as such, they partake of the flaws in our fallible imaginations."

For good or ill, contemporary philosophical reflection about technology has been deeply indebted to Martin Heidegger. His "The Question concerning Technology" (1954) offered a new interpretation of technology, its meaning in Western culture, and its role in contemporary human affairs by posing the question: What is like to live "in the midst" of a technological existence? Several philosophers (for example, Alfred Borgmann, Don Ihde, Hubert Dreyfus, Robert Scharff, and Michael E. Zimmerman) have explicated Heidegger's holistic interpretation of technology in order to explain the degree to which technology discloses Being and offers a kind of ontological truth. In this volume, Zimmerman considers Heidegger's philosophy of technology in light of posthumanism, focusing on Ray Kurzweil, a director of engineering at Google and leading trans/posthumanist futurist. Kurzweil, as we already noted, has predicted 2045 as the year of the Singularity and has hailed AI as the specter of human salvation. Looking at Kurzweil's futurism from the perspective of Heidegger's philosophy, Zimmerman explains what is problematic about posthumanism, as well as how to think positively and optimistically about its proposed human self-transformation. Posthumanism is problematic because it expresses the modern "techno-industrial nihilism" that fails to appreciate the degree to which the human, or *Dasein*, depends on the "clearing" (*die Lichtung*) which allows beings to manifest themselves and become intelligible. In its denial of death and its relentless pursuit of power, posthumanism utterly misses the depth and complexity of being human. When Heidegger's analysis of Being is applied to posthumanism, contemporary technoscience turns out to be not only a kind of "ontological blasphemy" because of its pursuit of infinite power but also an assault on what makes us specifically human, namely, our dependency, affectivity, receptivity, and temporality. Heidegger presciently suggested that the mechanization of the human will turn us all into "fungible raw material," the servants of technoscience rather than its makers. Yet, going beyond Heidegger, Zimmermann suggests that current research into the complexity of human consciousness may offer "a new clearing, one consistent with its own possibilities." That possibility is quite different from Kurzweil's understanding of AI because it takes into consideration what could possibly motivate a super AI to do something at a given time. Zimmerman's Heideggerian interpretation shows the internal contradiction in Kurzweil's futuristic vision, thus asking us to remain more skeptical about his presumably inevitable scenario.

Part II: Ethics and Politics of Envisioned Futures

If the essays of Part I explain how the anticipation of the future reflects the values and norms of the present, the essays of Part II tease out some of the ethical and political implications and conundrums of trans/posthumanist anticipated but imagined scenarios. Micha Brumlik suggests that a proper assessment of trans/posthumanism requires a philosophical anthropology that situates the human being in the order of things. Giovanni Pico della Mirandola (1463–1494) was the celebrated Renaissance thinker whose "Oration on the Dignity of Man" is considered by leading transhumanists to be the inspiration for contemporary transhumanism (Bostrom, 2005). In Germany, that philosophical anthropology thrived in the end of the 19th and early 20th centuries in the philosophy of Max Scheler, Helmuth Plessner, Ernst Cassirer, Hannah Arendt, and Hans Jonas, among others. From the perspective of philosophical anthropology, so Brumlik argues, transhumanism is neither innovative nor radical. Humans are by their very nature beings that do two things simultaneously: they extend their physical bodies through diverse technologies, and they endow these somatic extensions with symbolic meanings. Brumlik's anthropological perspective enables us to consider diverse technologies, including shelter, jewelry making, or weaponry, as technologies that are ultimately about reproduction. The proper way to engage transhumanism, therefore, is to question its conception of reproduction, which Habermas and others have defined as "liberal eugenics" (Habermas, 2003; Agar, 2004). Such designed and controlled reproduction, culminating in the cloning of humans, raises what Habermas called "species ethics," namely, ethics that pertains to humanity at large rather than to individuals. While speaking in the name of the freedom of individuals to choose, the program of controlled humanly engineered evolution curtails the freedom of future generations, making liberal eugenics ethically problematic. Brumlik is not utterly convinced by Habermas's objections to cloning, for example; instead, he argues that the engagement of transhumanism should not be undertaken in terms of species ethics but rather in terms of politics. Brumlik reminds us that "politics are always politics by humans and for humans" and that "political life is the highest way of life humans can achieve." But what kind of polity will it be in which carbon-based humans will coexist with augmented humans who have artificial internal organs, as well as with beings whose brains have been substituted by machines? Since human beings have always enhanced themselves by means of technology, the debate about transhumanism should not focus on the ethics of human enhancement, but rather on the inclusion of machines in political life. Interrogated from that perspective, transhumanism appears to be "trivial or empty." It is trivial since intelligent machines do not have "desires, needs, and abilities" that members of the biological species have and that anchor all political life, and it is empty because of thc possible existence of such machines has not been proven.

While Brumlik frames the examination of transhumanism in light of philosophical anthropological and political theory, Stefan Sorgner examines transhumanism in the context of Western discourse on the good life. Since transhumanism promises to increase human happiness, it behooves us to ask two questions: How do transhumanist philosophers understand human happiness, and what is the concept of the good that undergird their conception of happiness? The Western discourse on happiness began in ancient Greek philosophy that posited happiness as the ultimate good that all humans seek to attain, on the one hand, and to the claim that the ultimate good could only be achieved by those who are both virtuous and wise. Ancient Greek philosophy, thus, linked happiness, virtue, and knowledge as the main components of the pursuit of perfection (Annas, 1993), and that approach was adopted and adapted by the three monotheistic religions of the West: Judaism, Christianity, and Islam (Tirosh-Samuelson, 2003). Since transhumanism claims to generate not just "better people" but "perfect" people who will enjoy unlimited happiness, it is important to examine it in the context of Western reflections on perfection. Sorgner, who has written supportively about transhumanism (for example, Sorgner, 2009), argues that the "transhumanist discourse, however, is ambiguous because it employs various meanings of perfection" and, furthermore, that "transhumanist philosophers hold different and even incompatible notions of the good." The evidence for this critique comes from the writings of Nick Bostrom, the leading transhumanist philosopher, and Julian Savulescu, who is not a transhumanist but who is closely associated with the transhumanist project of human enhancement (cf. Savulescu & Bostrom, 2009). In Sorgner's analysis, these two thinkers reflect different conceptions of the good: whereas Bostrom serves as an example of the Renaissance ideal of the good life, Savulescu exemplifies what Sorgner calls "a common-sense account of the good." Sorgner's critical engagement with these two thinkers paves the way to his own alternative "radical[ly] pluralistic account of the good." Inspired by Nietzsche, a radical plurality of the good allows for a more diverse understanding of the good life that takes into account cultural differences, currently missing from the transhumanist discourse. Sorgner positions his alternative view as "metahumanism" that he locates between "transhumanism" and "posthumanism," as a way to promote human flourishing.

At the heart of the transhumanist vision of the ideal future is the unwillingness to accept death. If, for Dylan Thomas, "death shall have no dominion" expresses a human hope that captures the tragic human condition, for transhumanism this statement is to become a reality brought about by technology. Thorsten Moos, a Protestant Christian theologian, insightfully takes transhumanism to task by showing the degree to which it borrows from the Judeo-Christian tradition, while also secularizing and diminishing the depth of inherited motifs. Focusing on Nick Bostrom and Max More, two leading transhumanist philosophers, Moos argues that transhumanist thinkers "answer the problems of finitude and ambiguity by means of what might be called *technological denial*." Adopting what Grunwald labeled as "vision assessment," Moos analyzes the transhumanist vision of perfection, showing it to be both utopian and apocalyptic, notwithstanding the transhumanist claim that they are not concerned with utopia or paradise but only with "continuous progress." Moos situates the transhumanist discourse on death in light of religion and philosophical reflections on death and shows that "with the transhumanists, death has to be conceptualized basically not as an element of human nature in the normative sense of the word," because death (and, with it, illness and suffering) is simply not acceptable. The transhumanist crusade against death is, in fact, an endorsement of an "able-ism" that excludes the disabled and the "less-than-perfect." But transhumanism is not only utopian in its overcoming of death and suffering; it is also apocalyptic in its understanding of how perfection is to be attained. "Perfect" is a future state of affairs that can be accomplished either gradually through progressive realization (that is, the evolutionary model) or abruptly and catastrophically (the revolutionary model). The transhumanist "sense of urgency" and its preoccupation with the singularity illustrate the apocalyptic (or revolutionary) nature of the presumed evolutionary narrative from transhuman to posthuman. The apocalyptic mentality of transhumanism is inherently dualistic, ethically (good/evil), temporally (present/future), anthropologically (mind/body), and cognitively (mind/brain). In Moos's analysis, transhumanism adopts apocalyptic motifs and speaks very much like traditional religions, but without the reflective potential of traditional religions. Transhumanists "share the early Christians' expectation that the state of perfection is near," but they are afraid to die too soon, before they can truly enjoy it. To partake in the ideal future, transhumanists endorse cryonics, literally deep-freezing humans at death so as to overcome death. This literalist, embodied understanding of perfection and improvement exposes the poverty of transhumanism as a philosophical anthropology and as a way to deal with the meaning of being human.

While transhumanism challenges us to engage it philosophically, anthropologically, and ethically, it is not only a vision about human perfection; it is also a program of scientific and technological research. The contention of this volume is that all predictions about the future reflect the values and norms of the present. Franc Mali fully endorses this contention when he engages transhumanist visions of the future by asking why and how transhumanists want us to change. Cognizant of the fact that "social values shape technologies" and, conversely, that "technologies influence social values," Mali interrogates the interconnection between bioscience and technology with special attention to synthetic biology. Problematizing the questions of "safety and security, . . . human dignity and privacy, and the boundaries between the natural and the technological," Mali emphasizes the important role of experts in "anticipatory governance of such technologies." Emerging technologies make clear that existing models of policy making are outmoded (for example, decisionist and technocratic models) and that new models have to be identified. Mali's analysis focuses on the concept of *anticipatory governance*: unlike the eschatological predictions of transhumanist futurism, this vision of the future is "not interested in eschatological questions, and . . . operates without a grand vision concerning the destiny of humankind." We can indeed talk rationally and responsibly about policy making for and in the future, but there is no justification for technological determinism. Mali considers the challenges that face the various actors involved in anticipatory governance activities (for example, scientists, policy makers, regulators, industry stakeholders, and civil society activists) and argues for the need to act collaboratively, rationally, and prudently.

Imagining and anticipating the technoscientific future, then, is a social practice embedded in culture. The essay by Alfred Nordmann makes this point crystal clear when it states, echoing Heidegger, that "envisioned technology of the future is entirely social and not at all technical." Building on Taylor's and Jasanoff's analysis, Nordmann shows that "sociotechnical imaginaries do not merely valorize known technologies in certain ways but can form an idealized image of a technology that is different and better than any known technology." This insight is demonstrated historically in regard to specific technologies in the 19th century as well as in contemporary emerging technologies. A case in point is nanotechnology, which has generated "the hope—against our better judgment—for solutions to nearly all pressing problems." Nanotechnology is promoted as an example of "soft machines," but Nordmann argues that this construct is a category mistake because it blurs the organic and the mechanical. In agreement with Jean-Pierre Dupuy, Nordmann argues that nanotechnological research is contrary to the art of engineering and that it appears "to be dedicated to the paradoxical task of controlling the uncontrollable." The contemporary situation is particularly problematic because we can neither dismiss these unprecedented machines as "illusory and misleading" wishful thinking, nor can we "valorize" them "for their normative expectations of a better world." The technoscientific imaginary thus stands "in the horizon of hope and destiny, of what might be possible and what should be necessary." Nordmann invites us to recognize these social imaginaries for what they are: "symptoms perhaps of normative uncertainty and cultural instability."

Part III: Trans/posthumanism in Context

The third part of the volume showcases the four case studies that stood at the heart of the project's empirical and field research. Each case study explores the transhumanist imagination in a distinct cultural setting, drawing together multifarious settings, forms, and expressions of the transhumanist imagination. Each involved extensive documentary analysis and/or fieldwork in a specific area, and each employed somewhat different methods. The case studies were designed to speak to the project's cross-cutting theoretical questions, with the expectation that they would generate parallel findings that, when placed in conversation with one another, would speak to the larger questions of the project. The four studies have generated new insights, questions, and results.

Brice Laurent's essay examines a series of political experiments in European science policy undertaken in the name of perfecting European politics. These experiments reflect the European Commission's emergent focus on "responsible research and innovation" in agenda setting for innovation and governance in the European Union, a focus that emerged in part from a 2004 report on converging technologies. This report points to a (transhumanist) narrative of converging technologies as a moral imagination that informs U.S. innovation policy and rejects it as a vision of progress that contradicts European political norms. The idea of "responsible research and innovation" has since come to inform European initiatives in innovation and governance, including deliberative exercises designed to shape EU research agendas and, Laurent argues, to perform a European political identity. It rejects the sort of technological inevitability that is the hallmark of the transhumanist imagination and is, in certain ways, reflected in U.S. postures of governance. These practices have made collective imagination of technological futures a locus for engendering (and performing) a European politics of collective deliberation and solidarity. Thus,

Laurent observes that, whereas the U.S. constructions of progress treat imagination as matters of (expert) prediction, the EU Commission sees a moral imagination that, when married to a regime of innovation, threatens the integrity of European politics. Laurent's research offers insights about the ways differential constructions of responsibility figure in institutional practices of governance and deliberative politics that bear on the ways notions of moral pluralism and (secular) public reason are constructed in relation to governance of innovation.

Transhumanism is not only a vision about the transformation of society toward the ideal posthumanity, but a program that seeks to transform society in the present. The study of the Singularity University by Margarita Boenig-Liptsin and Benjamin Hurlbut offers a parallel analysis of American-based visions of technology-driven progress, though, in this case, rooted in a private sector institution. Founded in 2009 by Ray Kurzweil and the technology entrepreneur Peter Diamandis, the Singularity University (SU) is a Silicon Valley-based educational institution whose mission is to "'assemble, educate and inspire a new generation of leaders' capable of driving the development and use of 'exponential' and 'disruptive' technologies to 'address humanity's grand challenges.'" SU offers an annual graduate training program in which promising young innovator entrepreneurs are instructed in entrepreneurial strategies, creative techniques, and the social and ethical visions that the SU sees as critical for technological innovation.

Boenig-Liptsin and Hurlbut's study situates the transhumanist imagination in relation to more widely held visions of technological innovation and its relations-hip to modalities and repertoires of moral imagination. SU faculty members often declare that the Singularity University "is not about the Singularity and is not a university." Yet its project is to occupy the traditional role of the university as a locus for generating knowledge and technological capacity on the one hand and social and moral enlightenment on the other. Its project is to "innovate" justice by imagining and enacting technological transformation of human societies, environments, and bodies, while also disrupting and transcending what it sees as ossified, obsolete ideas and institutions. Though it is purportedly "not about the Singularity," its aim is to engender in its students the ability to see "exponentially," that is, to see the past and future in terms of the self-driven dynamics of technological change that portend inevitable materials, social and existential transformations of human life. These ideas are fundamental for transhumanism. Thus, the SU is not about the Singularity only insofar as it emphasizes more immediate technological futures. The logics of technological change, and the social, political and ethical postures they demand, are wholly consonant with the narrative of the Singularity. As such, the eschatological trappings of the transhumanist imagination are present in a subtle form: as an ethical posture that privileges technological innovation (and innovators) as agents of historical transformation, without explicitly invoking an eschatological endpoint. This ethical posture also entails a particular approach to social institutions, collective moral imaginations, and visions of governance. Put simply, insofar as human flourishing and social justice are reimagined as technological projects, innovation also becomes a project of governance—of assuming responsibility for the human good, including responsibility for imagining, articulating, and enacting that good. This is entirely consistent with transhumanism's emphasis on radical, individual agency, but it is also highly consonant with—and influential upon—an American vision of innovation. The SU is a space in which science and engineering expertise (of a sort that informs political and economic agendas and social discourse) and the evangelical and eschatological threads of transhumanism are unified—and difficult to separate. The case study, therefore, offers insights into the ways technological futures function as a locus of moral imagination, including in configuring moral authority in secular democratic politics, particularly in relation to notions of emancipation, justice, and progress. Furthermore, the case study draws attention not merely to ideas in the abstract but to their role in practices and institutions that actively configure priorities, norms, and postures of governance in contemporary "knowledge societies."

Nasser Zakariya's essay explores genealogical and critical relationships between transhumanist imaginaries and related, utopically minded constructions of technological determinism and the technological sublime. By following threads of transhumanism into historically prior traditions, he reveals linkages and disjunctions between present and past normative dimensions of scientific and technological imaginations. Narratives of technological determinism (namely, the notion that "technologies proceed in some sense according to their own logic") figure centrally in contemporary understandings of technology. As such, technological change becomes a frame through which historical change is narrated, and technology is made preeminent in accounts of social life. Zakariya draws particular attention to the role of the technological sublime in modern, secular formations. The experience of the technological sublime has played a constitutive role in the idea of technology and in engendering culturally shared imaginations of power, progress, and transcendence (Nye, 2006). Zakariya draws these phenomena up against Charles Taylor's analysis of the sublime as an outgrowth of secularization, constructed as a means to recover otherwise unavailable forms of fullness. In this light, transhumanism become an interesting site in which to see the "grammars of the secular" at work. He observes, "[I]f transhumanism is understood as a seeking of transcendence through some notion of exclusive humanism, its very terms seem to ... confute the secular/religious divide." How, then, do the imaginations of transcendence insipient in transhumanism stand in relation to other features of its cultural moment? Examining geneologies of transhumanism, scientific humanism, and related utopian projects, Zakariya observes that transhumanism refracts "the many active theorizations, hopes, and anxieties" associated with certain glamorous technological achievements of the 20th century. It is less a culmination than one of a multiplicity of utopian (and dystopian) imaginations of the technological futures that have informed imaginations of progress and transcendence in secular modernity. Thus, transhumanism offers a diagnostic for assessing conventional constructions of the secular, and their relationship to technology and to theological forms. Indeed, "the extent to which transhumanism can be understood through conventional *secular* notions suggests how much the varieties of *humanism*—like those of natural science—have enfolded comprehensive imaginations outstripping the apparently limited license of reason and have asserted reason's own authority to prove the inexplicability of its own reach." Attending to transhumanism's "oscillation" between secular and nonsecular keys illuminates how it projects its technological futures, but also the wider repertoire of imagination that renders technological futures simultaneously soteriological and secular.

Taken together, the essays by Laurent, Zakariya, and Boenig-Liptsin and Hurlbut interrogate the relationships (and disconnects) between critical studies of technology and of the role of the figures of "science" and "technology" in ideas and institutions of secular modernity. They represent an initial effort to leverage this project to explore the import of relationships between religion and innovation—actual and imagined— for institutions of secular modernity.

"Religion" is not an abstract entity but a term that covers ways of life shared by millions of people in numerous forms. John Evans, a sociologist of science and religion, examines attitudes of members of mainstream religious communities toward human enhancement. His essay provides a starting point to begin to tease apart the relationship between ideas of the role of technology in human life and notions of progress, transcendence, and hope that are present among religious communities. It is a fact of modern life that members of religious communities also exist within a wider, secular world that is shaped in powerful respects by technology, including by ideas of progress and the good that are articulated in the form of imagined technological futures. As a kind of hybrid of secular and religious motifs, transhumanism transgresses the secular/religious boundary. One outgrowth of Evans's research for this project is a quantitative, cross-national examination of forms of public "faith" in science and the implications for public responses to transhumanism (Evans, 2014). The quantitative study observed that faith in science as a source of meaning is generally increasing among younger populations but encounters resistance from religious communities. However, faith in science as a source of technological solutions to problems of human well-being—where it, in effect, sets aside its transcendent pretentions—enjoys increasing public "faith." Yet, as the other case studies suggest, this distinction between meaning and technological capacity is less clear cut that is generally appreciated. Indeed, Evans's study indicates that religious publics draw this distinction between "science" (as meaning) and "technology" (as capacity) to distinguish between those forms of authority that conflict with religion and those that are harmonious with it. Whereas the other studies have revealed putatively neutral technology to be an underappreciated site of moral imagination and meaning, Evans's essay suggests that the categorical distinctions that circulate in a common cultural vernacular reinforce these distinctions (and, one can speculate, the forms of trust, deference, and moral (dis)engagement that accompany them).

Evans's essay in this volume offers a qualitative study of attitudes of the religious public toward technologies of human enhancement, examining the feelings of the American religious public toward imagined human biotechnological futures. Evans observes that public attitudes toward technologies of human enhancement are highly consequential for ways technologies (whether imagined or realized) come to figure in social and political life. He demonstrates that common caricatures of religious attitudes toward technologies of human enhancement—caricatures that are often found in transhumanist writings—are inaccurate reflections of religious attitudes. Rather, he uncovers significant consonances between tranhumanist and Christian notions of redemption, of transcendence, and of the virtues of intervention in human suffering. Yet he discerns disjunctions primarily in notions of how technological capacity should be governed—according to what visions of the good, discerned through what means and by what authority. Evans's empirical findings point to several important insights. First, religious attitudes, at least among the public, are *not* at odds with innovation; thus, there is no basis for characterizing them as a regressive force. Second, where disagreement does emerge about appropriate forms of technological intervention, both (putatively secular) transhumanist views and religious views are already engaged with theological questions of meaning, purpose, and good, whether explicitly acknowledged or not. These insights simultaneously problematize the idea of religion as somehow outside and at odds with innovation, but also the idea that projects of technological innovation somehow sit unproblematically inside secular space, disengaged from the forms of meaning that figure in religious belief. Thus, the results of this case study also open critical questions about the constructions of secularism and religion that circulate around—and are informed by—technological visions and projects.

The volume concludes with more personal reflections of G. Pascal Zachary, who, for over a decade, has written on science and technology for the *Christian Science Monitor* and who has reflected on the cultural forces that undergird the pursuit of perfection in contemporary technology. Taking his own personal quest for enhanced consciousness by experimenting with LSD during the countercultural revolution of the 1960s, Zachary reflects on the similarities and differences between substance-based cognitive enhancement and the pursuit of enhanced cognition of artificial intelligence. Both projects are interested in improving the machinery of the mind, but computer-based programs appeared to avoid "the *unintended* consequences of ingesting drugs." Focusing on Vannevar Bush, "a scientific statesman and charismatic engineering leader," Zachary explains how he "anticipated the communal nature of mind expansion so that the proposition that the *collective intelligence* of the species might also grow through sharing between strangers made sense to him." Bush, as well as other futurists (for example, H. G. Wells), did not think about the future merely for themselves; rather, they articulated appealing narratives about the future that were implemented by others. The technological age was implemented because visionaries narrated the future in a way that invested a particular trajectory with meaning, but these scenarios are always contingent rather than deterministic or predictable. It is only in hindsight that things appear to us as "inevitable" and that the social reality they have put in place appears to us as something we could not live without. This is particularly true about the Internet and digital devices that crowd our life and make it harder for us to enjoy reflective time (Lightman, Sarewitz, & Desser, 2004). Zachary reminds us that "uncritical enthusiasm for technological change remains the signal response to the prospect of cognitive enhancement by the technoscientific elite" and that they have profound impact on the way we work, organize our leisure time, and care for the sick and the elderly. Digital technologies transform all aspects of our daily life and even Jaron Lanier, one of the early pioneers of the digital revolution, has recently admitted that "something started to go wrong with the digital revolution around the turn of the 21st century" (Lanier, 2010, 3). Underlying the project of mind expansion, be it through drugs or cognitive enhancement or computer networks, is always socially and economically driven, reflecting the logic of capitalism. As we reflect on the promise and perils of new technologies, we should never lose sight of embodied human existence in daily life experience.

Conclusion

The future is, in principle, unknowable and unpredictable. Yet humans have a propensity and perhaps also a deep psychological need to dream about and anticipate the future, not least to contend with the fear of death. In contemporary life, technological futures are a power and consequential site of such "forward dreaming." As a locus of utopian imagination, the technological vision is fundamentally different from prior utopian and dystopian narratives. First, such imaginations are grounded not in the unattainability of utopia as "no place," but in a relationship with the future. The in-principle reality of the technological future entails a different ethical relationship than that of the fundamentally unattainable utopian other-world. Second, imagined technological futures are simultaneously imaginations of human life. Where they take the human body as the object of technological transformation, they throw into question the nature and the meaning of the human in a fundamental sense. At the same time, they render the human future a question of what is technologically possible, in effect rendering questions of what human life is and what it should be subsidiary to questions of what the technological future has in store. For this reason, it behooves us to engage trans/posthumanism seriously. It is one such site of technological imagination. Furthermore, as the essays in this volume demonstrate, it is situated within—and refracts—some of the most widely held and least interrogated sensibilities of the contemporary moment. The effects of technological change today are felt in each and every aspect of human life: economics, politics, law, medicine, transportation, communication, culture, and education—the list goes on. Precisely because the shape of contemporary life is so profoundly informed by converging technologies, we must engage in a public, critical, and open debate about their significance, not least because our technological projects—actual and aspirational—are also sites of moral imagination. At stake in their making is not only the material products that emerge from an industrial pipeline, but the forms of meaning and moral imagination with which our built technological world is laden. Thus, the responsibility for imagining our technological future cannot be left to scientists, engineers, and innovators who would bring it about. Deliberation must encompass the complete repertoire of our humanistic capacities, for human life itself is at stake. At the level of scholarship, this responsibility embraces disciplines throughout the academic spectrum, including the humanities and the social sciences. Furthermore, such interdisciplinary engagement (for example, Tirosh-Samuelson & Mossman, 2012; Ranisch & Sorgner, 2014; and Mercer and Trothen, 2014) ought not be limited to academic experts but should engage all relevant social actors from public intellectuals to policy makers, members of religious communities, journalists, educators, and producers of culture. As Habermas has correctly noted, what is at stake today is the future of the human species, the continuity of culture, and the deepest values we hold dear (Habermas, 2003). Imagining the future entails a deep responsibility, and it is, in Hans Jonas's words (1984), the "imperative of responsibility" that should guide our actions. The task is not ours alone to complete, but neither are we free to desist from it.

References

Agar, N. (2004). *Liberal eugenics: In defense of human enhancement.* Oxford, England: Blackwell.

Annas, J. (1993). *The morality of happiness.* Oxford, England: Oxford University Press.

Badmington, N. (Ed.). (2000). *Posthumanism.* New York, NY: Routledge.

Badmington, N. (2004). *Alien chic: Posthumanism and the other within.* New York, NY: Routledge.

Barrat, J. (2013). *Our final invention: Artificial intelligence and the end of the human era.* New York, NY: St. Martin's Press.

Bauman, Z. (1993) *Modernity and ambivalence.* Oxford, England: Polity Press.

Beck, U. (1992). *Risk society: Towards a new modernity* (1st ed.). Newbury Park, CA: Sage Publications, Ltd.

Bell, D. (2000). Cybercultures rewriter. In D. Bell & B. M. Kennedy (Eds.), *The cybercultures reader* (2nd ed., pp. 1–9). London, England: Routledge.

Bell, D., & Kennedy, B. M. (Eds.). (2000). *The cybercultures reader* (2nd ed.). London, England: Routledge.

Bennett, G. (2015). *Technicians of human dignity.* New York, NY: Fordham University Press.

Bijker, W. E., Hughes, T. P., & Pinch, T. (1987) *The social construction of technological systems: New directions in the sociology and history of technology.* Cambridge, MA: The MIT Press.

Bloch, E. (1973 [1959]). *Das Prinzip Hoffnung.* Frankfurt, Germany: Suhrkamp. In English: *Principle of hope.* (N. Plaice, S Plaice, & P. Knight, Trans.). Cambridge, MA: MIT Press, 1986.

Bloch, E. (2000 [1923]). *The spirit of utopia.* Stanford, CA: Stanford University Press.

Bostrom, N. (2005). A history of transhumanist thought. *Journal of Evolution and Technology, 14* (1), 1–25.

Braidotti, R. (2013). *The posthuman.* Cambridge, England: Polity.

Buchanan, A., Brock, D. W., & Daniels, N. (2000). *From chance to choice: Genetics and justice.* Cambridge, England: Cambridge University Press.

Buchanan, A. E. (2011). *Beyond humanity: The ethics of biomedical enhancement.* Oxford, England: Oxford University Press.

Chu, T. (2014). *Human purpose and transhuman potential: A cosmic vision for our future evolution.* San Rafael, CA: Origin Press.

Clark, A. (1997). *Being there: Putting brain, body, and the world together again.* Cambridge, MA: MIT Press

Clark, A. (2003). *Natural-born cyborgs: Minds, technologies, and the future of human intelligence*. Oxford: Oxford University Press.
Clark, A. (2013 [2007]). Reinventing ourselves: The plasticity of embodiment, sensing and mind. In M. More & N. Vita-More (Eds.),*The transhumanist reader: Classical and contemporary essays on the science, technology and philosophy of the human future* (pp. 113–127). Originally published in *The Journal of Medicine and Philosophy, 32* (3) (2007).
De Grey, A. (2007). *Ending aging: The rejuvenation breakthroughs that could reverse human aging in our lifetime*. New York, NY: St. Martin's Press.
Eden, A., Moor, J. H., Soraker, J. H., & Steinhart, E. (Eds.). (2012). *Singularity hypotheses: A scientific and philosophical assessment*. Heidelberg, Germany: Springer.
Ellul, J. (1964). *The technological society*. New York, NY: Vintage Books.
Evans, J. H. (2014). Faith in science in global perspective: Implications for transhumanism. *Public Understanding of Science, 23* (7), 813–832.
Fukuyama, F. (Summer 1989). The end of history? *The National Interest*, pp. 3–18.
Fukuyama, F. (2002). *Our posthuman future: Consequences of the biotechnology revolution*. (1st Picador ed.). New York, NY: Picador.
Fukuyama, F. (2004). Transhumanism. *Foreign Policy, 144*, 42–43.
Garreau, J. (2004). *Radical evolution: The promise and peril of enhancing our minds, our bodies*. New York, NY: Doubleday.
Geraci, R. M. (2008). Apocalyptic AI: Religion and the promise of artificial intelligence."- *Journal of the American Academy of Religion*, 76 (1), 138–166.
Geraci, R. M. (2010). *Apocalyptic AI: Visions of heaven in robotics, artificial intelligence, and virtual reality*. Oxford, England: Oxford University Press.
Geraci, R. M. (2012). Video games and transhumanist inclination. *Zygon: Journal of Religion and Science, 47* (4), 735–756.
Goertzel, B. (2006). *The path to posthumanity: 21st-century technology and its radical implications for mind, society and reality*. Bethesda, MD: Academica Press.
Goertzel, B. (2007). *Advances in artificial general intelligence*. Washington, DC: IOS Press.
Gomel, E. (2014). *Science fiction, alien encounters and the ethics of posthumanism: Beyond the golden rule*. New York, NY: Palgrave Macmillan.
Gorski, P., Kim, D. K., Torpey, J., & Van Antwerpe, J. (Eds). (2012). *The post-secular in question: Religion in contemporary society*. New York: Social Science Research Council and New York University Press.
Graham, E. (2002). *Representations of the post/human: Monsters, aliens and others in popular culture*. Manchester, England: Manchester University Press.
Graham, E. (2004). Bioethics after posthumanism: Natural law, communicative ethics and the problem of self-design. *Ecotheology, 9* (2), 178–198.
Graham, E. (2007). Embodying technology, becoming our tools: Discussing the post/human." In J. Baxter (Ed.), *Wounds that heal: Imagination, theology and health* (pp. 125–148). London, England: SPCK.
Graham, E. (2013). *Between a rock and a hard place: Public theology in a post-secular age*. London, England: SCM Press.
Habermas, J. (1971). *Toward a rational society: Student protest, science, and politics*. (J. J. Shapiro, Trans.). Boston, MA: Beacon Press.
Habermas, J. (2003). *The future of human nature*. Cambridge, England: Polity.

Habermas, J. (2010). An awareness of what is missing. In J. Habermas et al., *An awareness of what is missing: Faith and reason in a post-secular age* (pp. 15–23). Cambridge, England: Polity.
Hansell, G., & Grassy, W. (Eds.) (2011). *H+: Transhumanism and its critics*. Philadelphia, PA: Metanexus Institute.
Hughes, T. P. (2005). *Human-built world: How to think about technology and culture*. Chicago, IL: University of Chicago Press.
Hurlbut, J. B. (2015a). Remembering the future: Science, law, and the legacy of Asilomar. In S. Jasanoff & S.-H. Kim (Eds.), *Dreamscapes of modernity: Sociotechical imaginaries and the fabrication of power* (pp. 126–151). Chicago, IL: University of Chicago Press.
Hurlbut, J. B. (2015b). Religion and public reason in the politics of biotechnology. *Notre Dame Journal of Law, Ethics and Public Policy, 29* (2), 423–452.
Hurlbut, J. B. (2015c).Technologies of imagination: Secularism, transhumanism, and the idiom of progress. In D. A. Yerxa (Ed.), *Religion and innovation: Antagonists or partners?* (chap. 4). London, England: Bloomsbury.
Huxley, J. (1963). The future of man: Evolutionary aspects. In G. Wolstenholme (Ed.), *Man and his future* (pp. 1–22). Boston, MA: Little Brown & Co.
Jasanoff, S. (2005). *Designs on nature: Science and democracy in Europe and the United States*. Princeton, NJ: Princeton University Press.
Jasanoff, S. (2015). Future imperfect: Science, technology, and the imaginations of modernity. In S. Jasanoff & S.-H. Kim (Eds.), *Dreamscapes of modernity: Sociotechical imaginaries and the fabrication of power* (pp. 1–47). Chicago: University Of Chicago Press.
Jonas, H. (1984). *The imperative of responsibility: In search of an ethics for the technological age*. Chicago, IL: University of Chicago Press.
Kroker, A. (2014). *Exit to the posthuman future*. Cambridge, England: Polity Press.
Kumar, K. (1991). *Utopianism*. Minneapolis, MN: University of Minnesota Press.
Kurzweil, R. (1999). *The age of spiritual machines: When computers exceed human intelligence*. New York, NY: Penguin Books.
Kurzweil, R. (2005). *The singularity is near: When humans transcend biology*. New York, NY: Viking.
Lanier, J. (2010). *You are not a gadget: A manifesto*. New York, NY: Vintage Books.
Latour, B. (1992). Where are the missing masses? In W. E. Bijker & J. Law (Eds.), *Shaping technology/building society: Studies in sociotechnical change* (pp. 225–258). Inside Technology. Cambridge, MA: MIT Press.
Latour, B. (1993). *We have never been modern*. Cambridge, MA: Harvard University Press.
Levin, Y. (2008). *Imagining the future: Science and American democracy*. New York, NY: Encounter Books.
Levitas, R. (2011). *The concept of Utopia*. Oxford, England: Peter Lang Oxford.
Lightman, A. P., Sarewitz, D., & Desser, C. (Eds.) (2004). *Living with the genie: Essays on technology and the quest for human mastery*. Washington, DC: Island Press.
Marcuse, H. (1991). *One-dimensional man: Studies in the ideology of advanced industrial society*. Boston, MA: Beacon Press.
Mercer, C., & Trothen, T. J. (Eds.) (2014). *Religion and transhumanism: The unknown future of human enhancement*. New York, NY: Praeger.
Minsky, M. (1986). *The society of mind*. New York, NY: Simon & Schuster.
Minsky, M. (2006). *The emotion machine: Commonsense thinking, artificial intelligence, and the future of the human mind*. New York: Simon and Schuster.

Moravec, H. (1988). *Mind children: The future of robot and human intelligence*. Cambridge, MA: Harvard University Press.
Moravec, H. (1999). *Robot: More machine to transcend mind*. Oxford, England: Oxford University Press.
More, M., & Vita-More, N. (2003). Principles of extropy, version 3.11.2003. Original version. The extroprian principles. *Extropy, 5* (5) (May 1990). Available at www.extropy.org/principles.htm.
More, M., & Vita-More, N. (Eds.) (2013). *The transhumanist reader: Classical and contemporary essays on the science, technology and philosophy of the human future*. Malden, MA: Wiley Blackwell.
Mitcham, C. (1994). *Thinking through technology: The path between engineering and philosophy*. Chicago, IL: University of Chicago Press.
Nayar, P. K. (2014). *Posthumanism*. Cambridge, England: Polity Press.
Nye, D. (2006). *Technology matters: Questions to live with*. Cambridge, MA: MIT Press.
Nynäs, P., Lassander, M., & Utriainen, T. (Eds). (2012). *Post-secular society*. New Brunswick, NJ: Transactions Publishers.
Prisco, G. (2013). Transcendent engineering. In M. More & N. Vita-More (Eds.), *The transhumanist reader: Classical and contemporary essays on the science, technology and philosophy of the human future* (pp. 234–242). Malden, MA: Wiley-Blackwell.
Rajan, K. S. R. 2006. *Biocapital: The constitution of postgenomic life*. Durham, NC: Duke University Press.
Ranisch, R., & Sorgner, S. L. (Eds.). (2014). *Post- and transhumanism: An introduction*. Frankfurt am Main, Germany: Peter Lang.
Sandberg, A. (2013). An overview of models of technological singularity. In M. More & N. Vita-More (Eds.), *The transhumanist reader: Classical and contemporary essays on the science, technology and philosophy of the human future* (pp. 376-394). Malden, MA: Wiley-Blackwell.
Sargent, L. T. (2010). *Utopianism: A very short introduction*. Oxford, England: Oxford University Press, 2010.
Savulescu, J., & Bostrom, N. (Eds.). (2009). *Human enhancement*. Oxford, England: Oxford University Press.
Scharff, R. C., & Dusek, V. (2005). *Philosophy of technology: The technological condition, an anthology*. Oxford, England: Blackwell Publishing.
Smith, C. 2010. *What is a person? Rethinking humanity, social life, and the moral good from the person up*. Chicago, IL: University Of Chicago Press.
Smith, M. R., & Marx, L. 1994. *Does technology drive history? The dilemma of technological determinism*. Cambridge, MA: MIT Press.
Solomon, M., & Bostrom, N. 2005. Transhumanist dream. *Foreign Policy, 146*, 4.
Sorgner, S. L. (2009). Nietzsche, the overhuman, and transhumanism. *Journal of Evolution and Technology, 20* (1), 29–42.
Taylor, C. (2003). *Modern social imaginaries*. Durham, NC: Duke University Press.
Tillich, P. (1971). Critique and justification of Utopia. In F. E. Manuel (Ed.), *Utopias and utopian thought* (pp. 296–309). London, England: Souvenir Press.
Tirosh-Samuelson, H. (2003). *Happiness in premodern Judaism: Virtue, knowledge and well-being*. Cincinnati, OH: Hebrew Union College Press.

Tirosh-Samuelson, H. (2014a). Religion. In R. Ranisch & S. L. Sorgner (Eds.), *Post- and transhumanism: An introduction* (pp. 49-71). Frankfurt, Germany: Peter Lang International Academic Publishers.
Tirosh-Samuelson, H. (2014b). Utopianism and eschatology: Judaism engages transhumanism." In C. Mercer & T. J. Trothen (Eds.), *Religion and transhumanism: The unknown future of human enhancement* (pp. 161–180). New York, NY: Praeger.
Tirosh-Samuelson, H., & Mossman, K. (Eds.) (2012). *Building better humans? Refocusing the debate on transhumanism*. Frankfurt, Germany: Peter Lang.
Transhumanism. (2012). Special issue of *Zygon: Journal of Religion and Science, 47* (5), 659–795. This issue consits of essays by Hava Tirosh-Samuelson, Robert Geraci, James Hughes, and Ron Cole-Turner.
Veblen, T. 1906. The place of science in modern civilization. *American Journal of Sociology, 11* (5), 585–609.
Vinge, V. (1993). The coming technological singularity. *Whole Earth Review,* (Winter). Reprinted in 2013 as "Technological singularity" in M. More & N. Vita-More (Eds.), *The transhumanist reader: Classical and contemporary essays on the science, technology and philosophy of the human future* (pp. 365–375) (Malden, MA: Wiley-Blackwell).
Winner, L. (1986). *The Whale and the reactor: A search for limits in an age of high technology.* Chicago, IL: University of Chicago Press.
Winner, L. (2005). Resistance is futile: The posthuman condition and its advocates. In H. W. Baillie & T. Casey (Eds.), *Is human nature obsolete? Genetics, bioengineering, and the future of the human condition* (pp. 385–411). Cambridge, MA: MIT Press.
Young, S. (2006). *Designer evolution: A transhumanist manifesto.* Amherst, NY: Prometheus Books.

Part I
Technological Imaginations

What Does the Debate on (Post)human Futures Tell Us?

1

Methodology of Hermeneutical Analysis and Vision Assessment

Armin Grunwald

1 Introduction and Overview

The debate on human enhancement and posthumanism refers to the concepts and expectations in the field of converging sciences and technologies aiming at interventions into the human mind and body (Roco & Bainbridge, 2002). This debate is among the major contemporary fields of intellectual engagement with future relations between humans and technology. Technology assessment (TA); science, technology, and society (STS) research; reflections in applied ethics; and ethical, legal, and social implication (ELSI) studies have been, and still are, conducted in this field at an early stage of development. However, all these activities are confronted with severe methodological difficulties concerning the lack of prospective knowledge required as a subject of analysis and assessment. It can easily be seen that there is extremely high uncertainty respecting openness about future enhancement technologies and their possible use in society, as well as about "soft" consequences of this use, such as possible changes in the self-image of humans and new relations between humans and technology. Instead of valid knowledge, there are visionary stories, imaginations, narratives, speculations, assumptions and propositions, expectations and concerns that currently form the basis for, for example, TA or ethical inquiry. This situation makes it difficult, perhaps impossible, to apply established approaches to assessing scientific and technological developments in the field of human enhancement (Sec. 2). The criticism of "speculative nanoethics" (Nordmann, 2007; Grunwald, 2010) is an excellent indicator of philosophical unease about this situation.

On the other side, orientation is required also in the field of human enhancement, for example, for shaping the agenda of funding agencies and research policies. This holds in particular for technology assessment aimed at providing knowledge and orientation for policy making (Grunwald, 2009). Thus, existing TA approaches have to be developed further, and, perhaps, new ones have to be invented and tested (cf. Nordmann, 2010; Grunwald, 2013a).

I will argue that a major role of technology assessment in very early stages of new and visionary developments is a *hermeneutical* one.[1] Based on a perspective that regards futures as social constructs (Sec. 3), I will recall the approach of vision assessment, which has been discussed elsewhere (Grunwald, 2007), and develop it further into the direction of a "hermeneutical futures analysis" (Sec. 4). The emerging novelties and challenges of human enhancement technologies have to be made tangible by making their content, their epistemic foundation, their normative elements, and their strategic constellation as transparent and clear as possible. This challenge leads to the postulate that the "hermeneutical side" of TA should be unfolded and developed further. Technology assessment must, on the one hand, uncover profound cultural, philosophical, and ethical issues. On the other, it must also look at the real-world constellation where the stories of the visionary futures are constructed and communicated by specific actors, as well as at the attitudes, concerns, and interests of the stakeholder groups and persons involved.

This methodological effort is intended, on the one hand, to help prepare society for coming debates by assessing ideas and visions with regard to social values, human rights, ethical criteria, and so forth. On the other, hermeneutical analysis of the current debates on posthumanism includes asking for the reasons why this debate has emerged in our contemporary situation and why it has been attracting so much attention for more than a decade now. Answers to these questions could help uncover and better understand today's specific challenges to the *conditio humana* and could thereby contribute to transparently clarifying our contemporary situation (Sec. 5). The possible emergence of an "enhancement society" based on trends toward social Darwinism and ever-increasing competition may be among the roots of uneasiness in the debate on human enhancement (Sec. 5).

---

1 This chapter builds on several publications of the author in this field, in particular on Grunwald, 2012a, 2013a, and 2013b. The added value consists in bringing together the proposal to distinguish between different modes of providing orientation by assessing and reflecting futures (2013b) with earlier work on vision assessment (2007; 2009; 2013a) in the field of new and emerging sciences and technologies, in particular on human enhancement.

2 The Debate on (Post)human Futures in the Light of the TA Experience

Technology assessment was introduced in the 1970s in the U.S. Congress as information resource and support for political decision making (Bimber, 1996). In recent years, its traditionally close relation to prospective research and policy advice (Sec. 2.2) has been challenged by debates on technovisionary futures (Sec. 2.3). While there is still a need for orientation (Sec. 2.1), the classical TA approach no longer works.

2.1 The Need for Orientation in the Field of Human Enhancement

While futuristic visions as in the field of human enhancement often appear somewhat fictitious in content, such visions can and will have real impact on scientific and public discussions (Grunwald, 2007). We must distinguish between the degree of facticity of the *content* of the visions and the fact that they are used in genuine communication processes *with their own dynamics*. Even a vision without any facticity at all can influence debates, opinion forming, acceptance, and even decision making. Visions of new science and technology can have a major impact on the way in which political and public debates about future technologies are currently conducted and will probably also have a great impact on the results of such debates—thereby considerably influencing the pathways to the future in at least two ways:

- Futuristic visions are able to change the way the world is perceived. The discussion on human enhancement, for instance, already alters the self-understanding of humans through the simple fact that it takes place (Grunwald, 2007). The *new conditio humana* designates a world in which there is no longer any ideal state of the physical and intellectual constitution of a healthy human being, but in which even this ideal state seems to be formable—independent of the questions whether and when the scientific and technical means for actually realizing a technical improvement of the human body, mind, and society will be available. The societal and public debate about the chances and risks of new technologies will revolve around these visions to a considerable extent, as was the case in the field of nanotechnology (cf. Brune et al., 2006) and as is currently the case in synthetic biology (Coenen, Hennen, & Link, 2009). Visions motivate and fuel public debate because of the impact these visions have on everyday life and on the future of areas of society such as the military, work, or health care and

because they are related, to some extent, to cultural patterns (DEEPEN, 2009). Negative visions and dystopias could mobilize resistance to specific technologies.
- Visions have a particularly great influence on the scientific agenda and, as a consequence, partly determine what knowledge will be available and applicable in the future (Dupuy, 2007). Directly or indirectly, they influence the views of researchers and, thus, ultimately also have a bearing on political support and research funding. Visions, therefore, influence decisions about the support and prioritization of scientific progress.

The factual importance and power of futuristic visions in the governance of knowledge and in public debate is a strong argument in favor of providing early policy advice in the fields of technovisionary sciences with a view to increasing reflexivity and transparency in these debates (Grunwald, 2013a). Policy makers and society should know more about these visions—they must be informed and "empowered" to deal constructively and reflectively with futuristic visions in processes of "anticipatory governance" and "responsible development."

This conclusion is supported by calls for a more democratic governance of science and technology (MASIS Expert Group, 2009). Futuristic visions contain a mixture of facts and values, allowing them to be used for ideological and interest-based purposes. Special consideration must, therefore, be given to the challenge of how democratic deliberation and public debate could be involved in shaping the future course of technovisionary sciences such as human enhancement, taking the described lack of knowledge and the so-called control dilemma (Collingridge, 1980) seriously. An open, democratic discussion of technovisionary sciences is a prerequisite for a constructive and legitimate approach to shaping the future research agenda, regulations, and research funding. The requirement for transparency with respect to future projections and the arguments, premises, and visions they comprise is indispensable (Grunwald, 2013a).

2.2 Impact orientation of technology assessment

It seems to be obvious to meet the described need for orientation with the available and established approaches. Technology assessment, in particular, should be mentioned here, since it was designed for exactly this purpose: to develop and communicate knowledge- and science-based orientation for dealing with the scientific-technical advance and its impacts on society (Grunwald, 2009). The impact perspective with special attention to unintended consequences of technology is characteristic for TA. This also applies to technology design approaches in the context of technology generation, like constructive technology assessment (CTA; cf. Rip, Misa, & Schot, 1995), where design aims to realize intended and prevent or minimize unintended impacts.

Impact orientation is inherent in central concepts of TA such as the *early warning* of technology-induced risks (Paschen & Petermann, 1992, p. 26), but also the *early detection* of the opportunities of technology in order to use them in the best possible way. Both concepts are about weighing up chances and risks, which are, in the end, only impact dimensions seen from different assessment perspectives (Bechmann, 2007). Technology assessment will contribute to the systematic expansion of foresight regarding the temporal and thematic impacts of scientific and technological progress instead of applying the principle of "trial and error," as was often done in the past. Therefore, research on the impacts of technology is the central analytical element of TA.

Technology assessment impact research must be *prospective* since it investigates the impacts of technologies that do not yet exist or will possibly never exist. Technology assessment can only contribute to design objectives like early detection or early warning if it works prospectively. However, the state of future knowledge in general and knowledge on technology impacts in particular is known to be precarious from an epistemic point of view. Since the possibilities of TA to develop reliable options for action in order to contribute to the realization of design intentions depend on the knowledge on impacts and its quality, this is one methodological focus of TA.

The earliest historical TA approach to solving this problem consisted in the use and further development of forecasting methods (mode 1 orientation following Grunwald, 2013b). The prognostic imagination of future developments was used to create a reliable framework that integrates pending decisions, for example, on the regulation or expansion of infrastructures, in the best possible way. While this approach often works for scientific prognoses, for example, in celestial mechanics or weather forecasts, the necessary conditions—causally closed systems—are generally not given in key societal fields of action and development including the way of dealing with scientific and technological progress. Since the course of the future depends on human decisions that might be influenced by the very impact statements (self-fulfilling and self-destroying prophecies), the problem of prognosis cannot even be solved by additional research. Rather than raising hopes for reliable prognoses, this results in considerable uncertainty about the impacts of technology and a, partly significant, diversity and divergence of corresponding technology futures (Grunwald, 2012a).

Instead of focusing on prognostic impact research, TA has developed possibilities for *orientation* by scenarios (mode 2 orientation following Grunwald, 2013b). Today, scenarios are established in many fields of TA (for example, sustainability analyses) as a standard concept for the systematic reflection on a future that is basically considered open and, therefore, not predictable. They are particularly common where there is a need for orientation on complex problems; where these problems are of overarching, societal importance; and where different knowledge, diverse opinions, and views need to be integrated (for example, Heinrichs et al., 2012). In this way, TA has learned to handle the openness of the future and the implied unpredictability in a constructive way.

2.3 The Debate on (Post)human Futures—beyond the Scope of TA

The necessary precondition of mode 2 orientation to be applicable is the existence of well-founded corridors of the envisaged future development. Frequently, those corridors are mirrored in sets of scenarios where often best-case and worst-case scenarios serve as the borderlines of "sensible" future developments. In the human enhancement debate, this precondition is not fulfilled: neither the mode 1 nor the mode 2 approach is applicable.

The human enhancement area is only one among other fields in this regard. In the past decade, there has been a considerable increase in visionary communication on future technologies and their impacts on society, for example, in the fields of nanotechnology, the converging technologies, synthetic biology, and climate engineering (Zülsdorf et al., 2011; Coenen & Simakowa, 2013). Visionary scientists and science managers have put forward far-ranging visions that have been discussed in science and the humanities and that also have been disseminated by mass media and have led to societal debates. These futuristic visions refer to a distant future, some decades ahead, and exhibit revolutionary aspects in terms of technology and of culture, human behavior, and individual and social issues. As a rule, little if any knowledge is available about how the respective technology is likely to develop, about the products that such development may spawn, and about the potential impact of using such products. Extremely high degrees of uncertainty respecting openness are, thus, involved, leading to severe controversies with regard not only to societal issues but also to the feasibility of the visionary technologies. Images of the future, then, sometimes range from paradise-like expectations to apocalyptic fears, without a reliable and agreeable possibility to rank the pro and con arguments in a transparent way (Grunwald, 2012a; 2012b, Ch. 10).

Frequently, TA aims at weighing risks against opportunities and providing an integrated, comprehensive, and balanced consideration. In the field of human enhancement and other technovisionary debates, however, this is not possible because it is not yet clear which possible technical developments should be considered as risks or as opportunities. On the contrary, the same technical issues can be taken as motivation to express fascination and euphoria, on the one hand, and as threat to humankind, on the other. Depending on different ethical positions, extremely contrary conclusions may be derived. Compare, for example, the position of Harris (2010) based on a liberalist view, with the position of Sandel (2007), characterized by a more communitarian stance. Thus, interpretations and assessments of the future prospects related to "converging technologies for improving human performance" (so the title of Roco & Bainbridge, 2002) show the *maximum imaginable disorientation*: they oscillate between expectations of paradise and of catastrophe (Grunwald, 2007, pp. 385ff.). If no approaches were available for assessing and scrutinizing diverging futures in a "rational" sense, then the seeming arbitrariness of the futures of human enhancement and their assessments would destroy any hope of gaining orientation by reflecting on future developments, as is the usual business of TA.

Thus, a classical TA-type approach will not work in this field simply because of the lack of knowledge about expectable consequences. It is absolutely not clear what the subject of an assessment should be (by the way, this was also the kernel of the criticism against "speculative nanoethics," Nordmann, 2007). One way of dealing with this situation could be, from a TA's point of view simple: Let's wait

- either until the debate has disappeared again because of its possibly nonrelevant content (in this case, the debate might well be conducted in philosophical seminars or in the "feature" sections of newspapers but would be lacking any practical relevance)
- or until more and better knowledge is available as subject to sound assessments as soon as the scientific and technological progress allows for more specific insight into paths of development, innovation possibilities, risks and opportunities, and so forth.

However, this strategy of waiting would contradict what has been said in Section 2.1—there is a need for orientation already today. Thus, we end up in a seemingly aporetic situation: orientation is needed but not achievable because of the lack of valid knowledge about innovation pathways, diffusion of enhancement technologies, and the consequences of their use. So it seems that the aim of providing this orientation though future communication is not achievable. When there is both a negative and a positive utopia, uncertainty and confusion are even increased. New tools for structuring, interpretation, criticism, rationalization, and assessment of visionary future communication are needed to overcome the described aporia.

3 Visionary Futures as Social Constructs

To continue along this path, it is advisable to recapitulate briefly the nature of future expectations, here especially of visionary futures. We make statements and forecasts on the future, simulate temporal developments and create scenarios, formulate expectations and fears, set goals, and consider plans for their realization. All this takes place in the medium of language (Kamlah, 1973) and is, thus, an element of the respective *present* time. Forecasters and visionary writers cannot break out of the present either, always making their predictions and projections on the basis of present knowledge, present assessments, and present values and interests. For this reason, we can talk about *possible* futures, about alternative possibilities of imagining the future, and explain why we expect something in the future. These are always *present pictures of the future*, but not pictures of what the future reality will be (Goodman, 1954; Picht, 1969).

If we talk about cyborgs or far-ranging human enhancements being possible in the future, we are not talking about whether these developments will "really" occur and what this occurrence would "really" imply but how we imagine *today* their possible manifestations in the future—and such imaginations differ greatly or diverge completely, as is not surprising in a pluralistic society (Grunwald, 2012a). Futures are, thus, always something contemporary and change with the changing attitudes, knowledge, diagnoses, hopes, and fears in each present. Futures are, thus, not something separate from the present, but a specific part of each present.

Futures do not exist *per se*, and they do not arise of their own accord. On the contrary, futures are "human-made" and constructed in a more or less complex manner. They are created by authors, teams, and institutes, who and which have interests, perspectives, diagnoses, expectations, and so forth. The shaping of futures is purposive action. Futures, regardless of whether they are forecasts, scenarios, plans, programs, visions, or speculative fears or expectations, are "produced" using a whole range of ingredients such as the available knowledge, value judgments, and suppositions. This constructed character of any future (that is, that it is the result of a construction process) is an essential point of departure for developing methods to extract orientation from them in cases when neither mode 1 orientation (by prediction) nor mode 2 orientation (by scenarios) would work (Grunwald, 2013b).

A discourse concerning content and quality of future statements is, thus, a discourse about the diverse components (knowledge, assumptions, extrapolations, values, and such) that are present in the respective positions on the projected developments and about the methodological approach to creation of the respective pictures of the future. A dispute about enhancement and posthuman futures, therefore, does not refer to the events projected to come about in a future present but to the reasons that the respective futures and visions are brought forward today. Thus, in taking seriously that visions of human enhancement are social constructs, we could learn something about the intentions of their authors, about their perception and discussion in scientific disciplines and in the public, about hidden hopes and expectations, but also fears and concerns by analyzing them.

In a nutshell, in the absence of the possibility of applying approaches allowing for mode 1 or mode 2 orientation, there is no possibility to use futures in the familiar way for orienting opinion forming and decision making. What remains to be done is performing a *hermeneutical turn*: focusing on what the visions of human enhancement might tell us about our present time and using them as a means of diagnosing our contemporary situation. However, this mode 3 approach (Grunwald, 2013b) describes a completely different mechanism of providing orientation compared to what we normally expect from considering and assessing futures by mode 1 and mode 2 approaches. The only orientation they can provide given their irreducible divergence is a hermeneutical insight to allow a better informed and reflected debate for preparing to make decisions and to take measures. It is a matter of reflexive clarification of the conditions under which people can act and decide today, taking into account divergent future perspectives. So, mode 3 orientation can only be understood as an offer to improve the conditions of an open, transparent, and democratic deliberation and negotiation facing strongly diverging pictures of the future.

4 Toward Hermeneutical Futures Analysis

New or further developed methodical approaches are needed to analyze and assess futuristic visions, especially their content, their cultural background, their degree of reality and probability, their feasibility, their normative aspects, and their impact on the public and political debate. Visions must be made the subject of prospective *hermeneutical* analysis in order to better understand their content and their strategic constellation. The primary issue is to clarify the *meaning* of the future projections: What is at issue; what rights might possibly be compromised; what images of humankind, nature, and technology are formed, and how do they change; what anthropological issues are involved; and what designs for society are implied in the projects for the future?

Thinking about these issues is obviously not aimed at direct policy action but is more about understanding what is at stake and issue in the debates on human enhancement—contributing to a "hermeneutics" of possibly changing elements of the *conditio humana* (Grunwald, 2007). In this way, hermeneutical reflection based on philosophical and social science methods such as discourse analysis can prepare the groundwork for anticipatory governance informed by applied ethics and technology assessment. Ultimately, this may promote democratic debate on scientific and technological progress by investigating alternative approaches to the future of humans and society with or without different technovisionary developments.

This "hermeneutics" of visions should address not only the cognitive but also the normative content and value dimension of the visionary communication, which are both culturally influenced. Hermeneutical analysis could result in better understanding of the origins and roots of the visions by uncovering underlying cultural elements that often form the background of normative attitudes and value assignments. An example of this type of analysis can be found in the DEEPEN project (DEEPEN, 2009; von Schomberg & Davies, 2010). One of the findings was that cultural narratives such as "opening Pandora's box" and "be careful what you wish for" form deep-ranging patterns of perception in the visionary public debates and concerns (on nanotechnology, in this case study).

However, the hermeneutical analysis of visions is not limited to a cultural and philosophical analysis of their contents and conditions. "Understanding" also includes knowledge about the contexts of action in which these visions were created and are communicated. To this end, a map of involved actors, including their institutional interdependencies, should be created using socioscientific means. Since future communication always has an intervening character (Grunwald, 2012b), it is also used for strategic purposes. Therefore, the knowledge of motives and interests of the actors involved is an essential momentum of a comprehensive understanding.

The expectation is that hermeneutical analysis in this interdisciplinary approach will help exploit futuristic visions to provide orientation. It might benefit from recent thoughts on vision assessment (Grunwald, 2009; Ferrari, Coenen, & Grunwald, 2012). Vision assessment should realize an early involvement of reflective analysis and prospective assessment in new fields of science and technology in order to meet the criticism that TA again and again came late. In R&D processes, it should provide accompanying analysis and reflection: a kind of "real-time TA" working simultaneously with the ongoing advances in science and technology (Guston & Sarewitz, 2002). In general, the public and political debate about future technologies could be informed by providing "meta-knowledge" about the visions under discussion: about their origins, their underlying premises and values, as well as about their societal impacts. Obviously, these thoughts can be considered as preparatory work for the hermeneutical futures analysis suggested here.

Another task of hermeneutical analysis could be to uncover the *biography of visions* (following Grunwald, 2013a). Futuristic visions are created and disseminated by authors, teams, scientists, and science managers or emerge from discourse within scientific communities. They are communicated via different channels, journals, networks, mass media, research applications, and so forth. Some of them, finding no resonance, will "die" within these communication processes, while others will "survive" and motivate actors and groups to subscribe to or oppose the visions—in either case, the story will continue. Only a few of the visions will find an audience via the mass media and will, therefore, be able to bring about "real" impact by influencing public debate and social perception or attitudes. Others may enter the political arena and result in political decisions, for example, about research funding, and may then disappear. The entire "life cycle" of futuristic technovisions, from construction to assessment and impact, thus raises a huge variety of research questions that can only be answered by giving interdisciplinary consideration to the item. The main objective would be to generate more knowledge about and greater insights into the social processes surrounding visionary futures, from their emergence and dissemination via different communication channels to their possible impact on decision making in the policy arena and other arenas of public communication and debate. Examining their historical roots (Coenen, 2010; 2014) will improve understanding. In this sense, we could regard futuristic visions as part of an ongoing societal and scientific communication process in which specific visions—for example, the molecular assembler (Drexler, 1986) or enhancement of the brain's capability—act as the necessary catalysts (Grunwald, 2012b).

So, in those cases where mode 1 orientation does not work out and even mode 2 orientation cannot be provided because of the lack of reliable knowledge, pictures of the future hardly tell us about futures in the sense of a present in the time to come, but rather *about us today*. If projections of the future are interpreted in a way that makes clear why we aggregate certain current ingredients to specific futures and argue dedicatedly about them, then we have learned something *explicitly* about ourselves. Mode 3 orientation, therefore, is about trying to learn something about ourselves, our societal practices, subliminal concerns, implicit hopes and fears from the diversity and divergence of futures. This form of orientation is modest. In the end, it consists of nothing more than improving the conditions for democratic debates and future decisions to make them more sophisticated, transparent, and open. Its motivation is based on the assumption that we are able to better understand ourselves, our societal debates, scientific positions, the different interests, hopes and fears, the often diverging perceptions and positions of societal actors—even beyond science—once we better understand the diversity and divergence of the concepts of the future (Grunwald, 2013b).

5 What Does the Debate on Posthuman Futures Tell Us?

Many of the visions on human enhancement are so speculative, both in terms of their feasibility and their impacts, that we have to ask ourselves why this debate between anxiety and enthusiasm has been carried on intensively for more than ten years now (relevant authors are, among others, Mihail Roco, William Bainbridge, Jean-Pierre Dupuy, Steve Fuller, Christopher Coenen, Ludwig Siep, Nick Bostrom, and John Harris). However, discussions of this range can be found wherever we deal with scientific and technological progress in general and are a manifestation of its far-reaching ambivalence. On the one hand, the application of technological progress to the human being itself instead of—as it used to be done—to its environment is a logical consequence of technological progress. From an emancipatory perspective, it is liberation from the constraints of nature, for example, regarding aging or sensory capabilities of human beings. Traditional evidence on human attributes becomes doubtful. The point that human beings are physiologically just the way they have become through evolution and that there are no alternatives is no longer valid.

On the other hand, the history of technologization is also accompanied by concerns about the loss of control and autonomy. Some well-known examples are

- the human being as part of a machinery created by himself; in order to make it work, he has to reduce himself to a functioning "cog in the wheel" (in the movie *Modern Times* by Charlie Chaplin, 1924);
- the superiority of an economic and technological system that takes on a life of its own over the individual (in Herbert Marcuse's "One-Dimensional Man," 1964 [Marcuse, 1991]);
- the antiquatedness of the human being compared to its technological creatures, as feared by Günther Anders already in 1956, with the inability of the human being to control technology being a consequence of this;
- the fear that the future does not need us human beings anymore since technology could take on a life of its own, develop autonomously, and would, therefore, no longer be dependent on us. Bill Joy (2000) argued that converging technologies could indeed lead to the possibility of "intelligent" artifacts. His concern was that humans might easily lose control over these armadas of artifacts. Instead of being servants to human needs, such artifacts could start acting independently of human orders and even gain control over the world.

The observation that the discussion about the "technological enhancement" of humans is part of a general debate on scientific and technological progress helps to understand its importance better; this is where hermeneutical futures analysis should begin. The many theoretical "what if..." statements, both expectations and fears, in these debates serve—this would be the hypothesis—*our self-assurance*. The question is where we are today in the light of the scientific and technological progress, what we expect of it or what we fear, what will or could be the next steps, and what we can do today to take part in the development of these steps instead of leaving them to their own blind dynamics. And the fact that all this is of a particularly controversial nature, given the relationship between humans and technology changing already today, might be one aspect to explain why this debate about future possibilities attracts so much interest despite its speculative nature.

There is also another aspect. The debate about speculative forms of technological human enhancement is so fascinating because we live in a society in which we can all too well imagine that enhancement technologies, if they would be available after all, would rapidly gain acceptance. Sandel (2007) nicely demonstrates how the American middle class is, to put it in my own words, obsessed with an "enhancement mania," especially in the field of child education or the "enhancement" of children for special purposes by ambitious parents who, for example, increase the growth of their children with drugs to make them better basketball players. It is the unrestricted competition of a performance-enhancing society with a never-ending spiral of more and more enhancements that boosts this thinking through competitive pressure.

The concept of constant self-observation for the purpose of enhancement is also increasingly incorporated into the individual's life. In sports and leisure, we can observe an increasing use of diagnostic tools to evaluate different body functions continuously —all with the aim of self-optimization, for example, to achieve certain quantitative goals with sportive activities. An above-average and constantly improved performance at work and during leisure time, a beautiful and strong body, and high-stress resistance become more and more important for many people, which is evidence of an environment where constant enhancement seems to be necessary to preserve the status quo and, thus, becomes a compulsion. Individual actors are part of a society in which competition is the prime driver of development. We live in a world of rankings and ratings and constant evaluations in which we need to present ourselves and beat our competitors, whether at work, in our love lives, or during an evening TV game show. Every success in the competition is only temporary and constantly threatened by "the others" who could catch up or overtake. In this way, the concept of competition and technological human enhancement are inextricably linked.

The idea that human enhancement could be linked with developments toward a performance-enhancing society, where enhancement technologies would be developed according to a market model and then spread, has hardly been mentioned to date. Instead, an individualistically narrowed perspective in the tradition of biological and medical ethics is prevailing. Protagonists of human enhancement regularly argue from a libertarian perspective: "A more mundane vision in a similar vein presents us with a society in which 'morphological freedom' and 'cognitive liberty' are core values.... In such a society, every individual would have the right to treat his or her own body as fully malleable object" (Coenen, Schuijff et al.2009, p. 43). While this position is, in a sense, "politically correct" and consistent with the principle of informed consent in medical ethics, it seems to be biased by restricting the argument to the individual level and refraining from taking the surrounding societal dimension into account. The libertarian argument regards individuals as monads who are free to make decisions on their own (Ferrari, Coenen & Grunwald, 2012). However, the societal perspective cannot be found there at all. Social constraints like competitive pressure or even the strengthening of social Darwinist tendencies, for example, in right-wing populist parties, cannot even be discussed in the concept of an individualistically narrowed ethics. I suppose that a hermeneutical analysis of the debate on human enhancement would reveal that it is not the fear of or the hope for a technologization of the human being that keeps the debate alive, but the sensitivities of today's society that are related to an unease about the spiral of competition and the further economization of all areas of life. Where competition turns from a source of creativity into a destructive constraint with the constant threat of doom, everyone is clutching at any straw; technological human enhancement could be a means of improving the competitiveness against rivals—at least temporarily. The debate on Ritalin, which is no longer new, is an indication of this development.

However, if the said connection between the enhancement debate and the unease about the current development of Western societies could be confirmed—and there are several arguments for this—this would say something about the direction of the societal debate that would *have to be* held about this: it would then not only be about ethical questions that might be answered in one direction or the other but also about the social system in which we live, for example, about the question of how much competition is beneficial for a human society to mobilize the creative potential of human beings without leading to unwanted and, finally, self-destructive endless increases of self-exploitation. In this case, the hermeneutical analysis would indeed have led to an essential finding. What remains is the task to follow up these hypotheses methodologically and empirically.

References

Anders, G. (1956). *Die Antiquiertheit des Menschen*. München, Germany: Beck.

Bechmann, G. (2007). Die Beschreibung der Zukunft als Chance oder Risiko? *Technikfolgenabschätzung—Theorie und Praxis, 16* (1), 24–31.

Bimber, B. A. (1996). *The politics of expertise in Congress: The rise and fall of the Office of Technology Assessment*. Albany, NY: State University of New York Press.

Brune, H., Ernst, H., Grunwald, A, Grünwald, W., Hofmann, H., Krug, H. , Janich, P., Mayor, M., Rathgeber, W., Schmid, G., Simon, U., Vogel, V., & Wrywa, D. (2006). *Nanotechnology — Perspectives and Assessment*. Berlin, Germany: Springer.

Coenen, C. (2010). Deliberating visions: The case of human enhancement in the discourse on nanotechnology and convergence. In M. Kaiser, M. Kurath, S. Maasen, & C. Rehmann-Sutter (Eds.), *Governing future technologies: Nanotechnology and the rise of an assessment* regime (pp. 73–88). Dordrecht, Netherlands: Springer.

Coenen, C. (2014). Transhumanism in emerging technoscience as a challenge for the humanities and technology assessment. *Teorija in praksa, 51* (5), 754–771.

Coenen, C., Hennen, L., & Link, H. J. (2009). The ethics of synthetic biology: Contours of an emerging discourse. *Technikfolgenabschätzung—Theorie und Praxis, 18* (2), 82–86.

Coenen, C., Schuijff, M., Smits, M., Klaassen, P., Hennen, L., Rader, M., & Wolbring, G. (2009). *Human enhancement study*. Science and Technology Options Assessment (IP/A/STOA/FWC/2005-28/SC35, 41 & 45). Brussels: European Parliament. Available from https://www.itas.kit.edu/downloads/etag_coua09a.pdf.

Coenen, C., & Simakova, E. (2013). STS policy interactions, technology assessment and the governance of technovisionary sciences. *Science, Technology & Innovation Studies, 9* (2), 3–20.

Collingridge, D. (1980). *The social control of technology*. New York, NY: St. Martin's Press.

DEEPEN. (2009). *Reconfiguring responsibility: Deepening debate on nanotechnology*. Available at http://dro.dur.ac.uk/6399/1/6399.pdf?DDD14+dgg1mbk.

Drexler, E. (1986). *Engines of creation: The coming era of nanotechnology*. Oxford, England: Oxford University Press.

Dupuy, J.-P. (2007). Complexity and uncertainty: A prudential approach to nanotechnology. In F. Allhoff, P. Lin, J. Moor, & J. Weckert. (Eds.). *Nanoethics: The ethical and social implications of nanotechnology* (pp. 119–132). Hoboken, NJ: Wiley-Interscience.

Ferrari, A., Coenen, C., & Grunwald, A. (2012). Visions and ethics in current discourse on human enhancement. *Nanoethics, 6* (3), 215–229.

Goodman, N. (1954). *Fact, fiction, and forecast*. London, England: Athelone Press.

Grunwald, A. (2007). Converging technologies: Visions, increased contingencies of the *conditio humana*, and search for orientation. *Futures, 39*, 380–392.

Grunwald, A. (2009). Technology assessment. In A. Meijers (Ed.), *Philosophy of technology and engineering sciences*. (Volume 9, pp. 1103–1146). Amsterdam, the Netherlands: Elsivier/North Holland.

Grunwald, A. (2010). From speculative nanoethics to explorative philosophy of nanotechnology. *NanoEthics*, 4 (2), 91–101.

Grunwald, A. (2012a). *Technikzukünfte als Medium von Zukunftsdebatten und Technikgestaltung*. Karlsruhe, Germany: KIT Scientific Publishing.

Grunwald, A. (2012b). *Responsible nanobiotechnology: Philosophy and ethics*. Singapore: Pan Stanford Publications Pte. Ltd.

Grunwald, A. (2013a). Techno-visionary sciences: Challenges to policy advice. *Science, Technology and Innovation Studies, 9* (2), 21–38.
Grunwald, A. (2013b). Modes of orientation provided by futures studies: Making sense of diversity and divergence. *European Journal of Futures Studies, 2* (1): 2-30.
Guston, D. H., & Sarewitz, D. (2002). Real-time technology assessment. *Technology in Society, 24* (1–2), 93–109.
Harris, J. (2010). *Enhancing evolution: The ethical case for making better people.* Princeton, NJ: Princeton University Press.
Heinrichs, D., Krellenberg, K., Hansjürgens, B., & Martínez, F. (Eds.) (2012). *Risk habitat megacity.* Heidelberg, Germany: Springer-Verlag.
Joy, B. (2000, April). Why the future does not need us. *Wired, 8* (4). Available at http://archive.wired.com/wired/archive/8.04/joy.html.
Kamlah, W. (1973). *Philosophische Anthropologie: Sprachkritische Grundlegung und Ethik.* Mannheim, Germany: Bibliographisches Institut.
Marcuse, H. (1991). *One-dimensional man: Studies in the ideology of advanced industrial society.* Boston, MA: Beacon Press
MASIS Expert Group. (2009). *Challenging futures of science in society.* EUR 24039. Brussels, Belgium: European Commission.
Nordmann, A. (2007). If and then: A critique of speculative nanoethics. *NanoEthics, 1*, 31–46
Nordmann, A. (2010). A forensics of wishing: Technology assessment in the age of technoscience. *Poiesis & Praxis: International Journal of Technology Assessment and Ethics of Science, 7* (1–2), 5–15.
Paschen, H., & Petermann, T. (1992). Technikfolgenabschätzung—ein strategisches Rahmenkonzept für die Analyse und Bewertung von Technikfolgen. In T. Petermann (Ed.), *Technikfolgen-Abschätzung als Technikforschung und Politikberatung* (pp. 19–42). Frankfurt, Germany: Campus.
Picht, G. (1971). *Prognose Utopie Planung.* Stuttgart, Germany: Klett.
Rip, A., Misa, T., & Schot, J. (Eds.) (1995). *Managing technology in society.* London, England: Pinter Publishers.
Roco, M.C., & Bainbridge, W. S. (Eds.) (2002). *Converging technologies for improving human performance.* Arlington, VA: National Science Foundation, Department of Commerce.
Sandel, M. (2007). *The case against perfection: Ethics in the age of genetic engineering.* Cambridge, MA: Belknap Press of Harvard University Press.
von Schomberg, R., & Davies, S. (Eds.) (2010). *Understanding public debate on nanotechnologies.* EUR 24169. Brussells, Belgium: European Commission.
Zülsdorf, T. B., Coenen, C., Ferrari, A., Fiedeler, U., Milburn, C., & Wienroth, M. (Eds.) (2011). *Quantum engagements: Social reflections of nanoscience and emerging technologies.* Heidelberg, Germany: AKA.

Manifestations of the Posthuman in the Postsecular Imagination

2

Elaine Graham

Have We Ever Been Human?

This paper sketches out the beginnings of a conversation between two contemporary concepts in critical theory: the posthuman (or post/human) and the postsecular. They have shared origins in the critique of the assumptions that shaped Western modernity, particularly the elevation of the human subject as autonomous and self-determining, distinguished from animals, nature, and machines but also uniquely rational and unconstrained by the bounds of superstition, tradition, and religion. The many contested visions of the posthuman (or post/human) condition preclude any assumption that we are merely talking about a successor species whose future trajectory and ultimate characteristics are already known, but require us to ask *whose version* of normative and exemplary humanity is assumed to be at stake.

If Bruno Latour's speculation that "we have never been modern" is a way of exposing the artifice of modernity, premised on a particular set of dualisms by which we separate the human from the nonhuman, the technological from the biological, artificial from natural, then so too the assertion that "we have never been secular" challenges the fixity of the boundaries between science and religion, the profane and the sacred and the modernist evacuation of faith from accepted conventions of public and moral reasoning. Both statements are designed to bring us up short and think again about the taken-for-grantedness of the boundaries by which supposedly fixed identities are defined. The postsecular is absolutely not about religious revival or triumphalism. In fact, it is deeply contested and contradictory. To consider society as postsecular is not to deny the real material and cultural changes that Western society (in particular) has experienced since the end of the twentieth century in relation to the fortunes of organized religion, but it does prompt us to consider, first, how concepts such as the "secular" and "religious" have been constructed and what happens when we think through their implications for understandings of subjectivity, public space, and moral reasoning. The concept of the postsecular represents highly novel challenges to prevailing understandings of identity and culture in a globalized world.

I want to explore, however, whether the "post" of the postsecular is best understood as working in a similar fashion to that which I once outlined in relation to the terminology of the posthuman: "as an interrogative marker, a critical cue, for questions concerning the authors, objects and political implications of appeals to [modernity, the secular] and "humanism" and to expose the "categorical instability" of such terms (Graham, 2002, pp. 36–37; Badmington, 2004).

For Stefan Herbrechter, critical posthumanism is essentially about contemplating the end of the human "without giving into apocalyptic mysticism or to new forms of spirituality and transcendence" (Herbrechter, 2013, p. 3). While he is warning against the invocation of quasi-mystical constructions of human uniqueness founded in spurious universals, his choice of language reminds us that religion, assumed to be surplus to Western modernity's requirements, is never far away insofar as the posthuman finds notions of the sacred and the transcendent everywhere within its discourse. But like many of the manifestations of religion within the postsecular, these invocations serve to challenge conventional demarcations of sacred/secular, religion/spirituality, or transcendence/immanence. They also overturn assumptions about the material world as devoid of spirit, and the spiritual as necessarily being disembodied, virtual, other-worldly, in the name of new ontologies and anthropologies of hybridity and re-enchantment. So we may characterize the posthuman as about "becoming-animal, becoming-earth and becoming-machine" (Braidotti, 2013, p. 66)—but is it also about "becoming divine" (Jantzen, 1998)?

The Posthuman Condition, or Being Human Has Never Come Naturally

The idea of the posthuman reflects the emergence over the past half-century of a technoscientific[1] culture in which, thanks to cybernetics, artificial intelligence, virtual reality, plastic surgery, gene therapies, and assisted reproduction, biological

1 "Technoscience" is intended to evoke the interpenetration of science and technology. Rather than being separate enterprises, scientific endeavor is conducted within certain material and political economic contexts and is always already a cultural practice. Similarly, technologies are more than tools or instruments but historically situated in networks of human labor and organizational systems (Haraway, 1991; Graham, 2002, p. 30).

humans are everywhere surrounded—and transformed—into mixtures of machine and organism. Technologies are now incorporated into our bodies as permanent, autonomously functioning devices, such as cochlear implants, intraocular lenses, heart pacemakers, or even artificial hearts; or they create immersive environments, such as cyberspace, that reshape our physical concepts of time, space, and distance. In a more prosaic sense, insofar as most of the population of the "first world" is completely dependent on technologies such as computer-mediated communications, broadcasting media, transport infrastructure, and smart gadgetry, technologies of many kinds now completely permeate everyday life.

Arguably, human evolution has always been driven in part by the emerging capacity for the manufacture and use of tools and artefacts, however rudimentary, but our relationship with our tools and technologies intensified with the construction of the world's first stored-memory computer in 1948 and the identification of DNA in 1953, as innovations in genetics, cybernetics, neurology, and information and communications technologies have transformed our personal and cultural lives (Graham, 2002). Yet consideration of the material implications of new technologies also gives rise to philosophical reflection as well: not only about the practical or social consequences of our increasing dependence on such technologies but also to what the ethical, political, and cultural implications are likely to be.

It means that what we call "nature" has been significantly reshaped by technology and that technology, in turn, has become assimilated into "nature" as a fully functioning component of organic life itself. Like many others, in my own work I have been concerned with the ways in which the digital and biotechnological age is rendering problematic taken-for-granted boundaries between humans, machines, and what we call "nature"—be that nonhuman animals or our environment—and, in turn, how it raises questions about how we—as humans, as builders and users of tools, artefacts, and technologies—think about ourselves (Graham, 2002). In that respect, the right question is anthropological, even ontological: as the theologian Philip Hefner has put it, "[T]he question can never be first of all 'what are we *doing* with our technology?' but it must be 'what are we *becoming* with our technology?'" (Hefner, 2003, p. 9).

For some, the technologization of everyday life represents an unwelcome encroachment upon human integrity and the immediacy of our encounter with reality. In the last century, the theologian Jacques Ellul spoke of the degradation of creative production to the routinization of mere "technique" (Ellul, 1965). Martin Heidegger (1993 [1954]) argued that technology has a particular way of "revealing" the world to its users that obscures all other possibilities than that of "standing-reserve"—essentially, an objectification of nature that reduces everything to a commodity.

For the contemporary philosopher Albert Borgmann, similarly, technologies have colonized everyday life to such an extent that they have become virtually invisible (1984, 2003). The conveniences of processed food, television, text messaging, and labor-saving gadgets have displaced more traditional skills and pastimes that may have involved more effort but that also constituted the essential fabric of daily life. By removing technologies from their human creators and from any sense of connectedness to their social or cultural contexts of production, Borgmann argues, we have emptied their accompanying activities of meaning. Technology, as the product of human labor, has itself become reified and now assumes a deterministic quality in which humans regard themselves as the objects of technological imperatives, rather than their subjects. The everyday familiarity of technologies blunts our sense of wonder at their complexity and dulls our curiosity toward their origins in the actual processes and relations of production, and their ubiquity serves as an ideological opiate to dull our moral and political sensibilities. The solution for Borgmann is to restore a sense of the origins of technologies in the human practices of fabrication and sociability, thereby returning them to the status of "focal things" (Borgmann, 2003, p. 27).

However, other accounts of the future are altogether more optimistic, regarding technologies as the benign tools of human self-actualization. Rather than envisaging the effacement of human agency or endangering our capacity to participate in authentic existence, these more technocratic analyses are single-mindedly positive about the ability of technologies to facilitate problem solving and advance human interests. Humans will continue to be the masters of their technological inventions, and this will enable them to dominate, subdue, and eventually transcend the forces of nonhuman nature as well. Technology enables us to fulfill and then ultimately transcend our humanity in favor of a brave new world of human evolution.

Technologies, long having been the means of enabling humanity to compensate for physical limitations by providing instruments of comfort and utility, now offer the opportunities to overcome the limitations of the flesh in more radical ways: no longer rectifying or repairing, but improving, embellishing, upgrading "nature."

The significance of the posthuman, therefore, lies not simply in examining the material impact of new technologies on human bodies, communications, family structures, reproduction, economy, and so on, but in the way those very technologies expose the discourses and values informing our assumptions about what it means to be human. In this respect, the posthuman might be best conceived as a kind of "thought experiment": an opportunity to think anew about the relationship between humans and our environments, artifacts and tools in a digital and biotechnological age: "Thus, the posthuman is not any one particular thing; it is an act of projection, of speculation about who we are as human beings, and who we might become" (Thweatt-Bates, 2012, p. 1).

What has been termed "critical posthumanism" (Simon, 2003), therefore, argues that digital, cybernetic, and genetic technologies have not only material but existential impact, in their ability to question the boundaries and categories by which we have always delineated "human nature."

Have We Ever Been Modern?

In preparing this paper, I found it helpful to return to the work of a social and cultural theorist who has proved influential in thinking about these issues: Bruno Latour (1993). In particular, his work locates the capacity of modern technoscience to unsettle the axioms of what it means to be human in the face of advanced digital, medical, and biotechnological procedures and systems.

On the one hand, Western modernity rests on the establishment of "distinct ontological zones" (Latour, 1993, p. 10) between species and categories: human/nonhuman, active/inert, culture/nature. This Latour terms a process of "purification." But the paradox is that the very fruits of modern science and technology that depend on the demarcation between human manipulation and transformation of its nonhuman others engenders, in its resulting products, a series of phenomena that precisely breach these boundaries. He characterizes this as a parallel—and contradictory—process of "translation" (Latour, 1993, pp. 10-11). Once we introduce a level of reflexivity or self-awareness regarding the coexistence of these two processes, he argues, we cease to consider ourselves truly "modern."

One of the chief artifacts of modernity and its twin processes of purification and translation, argues Latour, is, of course, the human subject itself or, perhaps more precisely, the discourse of humanism. The strong affinities between human and nonhuman nature, the malleability of genetic and digital technologies, the ubiquity of virtual and computer-mediated communications and their accompanying influence on everyday life, not least in taking for granted understandings of space, place, community, and embodiment means that "matter is not dialectically opposed to culture, nor to technological mediation" (Braidotti, 2013, p. 35). In other words, the biological and the technological, the material (or "real") and the virtual are coexistent and coevolving. The inability to disentangle everyday life from its (inter)dependence on or with advanced technologies renders the classical humanist subject obsolete.

This realization is at the heart of critical posthumanism. In a digital and biotechnological age, it is impossible to speak of a pristine, unadulterated "human nature" without considering the ways in which humans have always, as it were, coevolved with their tools and technologies. Latour (1993) argues that the picture we may like to have of the world as conveniently and neatly divided into organic and artificial, born and manufactured, nature and nurture, fact and value, is a fiction. More precisely, he says it is a fiction born of Western modernity: it allowed scientists to secularize nature, the better to dissect and objectify it; it allowed humanism to flourish; but it may now be functioning more as an ideology than a creative truth. For Latour, we have always been mixed up, coevolving with our tools, living a hybrid existence. What the rapid intensification of new technologies has done has simply exposed us to that understanding.

> [T]he deconstruction of the humanist subject coincides with a new posthumanist understanding of human and nonhuman subjectivity, as an integrated form of agency within diverse networks of information environments and nonhuman actors. (Herbrechter, 2013, p. 199)

Thus, in her essay "A Cyborg Manifesto," first published in the mid-1980s, Haraway (1991) famously argued that, in the face of new technologies, we have to realize—and celebrate—the accompanying erosion of many of the boundaries by which we have set the limits of normative or authentic human nature. For Haraway, the "blasphemy" of the cyborg rests in its refusal to observe the essentialist categories of species, gender, nature, and ontology, and in whom "the boundaries of a fatally transgressive world, ruled by the Subject and the Object, give way to the borderlands, inhabited by human and unhuman collectives" (1992, p. 328). The modernist construction of the self was contingent upon a particular phase of Western scientific inquiry and categorization; a recognition of the hybridity of the "posthuman condition" enables us to consider that the delineation of human nature is more a question of *boundaries* than of *essences* (Graham, 2002, pp. 33–37). Humanism privileges individualism and autonomy, whereas posthumanism sees subjectivity as a state of becoming amid networks of subjects and actors, human/nonhuman, tools and environments. You can't do "human" ahistorically either, as if "human" were one thing. "Human" requires an extraordinary congeries of partners. Humans, wherever you track them, are products of situated relationalities with organisms, tools, much else. We are quite a crowd, at all of our temporalities and materialities (which don't appear as containers for each other but as co-constituting verbs), including that of earth history and evolution (Gane, 2006, p. 146).

The cyborg deconstructs the dualism of human agent and inert tool; it absorbs the extent to which Western life in the late 20th- to early 21st centuries is complicit with technologies, that we cannot speak of "being human" without recognizing our dependence on machines, that increasingly it is difficult to draw definitive boundaries between species, that "humanism" as a category was always about "naturalizing" a construction. It is a refusal to fix or reify human nature independent of humanity's relationships with its environment, tools, and artifacts.

According to this account of human evolution and ontology, then, it is a mistake to assume that, sometime before the advent of biomedical, cybernetic, and genetic technologies, the boundaries between the human/nonhuman, natural/artificial, organic/technological were fixed and axiomatic and that to be human rested on a clear "ontological hygiene" of essential qualities (Graham, 2002). In fact, the boundaries have always been contested, and attempts to define the "human" in relation to the "nonhuman" are a work of exclusion, a denial of our entanglement, our complicity, with the world of our tools, technologies, and environments.

> For what does it mean to say that our context is *now* posthuman? What does it mean to say that the human is intertwined with non-humans *as never before*? When was the human *not* inextricably entwined in material, technological and informational networks? When was the human ever just "itself"? (Braun, 2004, p. 271)

Technologies are, thus, an integral part of our very human nature, or, in the words of Philip Hefner, a necessary means of "human becoming" (Hefner, 2003, p. 43). And this may be no more than a reminder that, via the processes of transforming or constructing the world, humanity cannot but refashion *itself* as integral elements of that world. "Human nature" ceases to be a fixed category and re-emerges as a constantly changing set of possibilities and configurations. The possibility of human evolution continuing through technologies is not to be rejected. Rather, it is possible to argue that a quintessential aspect of our very humanity is realized in and through our relationships with our tools and technologies.

My use of the *oblique* in the terminology of "post/human" (Graham, 2002) reflected a wish to introduce a kind of critical interval into debates about "the" posthuman (as monolithic or inevitable). It was intended to draw attention to the function and meaning of *both* terms in mutual interaction. I do not regard the "posthuman" as an inevitable condition, therefore, so much as an *interrogation* into the terms by which Western culture has defined what it means to be human (Graham, 2002, p.37).

Thus, talk of the posthuman or post/human is a way of tracing the fault lines by which we have differentiated organic from inorganic, nature from artifact, human from nonhuman; but now we recognize that these are conventions, rather than ontological distinctions, that determine the categories of human, nonhuman and "otherwise human" (Braidotti, 2013, p. 196). It represents a refusal to fix or reify human nature or essence independent of an account of humanity's coevolution with its environments, tools, and artifacts. It exposes the extent to which "being human" has never come naturally!

Yet there are further philosophical, not to mention theological, dimensions to this. In speaking about the ascendancy of modern humanism, Latour himself argues that it was premised on "the simultaneous birth of 'nonhumanity'—things, or objects, or beasts—*and* the equally strange beginning of a *crossed-out God*, relegated to the sidelines" (Latour, 1993, p. 13, my emphasis; see also pp. 32–35). And here we begin to get glimpses of the other element within the emergence of modernity: the birth of a discrete philosophy, or sphere of life, known as the "secular." The creation of an immutable, autonomous, self-actualizing humanity was as dependent on the suppression of the transcendent, the divine nonhuman, as it was on the creation of a binary opposition between the normatively human and its "others" in nature, the animal kingdom, or in the world of tools, technologies, and machines.

The Crossed-Out God: The Paradoxes of Postsecular Society

Enlightenment humanism often assumed an implicitly antireligious and secular basis, since it placed human self-determination and autonomy—freedom from dependence on the forces of tradition, superstition, and autocracy—at the heart of its ambitions. Humanism is premised on ideals of self-determination through pursuit of individual rights and liberation from irrational influences such as tradition, ignorance, or religion. Personal conscience and reason became the guiding sources of moral authority, unmediated by the power of monarch or church. "Profoundly secular in orientation, Humanism promotes respect for science and culture, against the authority of holy texts and religious dogma" (Braidotti, 2013, p. 29). So, if the hegemony of secular humanism is in question and the categorization of the human against a series of nonhuman others is exposed as artifice, then it may be no surprise to see that other shibboleths of modernity—namely, the separation of secular from sacred, reason from faith, human from divine—have also been thrown into doubt.

So an inspection of the origins of modernity and of humanism through the construction of certain material and discursive boundaries leads us to another frontier: that which demarcates the secular from the religious, the material from the metaphysical. In adopting the terminology of the "post-secular," I should not be misunderstood as attempting to make a philosophical or theological case for the existence of God or for advocating a return to premodern understandings of medical science, human rights, the humanity of women, the creation of the universe, or similar ideas. The whole point about the postsecular is that it does not simply represent a reversal of the processes of secularization that have taken place in the Western world since the European Enlightenment. Instead, while we can point to evidence of a "new visibility" of religion and religious pluralism within global politics and society, there is no sign of a diminishment in long-term institutional decline of the Christian churches, as well as the continuation of vigorous and principled resistance to the legitimacy of religious reasoning within political, legal, legislative, and moral debate (Harris, 2005; Dennet, 2007).

The revival of religion as a global political force from the late 1970s (in empirical and theoretical terms) engendered a wave of revisionist interpretations of the sociological secularization thesis. As it has emerged in contemporary debate within philosophy, theology, and social theory, the postsecular refers in particular to a revisionist understanding of classic secularization theory, understood broadly as a separation of church and state, the privatization of faith, and the general decreased importance of religion in public life. The conventional account of secularization sees religious decline—even extinction—as an inevitable consequence of modernization. Yet evidence suggests that this is not the case. In many of the most rapidly developing economies, such as Brazil, China, or India, religion continues to grow and to be a significant part of public life. A feature of the postsecular condition, thus, entails a shift in consciousness to allow a certain public recognition of religion, as in, for example, the interventions of faith-based activism within civil society but also in terms of recognition of religious identity and the legitimacy of religious reasoning in public debate (Beckford, 2012; Habermas, 2008; Dinham, 2012).

One manifestation of the postsecular is a new wave of philosophical thinking that seeks to retrieve religion from the margins of modernity, to reconsider it as a significant and fundamental dimension of dynamics of social cohesion and social change, as well as an inescapable cultural resource for the construction of identity, both civic and personal. No less a figure than Jürgen Habermas has been credited with bringing the term "postsecular" into mainstream philosophical debate, departing from modernity's separation of faith and reason and exhibiting a greater openness toward religious contributions to public discourse, so long as their arguments can be "translated" into terms intelligible to secular protagonists (Habermas, 2008, 2010). Habermas's intervention has, thus, sparked a series of debates about the viability of a secular public square and the conditions under which religious conviction might be mediated into a pluralistic, liberal polity (Beckford, 2012; Dillon, 2010; Sandal, 2012).

Other writers, such as José Casanova (1994), Talal Asad (2003), and Craig Calhoun (2010) have exposed the very constructedness of Western secularity. Grace Davie (2002) speaks of the "exceptionalism" of European secularization and questions the inevitability of religious decline in the face of global vitality. Slavoj Žižek's evocation of Christianity as the "fragile absolute" (2003) also captures a postsecular sensibility that does not attempt to deny Enlightenment critiques of religion or sociologists' empirical warnings of the very precariousness of organized religion, yet insists, against all the odds, that faith endures as a vital source of personal meaning and social capital. Rosi Braidotti (2008, 2013) and Judith Butler (2008) have both criticized the tendency of post-Enlightenment political theory—including much of Western feminism—to equate humanism and emancipation with secularism. The postsecular turn challenges such a bifurcation by revealing the cultural and political agency of women of faith and "makes manifest the notion that agency, or political subjectivity, can actually be conveyed through and supported by religious piety, and may even involve significant amounts of spirituality" (Braidotti, 2008, p. 2).

Yet this is not a simple return of the sacred, however, since there are some things that suggest our cultural landscape is still decidedly irreligious. Notwithstanding the persistence of religion and its bold return to the public stage, Charles Taylor insists that the conditions of belief have radically shifted "from a society in which it was virtually impossible not to believe in God, to one in which faith, even for the staunchest believer, is one human possibility among others" (Taylor, 2007, p. 3). Regardless of our own personal faith commitment, we have all crossed a secular Rubicon, never again to return to a world in which personal autonomy, reflexivity, and freedom of belief are not axiomatic.

Nor has such renewed philosophical currency of religion necessarily been matched by any corresponding level of institutional viability or of public understanding. The actual foothold of organized religion institutionally and in the public imagination remains tenuous and limited. Measures of active affiliation to traditional forms of Christianity continue to decline, especially among younger generations. Culturally, at least in the West, any direct experience of lived religion is a minority pursuit. Add to that the continuing prominence of the "New Atheists" such as Richard Dawkins (2008), Sam Harris (2005), and the late Christopher Hitchens (2007), who repudiate the suggestion that religion could be a legitimate or rational source of human understanding, and, in the popular mind, suspicion and scepticism are probably still the default positions.

It would seem, then, that we have a perfect storm. Religious institutions are fragile in the face of demographic decline and a preference for (personal) spirituality over (organized) religion; skeptics and critics of religion continue to question its very legitimacy as a respectable intellectual option, let alone a legitimate force in society; yet, in spite of it all, religion continues to be a significant source of social capital—not least as a remarkably potent mobilizing force for volunteers in civil society. Similarly, globally speaking, if religion is not the "cause" of political and cultural change, it nevertheless still exerts a powerful formative influence on personal identity, popular movements, nation building, geopolitical conflict, and humanitarian initiatives.

So we are confronted with a paradoxical, almost agonistic, situation of two simultaneous and seemingly incompatible developments: the attenuation of organized religion and the drift away from institutional affiliation, alongside the resurgence of religion as a global and local political and cultural force. Like Latour's analysis of modernity, then, it is only when we can no longer deny the coexistence of the forces both of resurgent religion and its "cultured despisers" that the solidities of humanism, modernity, and secularity begin to melt into thin air. For me, that is the essence of what we might term the postsecular and what makes it—to both advocates of secularization and religious orthodoxy alike—unprecedented and theoretically intriguing. The postsecular denotes not simply the end of the secularization thesis but its unpredictable afterlife, in which currents of disenchantment and re-enchantment coexist:

> No sooner had a thoroughly atheistic culture arrived on the scene ... than the deity himself was suddenly back on the agenda with a vengeance. ... The world is ... divided between those who believe too much and those who believe too little. (Eagleton, 2014, pp. 197–78)

While religious belief, practice, and identity have not been extinguished by modernity and although they have managed to survive (and in many contexts) prosper, therefore, I must stress once more that this is not a revival of previous forms but their mutation. Furthermore, the space religion/s occupies remains contested and agonistic: and it is this third space between the persistence of secular objections to public religion and the new acknowledgment of religious actors that may best be termed the "postsecular"—not just some kind of successor phase, but as something that troubles and unsettles our prevailing definitions of "secularity" in the first place, as well as leaving open the many, and often paradoxical, configurations of religion and public life that are now emerging (Graham, 2013). Indeed, the debate about "what follows the secular" mirrors much of critical posthumanist theory in subjecting the very category of "the secular" to renewed scrutiny. As the postcolonial theorist Talal Asad argues, the secular, like the human, has a history, a "genealogy" (2003, p. 192).

Have We Ever Been Secular?

Noting "the contemporary salience of religious movements around the globe" (Asad, 2003, p. 192), Talal Asad calls for a re-evaluation of the very premises of secularism via an examination of its very "genealogy." In particular, the rise of global Islam as a political force exposes the limitations of conventional narratives about secularization as pertaining merely to the eclipse of Western Christendom. Asad locates "the secular" as a discursive category, within an entire system of dualisms such as "belief and knowledge, reason and imagination, history and fiction, symbol and allegory, natural and supernatural, sacred and profane—binaries that pervade modern secular discourse, especially in its polemical mode" (2003, p. 23).

Asad's argument is that "secularism" constructs "religion" as its negated Other in order to establish its own coherence. The "secular" brands religion as a matter of belief relating to an ontological category of the "supernatural," whereas secularism by contrast deals with the natural and the social, in which the citizen is supreme public reality, and anything to do with transcendence or the nonmaterial is consigned to the private and the interior. Yet this categorization is historically and culturally contingent and emerged out of particular practices of reading the scriptures, discourses of religious experiences, and, of course, ways of configuring the relationship between church and state.

In similar vein, Terry Eagleton's thesis in *Culture and the Death of God* (2014) is, to evoke Latour, that we have *never* been secular. Eagleton's argument is that post-Enlightenment thought has repeatedly failed to exorcise its metaphysical foundations. Insofar as the Enlightenment was really an attack on the coupling of political sovereignty and ecclesiastical authority, particularly of an autocratic variety, it was never primarily atheistic but rather an attempt to ground its social and political values in the principles of natural, rather than revealed or creedal, religion. In fact, says Eagleton, the secularism of modernity was a misnomer. Rather, the history of modernity should be seen as an extended "rewriting of religious faith in secular terms" (Eagleton, 2014, p. 47).

In the secular, modern era, however, a series of intellectual movements or watchwords—reason, progress, the state, empiricism, and nature—assumed the role of God's surrogates. One of the illusions of modernity is, thus, its failure to see that the recession of religion from day-to-day consciousness obscures its extended afterlife in a range of apparently secular philosophies:

> The history of the modern age is among other things the search for a viceroy for God. Reason, Nature, *Geist*, culture, art, the sublime, the nation, the state, science, human-

> ity, Being, Society, the Other, desire, the life force and personal relations: all of these have acted from time to time as forms of displaced divinity. (Eagleton, 2014, p. 44)

Eagleton exposes all post-Enlightenment alternatives to religion as attenuated versions of religious worldviews that are incapable of exerting any moral leverage by being devoid of ultimate value. Any attempt to invest society, human knowledge, or morality with a sense of meaning fails because none of these surrogates is capable of transcending the conditions of its own generation: "As the rationalising process comes to infiltrate the cultural and religious spheres, as with the mechanistic world of Deism or the legalistic nature of some Protestant doctrine, these realms become less hospitable to questions of fundamental value, and thus less capable of underpinning political power" (Eagleton, 2014, p. 43).

It is the very artifice of any such ultimate values that is their ultimate undoing. Secular myths of nationhood, progress, the self, or the revolution cannot "be legislated into existence by philosophical fiat," according to Eagleton (2014, p. 61); these ideals never succeed in being more than "ersatz forms of religion" (p. 80). A universalist humanism may have replaced Christian revelation, but it betrays a deep continuity, not least in the way that the intelligentsia looked to organized religion to ensure that political reforms, having empowered the bourgeoisie, would not inflame the populace. It was, thus, in the interests of Enlightenment thinkers to promote the functional utility of religion as inculcating civic virtue and obedience in the populace: "religion is judged primarily in terms of its utility. It is acceptable only if it promotes the kind of morality one would still endorse without it" (Eagleton, 2014, p. 25).

If there is such a concept as the "postsecular," then, it must be recognized that secularization and the secular are in themselves already complex and diverse terms and have histories that enable us to see them not as fixed entities but as heuristic and conceptual frameworks that may now be in need of reconstruction. Since it was assumed that modernization is a universal process, advocates of secularization have assumed that secularization, too, would be a worldwide and globally uniform phenomenon. However, there is no reason not to anticipate that processes of modernization might be manifested differently in various parts of the world, depending on whether they originated in industrial and economic transformation, as in Western Europe, or were introduced as a result of imperialism, as in Africa, or by political elites, as in Japan and Turkey, and whether a society encountered capitalism in its early stages or in its advanced stages of production. Shmuel Eisenstadt has suggested, therefore, that it may be more appropriate to talk of "multiple modernities," all of which interact with indigenous cultures, including religious belief and practice, in different ways (2000, p. 593).

Posthumanism, Transcendence, and the Sacred

Despite the equation of modern science with secularism, it is intriguing to identify a number of ways in which religious discourses still imbue many of the considerations of what should constitute our post/human future. Even in a supposedly secular age, prevailing concepts of God and transcendence still impinge on the kinds of normative and exemplary models of divine and human nature that fuel our technological visions. This does not simply apply to those who regard technologies as responsible for the *disenchantment* of the world in the name of spiritual values (Borgmann, 2003) but also to those who deploy theological language to justify technologically driven human evolution. New technologies, it is said, will enable humanity to transcend physical limits, such as bodily finitude, illness, and mortality, or to transport their users to a higher plane of existence. Once more, this contradicts interpretations that regard science and religion as antipathetic, for here are new technologies as—in David Noble's words (1999)—*instruments of deliverance*: vehicles of ascent to higher planes, harnessing elemental powers, possessing the power to transport their users into a sacred realm of "transcendence," free of the encumbrances of the flesh.

The problem with many of the commentators on this phenomenon is that they work from a reified notion of "religion," centred on belief in a dispassionate, transcendent, disembodied, immutable God. What they seem to share is that they are shaped by appeals to a neo-Platonic ideal of "transcendence" or the drive to use human creative energies to abandon the physical world—as in the quest for outer space, or the wish to transform embodied experience into pure information, as in the human genome project, or characterization of cyberspace as the "heavenly city" free of the encumbrances of the flesh.

Cyberspace is envisaged as a "mythic community" (Bukatman, 1993, p. 151), the apotheosis of an ancient project to transcend the physical world in search of perfection and immortality: "[T]he subject is dissolved in the swirls of cybernetic information, but is at the same time further empowered through an extension of motility and spatial possession. Here, then, are the paradoxically simultaneous experiences of death and immortality that are fundamental to religious practice" (Bukatman, 1993, pp. 295–96).

The implications of this, however, are disturbing, not least because such a representation of humanity's religious quest through technoscientific endeavor rests on a distorted appropriation of the notion of "transcendence." It is taken to mean the final attainment of the Cartesian split between body and mind, the ultimate detachment of posthumanity from contingency, embodiment, finitude, and particularity:

> [C]yberspace approximates [to] the ideal of total knowledge—the knowledge not only of the angels, but the access to the omniscience that pervades the Enlightenment pursuit of the ultimate encyclopedia, Hegel's dialectical drive towards the absolute, Wagner's *Gesamtkunstwerk*. As such cyberspace offers the illusion of unlimited control, control beyond the limitations of the physical body. Cyberspace promises the sublime experience such that one can lose oneself as one appears to move through a vertiginous infinite at a speed which makes the heart beat quicker in a friction-free accumulation of knowledge. ... This is the ideal of total power. The lure of cyberspace lies in the endless opportunities it promises for new spaces, new sites, new sensations and new adventures. (Ward, 2000, pp. 250–251)

This perplexing re-emergence of religion, what David Noble (1999) calls the "religion of technology" expresses the notion that advanced technologies are a means by which the secular world of the "crossed-out God" will be re-enchanted. Yet the ways in which religion is evoked to support this project are also problematic, as critical dialogue between some of these tendencies and Christian theological anthropology will indicate. In light of the preceding discussion, we should adopt a heightened hermeneutics of suspicion regarding the apparent re-emergence of the sacred in such debates, and be prepared to see the categories of sacred/secular, transcendent/immanent, and human/divine as themselves undergoing critical reinterpretation.

This "technological sublime" actually rests, then, on an unexamined identification of "religion" with a flight from contingency and materialism and a spirituality that seeks fulfillment in union with a "transcendent" deity entirely removed from the immanence of this mortal world. The presumption is that the quest for technological advancement is at some level an expression of the *imago Dei* and that this entails a necessary mastery over creation, heedless of the fragility and interdependence of life—ambitions that have in the past been used as rationalizations for dominion over nonhuman nature and even colonized peoples. Its predilection for the qualities of detachment, omniscience, immutability, and incorporeality promote disdain for embodied contingency and foster technologies that are obsessed with cheating death, vulnerability, and finitude. To represent religion in this way, therefore, is to see it as providing the ideological underpinning for many of the dualisms that have fuelled modernity and technoscientific innovation (Jantzen, 1998).

In fact, this equation of spirituality and religion with other-worldly transcendence has been subjected to theological critique, which sees things differently, not least in its eschewal of a symbolic of transcendence premised on omnipotence, individualism, and immortality. Rather than regarding the immanent, embodied, material world as an impediment to genuine spirituality or—reminiscent of Gnostic worldviews—as the profane, flawed, pale reflection of the authentic divine world, this vision sees creation itself as the very realm of divine–human encounter:

"Whatever it is, salvation will be an affirmation of the essential finitude of human nature, not an escape from it" (Kelsey, 1998, p. 144).

Postsecular Posthumanisms

I have already discussed Donna Haraway's famous use of the cyborg figure to argue that technologies have become so ubiquitous that notions of purity of origin, of human uniqueness, are redundant. Better to think of ourselves as cyborgs, she says: a blend of the cybernetic and thc organic. That way, we are divested of the illusions of innocence, purity, detachment, and the pretence of transcendence—all dangerous facets of a kind of will to power. The cyborg delights in her technological complicity and recognizes the plasticity of categories of being. In the cyborg, Haraway advances a metaphor for post/human experience that resists seeing technology as a deterministic monolithic force or as a quick fix whose social and political implications are somebody else's concern. Ethically and existentially, it rejects solutions of either denial or mastery. It suggests that ethical and political engagement with our technologies proceeds not from a position of abstraction or purity but from a recognition of our affinities with and immersion in a highly technologized world (Graham, 2002; Garner, 2011).

Thus, Haraway eschews anthropological dualism in her characterization of the cyborg, and she also questions traditional binaries of spiritual and material, endorsing a more "sacramental" understanding, in which the symbolic or divine is always already instantiated in the material. Having never participated in the human categories of birth, gender, and embodiment, the cyborg *transcends* the constraints of such categories. Not for her the "innocence" of the Judeo-Christian myth of origins in an idyllic, prelapsarian Eden; the world has always been one of ambivalence and complicity.

> A cyborg body is no innocent; it was not born in a garden; it does not seek unitary identity and so generate antagonistic dualisms without end (or until the world ends). Intense pleasure in skill, machine skill, ceases to be a sin, but an aspect of embodiment. We can be responsible for machines; they do not dominate or threaten us. We are responsible for boundaries; we are they. (Haraway, 1991, p. 173)

But is Haraway plumping for a secular materialist over a spiritual or theo/alogical anthropology? It is certainly an affirmation of an embodied, hybrid experience in which post/humanity embraces its affinity with the material world and the panoply of its nonhuman, technological, bioinformatical "others." But is this a secular, a religious, or a postsecular sensibility? In declaiming, "I would rather be a cyborg than a goddess" (Haraway, 1991, p. 181) is Haraway intending to reject or to redirect the traditional premises of religion as equated with other-worldly, disembodied, antimaterial spirituality?

I have always understood Haraway's preference for a cyborg socialist feminism over a goddess feminism as the dismissal of the kind of appeal to technotranscendence that equates human evolution with mastery over nonhuman nature and victory over embodied contingency. Jeanine Thweatt-Bates insists, however, that it should also be read as a choice of technological culture over romanticized nature; we should read Haraway in this respect not as rejecting transcendence, so much as essentialism. Certainly, Haraway's cyborg writing exhibits a strong materialism and a "latent spirituality" that reflects the afterlife of Haraway's own Roman Catholic formation (Thweatt-Bates, 2012, pp. 82–83). Arguably, then, Haraway's cyborg represents yet another transgressive boundary crossing, of the assumed separation of disembodied soul or spirit from material, contingent, embodied reality.

For while Haraway's statement represents at first glance what Thweatt-Bates characterizes as a "double rejection of religious narrative" (2012, p. 30), her own roots in Roman Catholicism still endure. Haraway encourages her readers to "redeem" Jesus Christ as a "trickster" figure, his human/divine nature (and resistance, historically, on the part of Christian tradition to any attempt to split or bifurcate this hybridity) as a subversion of conventional humanist anthropology. It affirms immanence and materialism in keeping with a highly sacramental and incarnational apprehension of creation and redemption. Sacramental theology understands the created world of nature and culture—including human artifacts and activities—as the basis for the expression and revelation of divine grace; God's participation in history is affirmed via Christianity's perception of Jesus Christ as the human manifestation, or incarnation, of the divine. Put at its simplest, then, any religious experience or action is always already mediated through specific material conditions, such as language, artifact, ritual, or image. The material is always already invested with the symbolic, the profane with the sacred, human fabrication with divine redemption: "A genuinely human life, so the argument goes, would seek meaning and fulfilment in ordinary attachments and attainments conditioned by our finitude, vulnerability and bodily nature rather than attempting to renounce or overcome these characteristics and limitations" (McKenny, 2011, p. 178).

Rosi Braidotti, similarly, proposes a model of the posthuman that pushes at the boundaries of sacred and secular, transcendence and immanence, via an eco-feminist deconstruction of modernity's objectification of "nature." This particular postsecular take on a "posthuman feminism in the neo-vitalist mode" (Braidotti, 2008, p. 14) thus opts for a more integrated notion of all matter, including nature as infused by "spirit" and a relocation of transcendence not as negation but confirmation of, beyond, and within the immanent. Braidotti pursues an Spinozan, neovitalist philosophy in which the notion of transcendence or becoming divine through technology is less about achieving immortality or other-worldly existence than about the re-enchantment of matter itself.

This kind of re-enchantment of social theory has its parallels in other disciplines, as, for example, the resurgence of spirituality in social theory. Sometimes this draws for inspiration from First Nation, aboriginal, or indigenous belief systems (discussed in the work of the urban geographer Leonie Sandercock, 2006) which argue that the land is never mere property to be viewed as a commodity, only ascribed value through economic exchange. Rather, it is to be viewed as a living actor whose role as ground of being is vital for sustainable habitats and upon whom all life—human and nonhuman—is dependent. Whether it is expressed in explicitly spiritual or religious terms, this conception of the earth as sacred renders the environment as irreducible to human appropriation. Even though this is conceived more in terms of a kind of ecological panentheism than any kind of traditional theism (Jantzen, 1998), it introduces a "more-than-human" horizon to the ecology of the post/human: "A posthuman ethics for a non-unitary subject proposes an enlarged sense of inter-connection between self and others, including the non-human or "earth" others, by removing the obstacle of self-centred individualism" (Braidotti, 2013, p. 50). The posthuman is not the end of humanity per se, but rather a displacement of a certain hegemonic, taken-for-granted assumption of the human as normative. It "enforces the necessity to think again and to think harder about the status of the human … and the need to invent forms of ethical relations, norms and values worthy of the complexity of our times" (Braidotti, 2013, p. 186).

Transcendence is not premised on absence from this world in order to achieve consummation with the next—an essentially spatial definition founded on a premodern understanding of heaven "up there" and earth "down here"—but more in temporal terms, as reaching beyond the now and the given toward future, unrealized possibilities. This sort of transcendence describes a nonreductiveness and deep connectivity to experience that "does not depend on fusion with something beyond the world, but, rather, relies on the extent to which the world itself is beyond us, is enormous, and is, at least in the aspect of its beauty, whole" (Calhoun, 2012, p. 357).

Conclusion: Human, Nonhuman, and More-Than-Human

The postsecular is located, in a sense, "between a rock and a hard place" (Graham, 2013): confronted by the resilience of religion, spirituality, and the sacred on the one hand and by the political settlements and epistemological convictions of secularism, materialism, and humanism on the other. For Rosi Braidotti, there needs to be a way to steer between the Scylla of a "new belligerent discourses about the alleged superiority of the West ... expressed in terms of the legacy of secular Humanism" and the Charybdis of "post-secular practices of politicized religion" (2013, p. 36)—a third space, potentially, of qualified, contextual, but principled humanist politics and ethics that is capable of articulating alternative, less polarized accounts of what it means to be human. This is where critical posthumanism comes in, as a means of transcending the binary divisions toward an account of contemporary, global (in its political and ecological senses), postsecular visions of truth and obligation.

If the posthuman condition signals the return of the repressed Other of liberal humanism, then the postsecular reflects the recognition of the endurance of the sacred, often in unexpected ways. My own take on the postsecular is that it is a very long way from delineating any kind of religious revival; yet, as I have begun to indicate, it may provide a route away from Western modernist traditions of divine transcendence that bifurcate the spiritual and the material in favor of a future that acknowledges the affinities between the human, nonhuman, and more-than-human (in the form of re-enchanted realms of nature and cosmos) in a more integrated and responsible celebration of life in all its fullness and diversity.

References

Asad, T. (2003). *Formations of the secular: Christianity, Islam, modernity*. Stanford, CA: Stanford University Press.

Badmington, N. (Ed.) (2000). *Posthumanism*. London, England: Routledge.

Badmington, N. (2004). Post, oblique, human. *Theology & Sexuality*, 10 (2), 56–64.

Beckford, J. A. (2012). Public religions and the postsecular: Critical reflections. *Journal for the Scientific Study of Religion*, 51 (1), 1–19.

Berger, P. (1999). The desecularization of the world: A global overview. In P. Berger (Ed.), *The desecularization of the world: Resurgent religion and world politics* (pp. 1–18). Grand Rapids, MI: Wm. B. Eerdmans.

Borgmann, A. (1984). *Technology and the character of everyday life*. Chicago, IL: University of Chicago Press.

Borgmann, A. (2003). *Power failure: Christianity in the culture of technology*. Grand Rapids, MI: Brazos.
Bostrom, N. (2003). *The transhumanist FAQ: A general introduction*, version 3.0. Available at http://humanityplus.org/philosophy/transhumanist-faq/.
Bracke, S. (2008). Conjugating the modern/religious, conceptualizing female religious agency: Contours of a "post-secular" conjuncture. *Theory, Culture and Society*, 6 (2), 51–67.
Braidotti, R. (2008). In spite of the times: The postsecular turn in feminism. *Theory, Culture and Society*, 25 (6), 1–24.
Braidotti, R. (2013). *The posthuman*. Cambridge: Polity Press.
Braun, B. (2004). Querying posthumanisms. *Geoforum*, 35 (3), 269–273.
Bukatman, S. (1993). *Terminal identity: The virtual subject in post-modern science fiction*. Durham, NC: Duke University Press.
Butler, J. (2008). Sexual politics, torture, and secular time. *British Journal of Sociology*, 59 (1), 1–23.
Calhoun, C. (2010). Rethinking secularism. *Hedgehog Review*, 12 (3), 34–48.
Calhoun, C. (2012). Time, world, and secularism. In P.S. Gorki, D.K. Kim, J. Torpey, & J. Van Antwerpen (Eds.), *The Post-secular in question: Religion in contemporary society* (pp. 335–364). New York, NY: New York University Press.
Casanova, J. (1994). *Religions in the modern world*. Chicago, IL: University of Chicago Press.
Cooper, D. (1995). Technology: Liberation or enslavement? In R. Fellows (Ed.), *Philosophy and technology* (pp. 7–18). Cambridge, England: Cambridge University Press.
Davie, G. (2002). *Europe: The exceptional case. Parameters of faith in the modern world*. London, England: Darton, Longman and Todd.
Dawkins, R. (2008). *The God delusion*. Boston, MA: Houghton Mifflin.
Dennett, D. (2007). *Breaking the spell: Religion as a natural phenomenon*. London, England: Penguin.
Dillon, M. (2010). Can post-secular society tolerate religious differences? *Sociology of Religion*, 71 (2), 139–156.
Dinham, A. (2012). *Faith and social capital after the debt crisis*. London, England: Palgrave Macmillan.
Eagleton, T. (2014). *Culture and the death of God*. New Haven, CT: Yale University Press.
Eisenstadt, S. (2000). The reconstruction of religious arenas in the framework of "multiple modernities." *Millennium: Journal of International Studies*, 29 (3), 591–611.
Ellul, J. (1965). *The technological society*. (J. Wilkinson, Trans.) (R. Merton, Intro.). London, England: Jonathan Cape.
Gane, N. (2006). When we have never been human, what is to be done? An interview with Donna Haraway. *Theory, Culture and Society*, 23 (7-8), 135–158.
Garner, S. (2011). The hopeful cyborg. in R. Cole-Turner (Ed.), *Transhumanism and transcendence*, pp. 87–100. Washington, DC: Georgetown University Press.
Graham, E. (2002). *Representations of the post/human: Monsters, aliens and others in popular culture*. Manchester, England: Manchester University Press.
Graham, E.(2003). Frankensteins and cyborgs: Visions of the global future in an age of technology. *Studies in Christian Ethics*, 16 (1), 29–43.
Graham, E. (2013). *Between a rock and a hard place: Public theology in a post-secular age*. London, England: SCM Press.

Habermas, J. (2008). Religion in the public sphere: Cognitive presuppositions for the "public use of reason" by religious and secular citizens. In J. Habermas, (Ed.), *Between naturalism and religion: Philosophical essay* (pp. 114–147). London, England: Routledge.
Habermas, J. (2010). An awareness of what is missing. In J. Habermas et al., *An awareness of what is missing: Faith and reason in a post-secular age* (pp. 15–23). Cambridge, England: Polity.
Haraway, D. (1991). A cyborg manifesto: Science, technology, and socialist-feminism in the late twentieth century. In *Simians, cyborgs and women: The reinvention of nature* (pp. 149–182). London, England: Free Association Books.
Haraway, D. (1992). The promises of monsters: A regenerative politics for inappropriate/d others. In L. Grossberg, C. Nelson, & P. Treichler (Eds.). *Cultural studies* (pp. 295–337). New York, NY: Routledge.
Harris, S. (2005). *The end of faith: Religion, terror, and the future of reason*. London, England: Free Press.
Hayles, N. K. (1999). *How we became posthuman: Virtual bodies in cybernetics, literature and informatics*. Chicago, IL: Chicago University Press.
Hefner, P. (2003). *Technology and human becoming*. Minneapolis, MN: Fortress.
Heidegger, M. (1993 [1954]). The question of technology. In D.F. Krell (Ed.), *Basic writings* (pp. 307–342). London, England: Routledge.
Herbrechter, S. (2013). *Posthumanism: A critical analysis*. London, England: Bloomsbury.
Hitchens, C. (2007). *God is not great: How religion poisons everything*. New York, NY: Warner Twelve.
Humanity+ (2002). *Transhumanist FAQ 3.0*. Available at: http://humanityplus.org/philosophy/transhumanist-faq/.
Jantzen, G. M. (1998). *Becoming divine: Towards a feminist philosophy of religion*. Manchester, England: Manchester University Press.
Kelsey, D. (1998). Human being. In P. Hodgson & R. H. King (Eds.), *Christian theology: An introduction to its traditions and tasks* (pp. 167–193). London, England: SPCK.
Kurzweil, R. (1999). *The age of spiritual machines*. London, England: Orion.
Latour, B. (1993). *We have never been modern*. (C. Porter, trans.). Cambridge, MA: Harvard University Press.
Mazlish, B. (1993). *The fourth discontinuity: The co-evolution of humans and machines*. New Haven, CT: Yale University Press.
McKenny, G. (2011). Transcendence, technological enhancement, and Christian theology. In Ronald Cole-Turner (Ed.), *Transhumanism and transcendence: Christian hope in an age of technological enhancement* (pp. 177–192). Washington, DC: Georgetown University Press.
Noble, D. (1999). *The religion of technology: The divinity of man and the spirit of invention* (2nd ed.). New York, NY: Penguin.
Pew Forum on Religion & Public Life. (2012). *The global religious landscape: A report on the size and distribution of the world's major religious groups as of 2010*. Washington, DC: Pew Research Center.
Sandal, N. A. (2012). The clash of public theologies? Rethinking the concept of religion in global politics. *Alternatives: Global, Local, Political*, 37 (1), 66–83.
Sandercock, L. (2006). Spirituality and the urban professions: The paradox at the heart of planning. *Planning Theory and Practice*, 7 (1), 65–97.
Simon, B. (2003). Toward a critique of posthuman futures. *Cultural Critique*, 53, 1, 1–9.

Taylor, C. (2007). *A secular age*. Cambridge, MA: Belknapp Press of the Harvard University Press.
Thweatt-Bates, J. (2012). *Cyborg selves: A theological anthropology of the posthuman*. Burlington, VT: Ashgate.
Ward, G. (2000). *Cities of God*. London, England: Routledge.
Wolfe, C. (2010). *What is posthumanism?* Minneapolis, MN: University of Minnesota.
Žižek, S. (2003). *The puppet and the dwarf: The perverse core of Christianity*. Cambridge, MA: MIT Press.

Perfecting the Human: Posthuman Imaginaries and Technologies of Reason

3

Sheila Jasanoff

Religious or secular, humanity's dreams of the future have always been posthuman. As long as people have tried to fathom the reasons for sorrow and evil, they have also looked for ways of transcending the bitterness of being human. Take, for instance, the myth of Icarus that I encountered in early childhood. I grew up speaking Bengali as my mother tongue and learned English only as a second language. I must not have been quite six when I read that story in an English lesson with my father, and it fascinated me. The youthful, doomed flier was the subject of the first sentences of any literary quality I remember learning on the way to becoming bilingual: "Icarus flew too near the sun. The heat of the sun melted the wax and Icarus fell into the sea and was drowned."

The dream of escaping the burdens of the human condition is older than old. What gives the contemporary discourse on posthumanism, or transhumanism,[1] its novelty is not simply talk of breaking the fetters of mortality or the limits of human cognition. The newness, some would say, lies in blurring the boundaries between technology and the human. The blissful life everlasting promised by the great world religions seems almost attainable through the ingenious mastering of nature. In the posthuman condition as currently conceived, we expect to become as indomitable and indestructible, and as interlinked, as our most empowering machines: blood, bones, and nerves mixing with steel, wire, and circuitry; the natural biochemistry of our DNA reprogrammed with molecular precision; and individual minds vastly exceeding their grasp through as yet unimagined feats of computer-assisted connectivity. Yet, as in the story of the master architect Daedalus

---

1 I use the terms "posthumanism" and "transhumanism" interchangeably in this chapter, although the terms have somewhat different nuances. Posthumanism implies an end to humanism as it is currently experienced and the beginning of an altered state of human affairs. This reading accords well with some of the utopian imaginaries discussed here.

and his fatally adventurous son, the technological has arguably always been part of the posthuman dream. It is our understanding of what being human means that has changed along with our technological achievements—and with those changes also our imagination of what lies beyond, or after, humanness as we know it.

Transcending the human condition is partly about physical escape from death and other chafing limits on human abilities, such as forgetfulness, pain, and disability, but transcendence also has a salvationary appeal. We seek not only physical but moral improvement. Our social and cultural efforts, individual or collective, could always be more perfect, like the "more perfect union" envisaged by the framers of the U.S. Constitution. I once heard the editor of a noted science journal say that he saw nothing wrong with human enhancement if it would allow him, a talented amateur pianist, to play a difficult sonata nine times right and only once less perfectly, in place of his ordinary record of nine times imperfectly and only one time as his ear wanted to hear it. The desire here was not just to align instrument and muscles more smoothly, a feat one might hope to achieve with practice; it was about incorporating an ideal of musicality into one's body. With science and technology as ready servants, that wish for perfection comes within reach: in the modern posthuman imaginary, what should be can be.

There is a paradox here, however. The technologies with which we hope to better ourselves are our own creations, and, as such, they partake of the flaws in our fallible imaginations. It is not only that even the best technological designs are never failsafe and do not always behave as expected. More to the point, technologies articulate normative visions of what life is and how it should be lived. They are, in effect, always already embedded in historically and culturally situated sociotechnical imaginaries (Jasanoff & Kim, 2009, 2015). They, therefore, incorporate a period's ethical presuppositions and its social and cultural blindnesses as well as its acknowledged hopes and fears. Any attempt to integrate the human and the technological so as to achieve perfection should take account of those mutual embeddings. A question for today's futurists, then, is whether we, sitting here in the early 21st century, can think about perfectibility, enhancement, or the posthuman condition with descriptions of technology that are thick enough to reflect its limits as well as its possibilities.

Circulating concepts like posthuman or transhuman invite philosophers to get precise, to pin down the exact meanings of "post" and "trans" when appended to the term "human." The methodological bent of science and technology studies (STS) is more modest but also more messy. Science and technology studies revels in the groundedness of practice, following around concepts no less than people and things, not only to map the contexts in which ideas germinate but also the effects they achieve while circulating. To get a better handle on posthumanism, STS therefore asks what the concept means in practice. In turn, that question leads to others: to whom does posthumanism have meaning, and why does anyone want to go beyond whatever humanness implies? And if transcending humanity is itself a sociotechnical imaginary, one must also ask what kinds of enhancement are implicated in posthumanist discourse and what visions of welfare and the public good that discourse encodes. The paradox of perfecting human nature with technologies born of our all-too-human imperfection needs to be confronted with analytic tools that do justice to the enmeshing of the technological, the social, and the moral.

Nowhere is that entanglement more plainly visible than in projects of technologically aided nation building in the high-modern period. Those efforts reveal much about what social and political leaders of the recent past hoped to gain from technology, which dimensions of the human condition they sought to alleviate, and why, in most cases, their hopes of liberation ran aground. In this essay, I therefore begin by looking at the role of modern nation-states in articulating ideas that we might today regard as posthuman, teasing apart the elements of the public good that were built into that era's big emancipatory promises.

Following a survey of four such projects that each failed, I turn a comparative lens on recent national engagements with nuclear power to show how enduring political and cultural commitments shape the imagined futures toward which state-sponsored technologies are directed. I then ask how posthuman visions look not from the standpoint of grand state enterprises but as viewed through people's everyday practices of living with technologies. I conclude with reflections on the role of critical scholarship in exposing the normative implications of dominant posthuman discourses, thereby paving the way for potentially more inclusive conversations about what should lie beyond today's perceived horizons of human capability.

Imaginaries of High-Modernism

Modernity's detractors are legion. Ever since Max Weber deplored the rise of bureaucratic rationality and the disenchantment of nature (Weber, 2004 [1918]), leading social thinkers have taken aim at the encroaching, rationalizing machinery of industrial modernity that hollowed out the spirit and reached a horrific climax in the ovens of Auschwitz. The roster of critics reads like a who's who of twentieth-century European social thought: Zygmunt Bauman (1991), Jacques Ellul (1964 [1954]), Michel Foucault (1979), Jürgen Habermas (1970), Martin Heidegger (2008 [1954]), Herbert Marcuse (1964) and the Frankfurt School, and James Scott (1998).

Against that weight of critique, the turn toward refiguring technology as release and salvation feels fresh and new, a cool breeze blowing into stuffy, overheated rooms inhabited by the dead voices of a predigital, presilicon age. The plasticity and expansiveness of information and the internet of things seem to render ineffectual and obsolete the imagined iron cages of an earlier generation. Technology today has resurfaced as the paramount instrument of agency, empowering each individual to tear down the imprisoning walls of structure, whether institutional or bodily, as imagined by my amateur pianist's yearning for artificially enhanced musicality. It is hardly surprising that, in this new age, so many scholars have reacted with enthusiasm to actor network theory (ANT), which celebrates the individual idea bearer's power to remake the world. In an intellectual environment too long dominated (some think) by a suffocating emphasis on the structures of entrenched power, the pendulum swing toward pure agency offers a refreshing sense of control, renewal, and possibility. Unsurprising, too, in the current climate of posthuman thinking is the appeal of Bruno Latour's siren song of integrating human and nonhuman, material and social, inanimate and political into a veritable, fully desegregated parliament of things (Latour, 1993; see also Latour, 2012).

Yet it is important to recall that, well before the digital era, at the pinnacle of what we now call high-modernism, the world was also full of dreams of posthumanism, in the sense of imagined release from common human ailments, by technological means. Indeed, the high-modernism of the 1950s was at its core an escapist vision, albeit one of collective rather than individual transcendence. This was a time of confidence in grand state planning supported by technological advances that would liberate humanity from the sufferings of the poor and disadvantaged. Its apogee was the Green Revolution, that great founding myth of technoscientific progress: the Borlaug-ization[2] of the world that banished hunger and turned famine-ridden societies, such as the Indian subcontinent, into lands of plenty enriched by exportable surpluses of grain.

I want to illustrate modernism's posthuman commitments with two small mid-century texts that, though unconnected and somewhat randomly chosen, capture several of the most common themes of high-modernist posthumanism. The first is

---

2 In coining the term "Borlaug-ization," I implicitly refer to Bruno Latour's famous 1998 work *The Pasteurization of France*, widely regarded as the most important and extensive elaboration of actor network theory developed by Michel Callon and Latour himself at the École des Mines in Paris. The fact that there is no comparable ANT study of Borlaug may speak to the power of the Green Revolution's mythic status; it seems not to lend itself to that kind of academic demystification. Instead a much more political debate continues to center on the revolution's distributive impacts and whether the costs to the losers were worth the benefits to the winners.

"Children's Creed," a 1950s work by a little-known poet named Dorothy Roigt that the United Nations Women's Guild, which my mother loyally served from 1956 until the early years of this century, adopted as its signature text:

> There shall be peace on earth, but not until
> All children daily eat their fill,
> Go warmly clad against the winter wind
> And learn their lessons with a tranquil mind.
> And thus released from hunger, fear, and need,
> Regardless of their color, race, or creed,
> Look upward smiling to the skies,
> Their faith in life reflected in their eyes.[3]

The second, better known text derives from a speech by the financier and public official Lewis Strauss, architect of the U.S. "atoms for peace" program, who became one of the first members of the Atomic Energy Commission:

> It is not too much to expect that our children will enjoy in their homes electrical energy too cheap to meter, will know of great periodic regional famines in the world only as matters of history, will travel effortlessly over the seas and under them, and through the air with a minimum of danger and at great speeds and will experience a life span far longer than ours as disease yields and man comes to understand what causes him to age (Strauss, 1954, p. 5).

These two texts, one an obscure poem, the other a widely circulated message of postwar technocratic confidence, converge in their articulation of key ideas that were constitutive of high-modernism's dreams of release from human needs and wants: progress, efficiency, accessibility, eradication, and transparency. Each was perceived as a desirable feature of good societies as well as of good technoscience, and each was applauded as a worthy end in itself. In the postwar imaginary, there was no friction between the advancement of science and technology and the betterment of society.

Progress for midcentury moderns meant unending growth and unlimited development propelled by a seemingly unshakeable consensus about the nature of public goods and the feasibility of policy solutions. Social progress, moreover, was yoked to progress in science and technology, as if (despite Hiroshima and the Holocaust) the two could not be disentangled. A sense of limitless promise in

3 The poem underwent several modifications over the years to keep it politically correct. In the last line, for example, the original "faith in God" was replaced by "faith in Man" and eventually "faith in life."

both science and society marked U.S. adviser Vannevar Bush's famed report to President Franklin D. Roosevelt seeking institutionalized governmental support for "science—the endless frontier" (Bush, 1945). In those same years, the United Nations began channeling its development efforts through a Technical Assistance Board, predating by more than a decade the founding of the UN Development Program in 1966.

Efficiency, too, carried a liberating ring for the high-modernists, especially in the New World's can-do commitment to technologies of mass production. In a virtuous economic system, efficiency meant allowing more people to do the things they wanted to do, in less time, at less cost, with less friction and less waste. This was the utilitarian world of Fordism and Taylorism on the U.S. factory floor, but efficiency in America marched hand in hand with thoughts of individual freedom and plenty, rather than with European anxieties about massified discipline and subjugation. A popular 1948 autobiographical novel by Frank Bunker Gilbreth Jr. and Ernestine Gilbreth Carey, *Cheaper by the Dozen*, captured the American spirit. It was a comic recollection of the child-rearing adventures of the authors' parents, the efficiency experts Frank Bunker Gilbreth Sr. and Lillian Moller Gilbreth.

Accessibility, a cardinal virtue of democracy, was likewise very much part of the modernist paradigm. Lewis Strauss, as noted earlier, famously proclaimed that the nuclear age would bring "electrical energy too cheap to meter." With energy costs driven to near-zero, Strauss and men of his era presumed that less developed societies would gain the power to pull themselves up to industrial country standards with fewer demands of cost and infrastructure. Contrary to Langdon Winner's (1986) characterization of nuclear power as intrinsically undemocratic or Gabrielle Hecht's (2012) study of the African bodies sacrificed to uranium mining, early proponents of nuclearity imagined the splitting and harnessing of the atom as a globally democratizing force.

Eradication in the context of midcentury posthumanism meant the elimination of ancient, acknowledged evils—hunger, disease, aging, and manifest physical barriers to mobility. These were the ills that Prince Siddhartha encountered outside his carefully guarded palace compound, ills that set him on the road to asceticism and nirvana. Now, as both Roigt's poem and Strauss's promises testify, those natural evils were recast as eradicable, not through inwardness and spiritual retreat but through all-out assault on nature's errors and malfunctions, aided by the technological armamentarium of the American century.

Transparency, the ability to look through opaque barriers that block understanding, was a fifth salient, if usually implicit, element of high-modernism's imaginary of the posthuman condition. James Scott (1998), following Michel Foucault, identified "legibility" as a pathology of the high-modernist paradigm, designed to extend the state's powers of observation and control into the furthest reaches of governed societies. Such confidence, he rightly argued, was often misplaced, and many grand development schemes failed, as prescient thinkers might have predicted. But transparency had a positive valence that Scott's Foucauldian disdain for the eye of power failed to take into account. In the imagination of modernist reformers, transparency allowed power to be seen as well as to see. Envisaged in this way, transparency retains a positive valence in contemporary discourses on good technologies (for example, open source and open code) as well as on good government.

Fractures in the Frame of Development

If "big" and "built" were common themes in the mantra of high-modern posthumanism, not everyone bought in, even in the decades of greatest postwar optimism. There is a famous image of Jawaharlal Nehru sitting with Mahatma Gandhi, famous in part because even in their attire—Nehru all in starched white to his peaked cap, Gandhi bare-chested in his traditional loincloth—these nation builders represented such different visions of how to achieve the very kinds of release that they hoped independence would bring to India. Both presumably were dressed in cloth handloomed in India, as the practices of weaving were repatriated back from Britain's Lancastrian mills. But Gandhi's vision of *khadi* looked different from Nehru's. Gandhi stood for the small-scale, the rural, the village-based, an embodiment of E. F. Schumacher's (1973) principle "small is beautiful," whereas Nehru envisioned modern India on a grander Western model. Energy was central to Nehru's project, and dam building was his answer. The sheer physicality of those constructs called to this leader at the dawn of Indian independence. At the dedication of one iconic dam in October 1963, Nehru proclaimed, "Bhakra Nangal Project is something tremendous, something stupendous, something which shakes you up when you see it. Bhakra, the new temple of resurgent India, is the symbol of India's progress." Dams in place of temples: it was an image that would define the early years of India's development.

I grew up in the halo of that vision. My father, Sudhir Sen, was the first chief executive officer of the Damodar Valley Corporation (DVC), the eastern Indian version of the Tennessee Valley Authority. Our family holidays often included tours through multiple DVC construction sites: Bokaro, Durgapur, Mython, Panchet Hill. As a child, I reveled in the excitement of those site visits, from the huge, dusty dams rising up from vast earthworks and the reservoirs being dug behind them to the neat, well-built houses into which displaced villagers would be relocated. Long years later, after I became involved with STS, I hesitated to interrogate my father about his experiences of that period, perhaps unconsciously fearing to hear uncomfortable truths; but his writings indicate that, although he was caught up in a particular ideology of modernization, he was not sanguine about nationhood conceived as a gigantic engineering project:

> At the dawn of independence India relied wistfully on her high dam builders. During this TVA phase of India's economic development, a well-known Indian engineer used to proclaim off and on that he was going to build the highest dam in the world, suggesting implicitly a new yardstick for measuring national greatness—the height of the dam and the millions of cubic yards of concrete poured. That many engineers in India, if left to themselves, like to build monuments to themselves, regardless of the time and cost involved, is a commonplace of history. India had yet to discover this (Sen, 1974, p. 86).

India's independence on August 15, 1947, occurred in a period of high anticipation, when the end of a great war and the start of decolonization linked the victorious United States and the newly independent nations of the world in a shared faith in development and progress. But many of those bright hopes proved illusory. What failed in the modernist vision, we can say in retrospect, was not so much our ability to extend human faculties by technological means, but rather our *social capacity* to match ends and means to produce shared and achievable visions of a good society.

Four brief case studies, each representing a big technology and each dedicated to eradicating one of the specific constraints on human liberty mentioned in Roigt's poem—fear, hunger, need—can be read as cautionary tales about the dangers of emancipatory projects based on imperfect knowledge and unacknowledged value differences, coupled to governance schemes that ordered liberation from on high while neglecting the forms of life of those being governed. These are the stories of the bomb, Bhopal, Brasilia, and Biosphere 2.

First, the atomic bomb. This terrifying technology, which initiated the nuclear age, the Cold War, and the superpowers' unbridled arms race, exposed deep fissures in the very idea of progress, even though atom-splitting was at first conceived as the ultimate weapon against totalitarian terror and eventually reconceived as an instrument of peace. It brought the threat of annihilation that, on its first manifestation at the Trinity test site in New Mexico, shook J. Robert Oppenheimer into quoting from the Bhagavad Gita: "I am become death, destroyer of worlds." As the only country that used this ultimate weapon in war, the United States spent decades creating, and selling, a countervailing myth of the peaceful atom. Despite vigorous attempts to contain public fears, however, the bomb heightened a sense of helpless dread in all who watched the superpowers' mad stockpiling of devices whose only foreseeable end use was the total destruction of life on earth. The eventual stalling of nuclear power in both the United States and Germany showed how difficult it was for two of the world's most technologically advanced nations to maintain a credible split between atoms for peace and atoms for war (Jasanoff and Kim, 2009).

The second is the story of Bhopal, the 1984 toxic leak of methyl isocyanate gas at a Union Carbide pesticide plant in India that killed thousands and injured hundreds of thousands (Jasanoff, 2014). India imported this technology as part of the newly independent nation's effort to banish the specter of mass starvation and to support, with chemical fertilizers, its investment in the grains of the Green Revolution. The disaster stands as a powerful reminder that the effectiveness of a technology crucially depends on its social supports and that transferring the material components of technological systems without their social foundations can lead to terrible consequences. The accident revealed massive inadequacies in risk assessment, regulation, and hazard communication as the U.S. chemical industry's products reached out to global markets. But as the 20th century's most lethal industrial accident, the Bhopal disaster also underlined profound asymmetries of power that gave the lie to the redemptive message of development. Union Carbide and the U.S. legal system, not the victims nor the government that represented them, controlled the terms of the official settlement. Thirty years after the catastrophe, Bhopal's victims continued to feel doubly betrayed, by Carbide and by the Indian state.

The third project is Brasilia, an imagined capital city willed into being in the middle of nowhere by Juscelino Kubitschek, president of another resurgent nation-state wishing to demonstrate its power and prowess through an escapist leap from poverty into modernity. The core of the city is shaped like a bird, or perhaps an airplane, built around a gigantic cross of highways, signifying release and transcendence through their very imprint on space. Partnering with Oscar Niemeyer, the famed Brazilian architect, Kubitschek opted for a spare utopian design eradicating the messiness of ordinary urban life: no slums, no congestion, no well-trafficked commercial roads, not even a normal public square where people could gather in great, convivial crowds (Scott, 1998, pp. 121–125).

Yet, the real Brasilia with its millions of residents has grown up relatively unplanned in the quadrants defined by the bird's body and wings, while the planned inner core remains a thinly populated, monumental, architectural space, cut off from the hum and bustle of thriving inner city life. The dialectic between planned and unplanned began early. When I visited Brasilia in 2008, my sociologist guide told me that Niemeyer, a man of impeccable leftist credentials, had not calculated where the migrant workers would live at the end of the laboring day, a point Scott also emphasizes. They set up temporary squatter settlements that eventually came to serve as nuclei for Brasilia's peripheral townships. Initially reproducing Brazil's stark class distinctions between rich and poor (Scott, 1998, pp. 117–130), those peripheral cities now offer just those forms of urban congestion and vibrancy that Kubitschek and his architect of choice set out to eliminate from Brasilia's core. Life, it seems, remained unruly in a way that the city planners did not expect. Urban ecology proved more resilient than the grids of modernity that Scott so cogently critiqued in his denunciation of Brasilia's imposed legibility.

On a much smaller scale, and with no tragic consequences for human life, the Biosphere 2 project also illustrates the triumph of ecology, human and natural, over a planning mentality that sought to produce a self-sufficient machine for human and planetary survival. A group of unlikely private actors came together in the 1980s to plan and build a closed, mini-model of earth designed to support human life with no external inputs or interventions. It was a prototype for an escape hatch if all planetary systems failed and humans needed to travel elsewhere in the galaxy. Inspiration came from a counterculture group headed by engineer and entrepreneur John Allen and his small team of followers; funding came largely from Ed Bass, a Texas billionaire and environmentalist who had funded several previous utopian projects with Allen's group. Together, they backed the construction of a futuristic greenhouse in the appropriately named town of Oracle, Arizona, hoping it would function as a microcosm for the eight crew members who intended to take up residence in its tightly sealed environment for two years.

In its juxtaposition of pride and ignorance, Biosphere 2 is especially reminiscent of the myth of Icarus. It was an overly ambitious flight on still untested wings, a utopian dream that turned dystopian for the dreamers (Reider, 2009). The coalition of visionaries who built the project neglected the mundane complexities of the biosphere they were mimicking. The internal environment of Biosphere 2 quickly proved deadly to many of the imported plants and animals and near disastrous even for the humans. Red ants multiplied and birds died, while invisible bacteria in the soil sucked out the oxygen needed by the human inhabitants. In the end, emaciated and ill, the Biospherians barely completed their self-imposed two-year exile. Tellingly for today's posthuman visionaries, psychological strains on the participants proved even more disruptive than physical ones, and the once single-minded team splintered into two groups unable even to speak to one other.

Theorizing Posthumanism

Thus far, I have argued that high-modern posthumanism failed in its efforts to transform and redeem suffering human lives because the grand planners did not fully comprehend the very conditions of life that they sought to transcend. The promises of collective improvement through technology have repeatedly been frustrated not merely through mischance or design errors but because planners overestimated their own capacity to define and enforce the collective good: to turn weapons of mass destruction into credible instruments of peace and plenty; to transfer technologies without heeding their social contexts; to empty urban space of the comforting untidiness of lived lives; and to presume that an artificial microcosm would function more perfectly than the larger natural and social systems it mimicked. To help us move beyond these individual case studies, I would now like to situate posthumanism in a theoretical setting defined by the concept of sociotechnical imaginaries. Part of the reason for this move is to explore why, in spite of bows to such seemingly universal notions as enlightenment or progress, technology so often moves along different trajectories when it is embedded in different cultures and societies. The turn to theory also serves a normative purpose, allowing us to ask not what posthumanism is but what are the conditions of possibility that shape a period's or a culture's particular aspirations to reach beyond the human. By making these precursors visible, we can promote a more explicit accountability for the kinds of futures that today's posthumanists of all stripes would like to project onto our collective futures.

Sociotechnical imaginaries have been defined as "collectively held, institutionally stabilized, and publicly performed visions of desirable futures, animated by shared understandings of forms of social life and social order, attainable through and supportive of advances in science and technology" (Jasanoff and Kim, 2015). That imaginaries are "collectively held" is of key importance. To achieve effects, an idea such as posthumanism cannot simply be the brainchild of a single individual or small group. Such actors may articulate what STS scholar Stephen Hilgartner (2015) calls "vanguard visions," but those must reach beyond individual minds and sufficiently permeate through society to motivate collective projects for making the visions practicable. Imaginaries, too, are "institutionally stabilized," in that powerful institutions commit to achieving and sustaining them. They are "publicly performed," that is, they are accompanied by concrete actions that have the power to satisfy supporters and persuade new adherents. The "shared understandings" that constitute sociotechnical imaginaries, moreover, are profoundly normative: these are desired or desirable forms of social life and social order that advocates want themselves and others to adopt. Imaginaries are *futurescapes* that collectives actively wish to embrace.

The framework of sociotechnical imaginaries allows us to think about the deep structures of society in ways that make visible ordinarily invisible connections among the material, the social, and the normative. The Karlsruhe region, for example, where this book was partly conceived, houses both a nuclear power plant and the German Constitutional Court. The former belongs to technoscience, most would say, and the latter to society. Yet, when examined under the heading of sociotechnical imaginaries, one can see that both are engaged in constituting Germany's social order through instruments that, though different in kind, are interactive, mutually supportive, and, hence, in my terms coproductionist (Jasanoff, 2004).

Because of its promise of unending energy, its ties to security and warfare, and its potential for catastrophic accidents, nuclear power emerged everywhere as a constitutional technology, implicated in realizing the kinds of futures that nations imagine for themselves. When adopted as part of a national energy portfolio, however, nuclear power plants have to interact with a wide array of social institutions, such as political parties, environmental movements, expert committees, legislatures, and constitutional courts. How a nation deals with nuclear power thus offers a window into its dominant sociotechnical imaginaries; the particularities of that imagination, moreover, emerge most forcefully when set beside the constitutive elements of other nations' assumptions and approaches (Jasanoff and Kim, 2013). Here I briefly compare three cases: the United States, Germany, and South Korea. In each, nuclear power has been slotted into place differently, reflecting basic divergences in the constitutional orders that cradle national technological choices.

In United States, nuclear power remains very much a part of the national imaginary of a 21st-century energy policy, even though no new nuclear plant sites have been approved since the 1979 Three Mile Island accident. Unlike in Germany, the *de facto* U.S. abandonment of nuclear power rests on tacit, not formal, policy. The nuclear industry concluded decades ago that it could overcome neither political opposition nor economic barriers to deliver cost-effective electricity. But this was a corporate decision based on one era's socioeconomic calculations, hence, both provisional and reversible. As demand for nonfossil fuel energy has risen in response to climate change, the U.S. nuclear industry has found it possible to contemplate a comeback.

Faced with the same threat, Germany is once again in the grip of a utopian posthuman imaginary, the *Energiewende*, or energy transition, that will bring about the kind of economically, environmentally, and politically sustainable future that Germans yearn for. The nation's future depends on satisfying its energy needs without sacrificing its economic and social well-being. In turn, that goal demands that the power-generating potential of any fuel source, including the atom, be fitted to cultural expectations concerning the distribution of public and private risks and responsibilities. Nuclear debates have long been a site in which Germany confronted what kind of nation it wishes to be (Nelkin and Pollak, 1981). The answer for the moment is nuclear-free: under current policy, existing nuclear plants will be phased out, and the nuclear option will not be a part of the German *Energiewende*.

South Korea offers yet a third weighing of the benefits and risks of nuclear power, driven by a different imaginary of nation building through technology. Public commitment to the principle of developmental nationalism has meant that nuclear power enjoys a high degree of public support, as evidence of South Korea's ability to catch up and compete with the world's most technologically developed nations. That commitment to progress through technology ensured, for example, that the Chernobyl accident, which inflamed German risk consciousness, had relatively little impact on Korean public opinion or public policy. Indeed, unlike both Germany and the United States, South Korea has become an exporter of nuclear plants and expertise to other nations seeking to enter the club of nuclear power production (Jasanoff and Kim, 2009, 2013).

We see, then, that, in deciding how to deal with nuclear power, three technologically advanced nations have opted for three different paths based on their collective visions of what goods are to be achieved, how risks and losses will be distributed, and who decides such issues in each society. Table 1 offers a necessarily simplified picture of the nuclear imaginary in each country, capturing the salient differences. These are summarized below along five axes: policy focus, public framing, political stakes, controversies, and closures. Each deserves closer analysis.

	United States	Germany	South Korea
Policy Focus	Controlling radiation	Increasing transparency; ensuring responsibility	Building national technological capacity
Public Framing	Runaway accidents; catastrophic harm	Irresponsible and catastrophic harm	National independence; energy security
Political Stakes	Governance of technology	Governance of state-society relations	Governance of nation
Controversies	Scientific risk assessments	Lack of state and expert accountability	Political inclusion and participation
Closures	Court cases; NIMBY actions	Political mobilization; constitutional judgments; legislation	Administrative action; economic incentives

Tab. 1 Comparison of Nuclear Imaginaries—United States, South Korea, Germany

To begin with, policy responses in the three countries have followed distinctive trajectories. Over decades, the United States has invested the greatest resources in scientific studies of risk assessment and physical containment; Germany has devoted primary attention to political deliberation, in and out of the formal chambers of government; and South Korea has concentrated on building national capacity in nuclear engineering. These variations, in turn, stem from divergent framings of nuclear benefits and risks and how they should be allocated, managed, and controlled. The boundaries between public and private responsibility, in particular, rest in each nation on disparate foundations. These include not only the formal constitutional principles by which states manage internal relationships across governmental institutions but also their understandings of what is due to citizens when the state embarks on shaping the future through big technoscientific projects such as nuclear power.

In framing its policies for nuclear power, U.S. policy makers split questions of benefit from those of risk early in the game, tacitly determining that benefits were rightfully private and should be left to the market, whereas risks were to be socialized in the national interest. The atoms for peace program privatized power generation, but the Price-Anderson Act of 1957 provided insurance cover to shield the private sector against catastrophic risk. Leaving technology development to the market thus necessitated an assumption of risk on the part of the state—but only after harm occurs, not through an *ex ante* determination of how much risk is acceptable to the public. Germany chose the opposite course. Nuclear power was seen from the start as centering on the state's responsibility to ensure safety and security for citizens. Therefore, even though the plants are privately operated, it is up to the Bundestag to see to it that nuclear technology meets German citizens' expectations of responsible governance, including freedom from unmanageable risk. In South Korea, nuclear power remains a partially state-owned enterprise, even after economic restructuring that moved the country toward a more neoliberal model in the later years of the 20th century. Protest efforts by antinuclear citizens have had little policy impact.

What is at stake politically in developing nuclear power? Again, the answers vary across nation-states. In America, the threat of a runaway accident looms largest, and communities have mobilized against nuclear power in classic displays of the "not-in-my-backyard" syndrome. In Germany, though catastrophic harm is certainly an issue, the possibility that no one would be responsible if such a breakdown happens is of greater concern. Whereas the physical technologies of containment receive the most attention in America, Germans have focused more on holding both industry and the state accountable through law. In South Korea, energy insecurity and dependence on imports are seen as far more significant threats than the possibility of a runaway accident. Summing up, we can say that political stakes in each country arise from different understandings of which future needs to be governed: the future of technology in the United States, of state-society relations in Germany, and of national competitiveness in South Korea.

Nuclear controversies in the three countries have tracked these initial framings and political stakes. In the United States, debates focused primarily on risks to human and environmental health and safety, with repeated challenges to regulatory agencies' discretion in setting industry standards. Disputes were worked out case by case, in common-law fashion, with the Supreme Court periodically stepping in as arbiter of last resort. This litigious strategy hindered consensus building, instead fracturing debate along interest-group lines and generating frequent, protracted scientific debate.

One notorious consequence of the U.S. approach is the long-planned national depository for spent nuclear fuel at Nevada's Yucca Mountain. Dubbed "the most studied real estate on the planet" in a 2006 U.S. Senate committee report,[4] the site has consumed billions of dollars in preparatory studies, as well as untold person-years of lobbying and advocacy, all without a final resolution. If technoscientific projects sometimes serve as instruments for building national consciousness (Ezrahi, 1990), then Yucca Mountain represents a notable failure of this strategy. As the world's first mover toward nuclearity, the United States has notably fallen short in the effort to build national unity around the imaginary of atoms for peace.

Closure mechanisms in each country, finally, correspond to the variations in policy and politics described above. With litigation as the preferred mode of problem solving, U.S. controversies often end with Supreme Court judgments, typically ratifying the nation's official pronuclear policy. Unofficially, however, the United States has maintained a nuclear moratorium for decades, reflecting unresolved public anxieties about plant safety and state responsibility. Those concerns parallel public questioning in Germany where, however, the resolve against nuclear power rests on official decisions by successive administrations. Most recently, Japan's Fukushima disaster solidified Germany's antinuclear stance, ensuring that nuclear power will not be part of the energy transition imaginary. In marked contrast, the South Korean government has successfully used managerial authority and economic incentives to persuade citizens of the continued viability of nuclear power. Incentives work in Korea not by virtue of the funding alone but because much of

---

4 U.S. Senate, Committee on Environment and Public Works, Staff report to Senator James Inhofe, Washington, DC, March 2006, http://www.epw.senate.gov/repwhitepapers/YuccaMountainEPWReport.pdf.

the country firmly buys into the imaginary of national independence and success through technology (Jasanoff and Kim, 2009).

The United States, Germany, and South Korea are all constitutional democracies, but, in the case of nuclear power, observed through the lens of sociotechnical imaginaries, we see that their ways of confronting the future could hardly be more different. Formally, their policy styles continue to diverge, with the United States still pluralist and less subject to party discipline, Germany still significantly corporatist, and South Korea still statist in spite of democratization and neoliberalization. More fundamentally, though, each nation resorts in moments of crisis to well-institutionalized strategies for imagining or reimagining collective futures. What can we say, however, about posthuman imaginings in an era when technologies of future making—through the convergence of nano, bio, info, and cognotechnologies—seem more decentralized, more amenable to personal agency and choice, and far less subject to state control than nuclear power?

Postmodern Posthumanism

In the era of Google, we can almost dispense with social psychology. Instead of an abstruse disquisition by a current-day Carl Jung, we can turn to Google images and find out what our collective unconscious is thinking on any given subject. If one searches under "transhumanism," for example, one finds images that fall into several recurrent genres. One is of androgynous figures, speaking to the fact that sexuality, an important definer of the transhumanist movement, is seen as plastic and malleable, like consciousness itself. Apart from androgynous figures, the single most frequent image is of the human head with various kinds of enhancements built onto it. The diagnostic instrument of Google images, a visual collectivization of what people imagine, suggests that the computer-human interface, with its focus on engineering individual cognitive capacity, is the chief driver of contemporary thinking about transhumanism. Interestingly, this imagery leaves little room for other forms of connectivity that, arguably, are already affecting the human ability to reach beyond the self and forge connection with minds and collectives beyond one's own. Social networking technologies, for example, do not seem to have much significance for transhumanists. Facebook, Twitter, and indeed Google itself are absent from imagery that centers on connecting humans with nonhumans. Equally, the deliberative technologies through which we decide who we are as social beings and what constitutes the public good do not appear in pictorial representations of transhumanism.

These omissions are all the more puzzling because, in this technology-saturated age, machines and human artifice mediate all social activity. Take voting, for example. The same Google search that connects transhumanism with computer-enhanced brains displays many images of new kinds of voting machines with properties designed to encourage greater electoral participation. The U.S. National Institute for Science and Technology is involved in designing a voting machine that will make it easier for elderly, disabled, or simply busy people to cast their votes without needing to leave their homes, thereby sustaining a key form of connection between citizens and their imagined national communities. Yet, sociotechnically mediated behavioral spaces, such as real or virtual polling places, or indeed sidewalks crowded with texting teenagers unconscious of pedestrian traffic around them, apparently fall outside conventional imaginaries of the posthuman.

Legal institutions likewise do not feature prominently in discussions of posthumanism, except perhaps in the form of a presumed "law lag" that erects unreasonable obstacles against the achievement of technologically assisted transcendence. Yet, surely the law is one of the foremost technologies with which we seek to overcome the burdens of the human condition, to become in that sense posthuman. The U.S. Supreme Court has repeatedly ruled, for good and for ill, on what kind of collective entity the United States should strive to be—striking down separate but equal education for blacks and whites, affirming women's right to an abortion, approving adult patients' right to die, upholding gay marriage, recognizing corporations as persons in the political process, and refusing limits on gun ownership. These decisions grow from a technology no less intricate than the world's most sophisticated computers. Behind even the most trivial legal decision lies a thousand-year tradition of textual interpretation, with specialized techniques of reading and writing, and practices of advocacy designed not only to help articulate a society's norms but also to persuade those whom it touches that the law is fair, objective, and just. It is curious that so powerful a technology of future making should become invisible to people whose imagination of posthumanism derives mainly from human-computer interactions.

That neglect of deliberative capacity and its moral implications has troubled science fiction authors for decades. The classic 1956 film *Forbidden Planet* offers one example. Space travelers from earth arrive on a planet that bears the relics of an ancient civilization, but the makers of that civilization have vanished, leaving a mystery behind. Ultimately, the film offers a Freudian explanation of what happened, and it is a classic tale of overreaching. The Krell who inhabited the forbidden planet became progressively more knowledgeable and more powerful. But enhancing their cognitive capacity without attending to moral self-governance allowed their id, or uncontrollably evil self, to acquire disproportionate strength; with no superego to control them, their hubris ensured their annihilation.

Our own failure to take note of humankind's highly developed technologies of deliberation when thinking about posthumanism threatens to lead into a similar trap. The visioneering of the information age is not attuned to building a robust collective—a coherent "we"—for whom posthumanism makes sense as a moral or political commitment. Fixated on the newly imagined plasticity of the body, the consumerist posthumanism of the "selfie" age pays little attention to age-old practices of building collectives—imagined communities in Benedict Anderson's (1983) terms—that are bigger than the sum of their parts because they occupy a shared normative space. To think more seriously about a *common* posthuman future, it is, therefore, essential to engage in what we may call "reasoneering," that is, asking explicitly (as Germans arguably have done with nuclear power) what are the social ends toward which we should direct our technologies. Whether societies proceed through the classic instruments of deliberative democracy or by other means, a posthumanism that never questions what humankind hopes to achieve through technological advancement, as community, nation, or species, may turn out to be a future that not many will enjoy inhabiting.

How, then, should academic analysts of posthumanism seek to influence this open-ended promissory discourse? Scholars of STS would argue for a symmetrical approach, taking equally seriously the visions of ordinary people and those with special resources or know-how to move technology's frontiers. The right to imagine is, after all, a common human right, and empirical work on lay understandings and aspirations of transcendence can bring to the fore deeper visions of the public good than the prolific statements of a few vanguard visioneers. Let us look at a few headlines from a day's offerings in the *New York Times*, newspaper of record in a country well known for its technoscientific enthusiasms, to get a sense of the forms of liberation that people living in high-tech societies actually crave:

1. Coveting not a corner office, but time at home (Rampell, 2013, July 8).
2. Privacy group to ask Supreme Court to stop N.S.A.'s phone spying program (Risen, 2013, July 8).
3. Taking on rules to ease Sikh's path to the army (Dao, 2013, July 8).
4. A new tool aims to help Facebook users dig deep (Goel, 2013, July 8).
5. In Europe greener transit on existing infrastructure (Gies, 2013, July 8).
6. A deadly derailment in Quebec underlines oil debate (Austen, 2013, July 8).
7. Beyond the finish line (Rohan, 2013, July 8).

The most striking feature of these headlines, and the stories behind them, is the thorough entanglement of the technological and the social, the modern and the traditional. In much that is newsworthy about the contemporary world, the technologies of deliberation, social solidarity, and cohesion occupy the same narrative spaces as the technologies of release, escape, enhancement, and transcendence that are thought to define the posthuman. High-tech living seems, if anything, to have strengthened the human desire for social connection, the need for compassion, and the reaffirmation of memory—while people continue to acquire and enjoy the playthings of modernity. In the mixing up of people and things and values, these stories bear out Bruno Latour's (1993) observation that "we have never been modern," but deny his associated claim that modernity has imposed a constitutional division between the natural and the social. Hybridity, acknowledged and embraced, seems instead the desired order of the day.

To judge by the first story, what women want is not so much to achieve positions of power and influence (the "corner office") by "leaning in" (Sandberg, 2013), but rather to enjoy the pleasures of mothering, even if that choice prejudices one's career at the uppermost reaches of Silicon Valley. The featured mother in the story "finds climbing a career ladder less of a concern than finding a position that offers paid sick leave, flexible scheduling or even the opportunity to work fewer hours" (Rampell, 2013, p. A1). To be sure, the nursery beside the corner office remains an attainable goal for some women, such as Marissa Mayer, who, as CEO of Yahoo, built a nursery for her young son next to her own office. That story raises questions about equality and inequality in a company where the female CEO demands that people must show up at the workplace, not work from home, but where few other mothers have the clout to order their nurseries to be placed next to their offices (Carlson, 2014).

Several of the headlines indicate that, in public consciousness, technology cannot be easily divorced from institutional power. Confronted by Edward Snowden's disclosures about the National Security Agency (NSA), civil liberties groups turned to the Supreme Court to assert their privacy rights. The law served as the vehicle of choice for citizens wishing to live free from the state of total surveillance imagined by the NSA. An American Sikh soldier's attempts to reform the U.S. Army's ban on wearing his religion's mandated beard and turban points to a way of mediating between private faith and the secular public sphere different from the strict separation mandated by the rationalizing instincts of the world's most technologically sophisticated military.

Other stories expose the enduring influence of the past on any attempts at future making. Facebook's new tool is an enabler of social memory beyond everyday human capacities for recall: "designed to make it easier to find that lost photo or restaurant recommendation and unearth other information buried within your social network" (Goel, 2013, p. 81). Europe's green transit systems follow older structural lines, just as coal, oil, and electricity delivery systems in the United

States successively built on earlier infrastructures (Jones, 2014). The derailment in Quebec not only realizes the tragic consequences latent in hazardous technological developments but underscores their place in ongoing social debates about energy futures. And the most cryptic header, "Beyond the Finish Line," tells the story of a victim of the Boston Marathon bombings of April 15, 2013, who was caught up in a horrific event that pulled together ancient religious hatreds, modern social alienation, and technologies of both destruction and healing in a postmodern collage of superhuman skill, endurance, and survival.

These signals from the complex real world suggest that, as scholars and citizens, we need to resurrect all of the tried-and-true discourses of humanism if we want to figure out what kinds of humans we should be. Relevant theoretical discourses exist across many academic disciplines, along with their counterdiscourses, which are also worth noting. Anthropology more than most fields investigates collective imaginaries and their role in shaping human subjectivity (Anderson, 1983; Appadurai, 1990). Law offers the frameworks of human rights and the rule of law, along with critiques of those ideas when they become tools of governance rather than simply aspirational goals (Kennedy, 2009). In sociology, Pierre Bourdieu (1990) and his followers developed the notion of the *habitus*, structured economies of worth and exchange that people construct wherever there is such a thing as a society (Boltanski and Thévenot, 2006). Environmentalism in the late 20th century offered the ambiguous yet expansive concept of sustainability, which links together the fate of humans and the planet into a single common future (WCED, 1987). And recent theorizing in STS has grappled with questions at the heart of knowledge-power formations, with work on coproduction (Jasanoff, 2004), institutional reflexivity, sociotechnical imaginaries (Jasanoff and Kim, 2015), and bioconstitutionalism as examples.

To end on a comparative note, if the posthuman refers to a redemptive future for all of humanity, one may well ask who the "we" is that is entitled to imagine the unfolding of posthumanism. The German expression *andere Länder, andere Sitten* (other countries, other customs) calls attention to the cultural multiplicity of the human collective. The saying is sometimes uttered ironically, with an arched eyebrow of tolerant amusement. But comparison is also a technology of humility that invites one to consider the constructed foundations of one's own seemingly natural and rational beliefs (Jasanoff, 2003). Many normative questions that arise in connection with posthumanism—which boundaries are fundamental, which can be transgressed without fear—yield different answers when one locates oneself in different traditions of thought. The human-animal distinction, for example, carries one set of resonances in a religious context that sees the human as created in God's image and an altogether different set in a culture of more than a billion people in which a trans-species, elephant-headed deity embodies the divine stewardship of human well-being. Modernity's vaunted separation of the secular from the religious is likewise not a universal given. Indian notions of law, for example, call to mind an image, the top of an Ashokan pillar, that appears on the nation's coinage as a symbol in everyday exchange. Those pillars inscribed with the laws of the state were erected by the emperor Ashoka in the third century BCE after his conversion to Buddhism. But they were surmounted by the lion of power and wheel of *dharma*, the Sanskrit word for religious righteousness. Secular and religious were bound together in a way that is, at the very least, different from the norm in Western Enlightenment societies. All this suggests that, as we contemplate the prescriptive aspects of posthumanism, we need to say which human, located where in time and space, what attributes of that human need enhancement, and how to respect the different localizations of the varied, blemished humanisms we may wish to transcend.

Even in our technologically interconnected, increasingly globalized world, human beings still come from societies in which imagination and understanding are occupied by different nature-culture settlements and by different histories of origins, purposes, and the general good. In asking how the assuredly remarkable technological instruments we have at our disposal can lead us to new forms of being posthuman, it would be foolhardy not to take account of the different places we come from, each with its own ways of reflecting on what it means to be human and how to achieve the good during our fallible, finite occupation of this earth.

References

Anderson, B. (1983). *Imagined communities: Reflections on the origin and spread of nationalism.* London, England: Verso.

Appadurai, A. (1990). Disjuncture and difference in the global cultural economy. *Theory, Culture & Society, 7* (2), 295–310.

Austen, I. (2013, July 8). Deadly derailment in Quebec underlines oil debate. *New York Times*, p. A4.

Bauman, Z. (1991). *Modernity and ambivalence.* Ithaca, NY: Cornell University Press.

Boltanski, L., & Thévenot, L. (2006). *On justification: Economies of worth.* Princeton, NJ: Princeton University Press.

Bourdieu, P. (1990). *The logic of practice.* (R. Nice, Trans.). Palo Alto, CA: Stanford University Press.

Bush, V. (1945). *Science: The endless frontier.* Washington, DC: Government Printing Office.

Carlson, N. (2014, December 21). What happened when Marissa Mayer tried to be Steve Jobs. *New York Times*, p. MM22.

Dao, J. (2013, July 8). Taking on rules to ease Sikhs' path to the army. *New York Times*, p. A9.

Ellul, J. (1964 [1954]). *The technological society.* New York: Alfred A. Knopf.
Ezrahi, Y. (1990). *The descent of Icarus.* Cambridge, MA: Harvard University Press.
Foucault, M. (1979). *Discipline and punish.* New York: Vintage.
Gies, E. (2013, July 8). In Europe, greener transit on existing infrastructure. *New York Times*, p. B4.
Gilbreth, F. B., Jr., & Carey, E. G. (1948). *Cheaper by the dozen.* Boston, MA: Thomas Y. Crowell.
Goel, V. (2013, July 8). A new tool aims to help Facebook users dig deep. *New York Times*, p. B1.
Habermas, J. (1970). *Toward a rational society: Student protest, science, and politics.* (Jeremy J. Shapiro, Trans.) Boston, MA: Beacon.
Hecht, G. (2012). *Being nuclear: Africans and the global uranium trade.* Cambridge, MA: MIT Press.
Heidegger, M. (2008 [1954]). The question concerning technology. In D. F. Krell (Ed.), *Basic Writings* by Martin Heidegger (pp. 307–341). New York: Harper Perennial Modern Thought.
Hilgartner, S. (2015). Capturing the imaginary: Vanguards, visions, and the synthetic biology revolution. In S. Hilgartner, C. Miller, & R. Hagendijk (Eds.), *Science and democracy: Knowledge as wealth and power in the biosciences and beyond.* Abingdon, England: Routledge.
Jasanoff, S. (2003). Technologies of humility: Citizen participation in governing science. *Minerva, 41* (3), 223–244.
Jasanoff, S. (Ed.). (2004). *States of knowledge: The co-production of science and social order.* London, England: Routledge.
Jasanoff, S. (2014). Vulnerability and development: Bhopal's lasting legacy. In A. Hommels, J. Mesman, & W. E. Bijker (Eds.), *Vulnerability in technological cultures: New directions in research and governance* (pp. 89–108). Cambridge, MA: MIT Press.
Jasanoff, S., & Kim, S.-H. (2009). Containing the atom: Sociotechnical imaginaries and nuclear regulation in the U.S. and South Korea. *Minerva, 47* (2), 119–146.
Jasanoff, S., & Kim, S.-H. (2013). Sociotechnical imaginaries and national energy policies. *Science as Culture, 22* (2),189–196.
Jasanoff, S.,& Kim, S.-H. (Eds.) (2015). *Dreamscapes of modernity: Sociotechnical imaginaries and the fabrication of power.* Chicago, IL: University of Chicago Press.
Jones, C. F. (2014). *Routes of power: Energy and modern America.* Cambridge, MA: Harvard University Press.
Kennedy, D. (2009). *The rights of spring: A memoir of innocence abroad.* Princeton, NJ: Princeton University Press.
Latour, B. (1988). *The pasteurization of France.* (A. Sheridan & J. Law, Trans.) Cambridge, MA: Harvard University Press.
Latour, Bruno. (1993). *We have never been modern.* Cambridge, MA: Harvard University Press.
Latour, Bruno. (2012, Winter). Love your monsters: Why we must care for our technologies as we do our children. *The Breakthrough.* Available at http://thebreakthrough.org.
Marcuse, H. (1964). *One-dimensional man.* London, England: Routledge and Kegan Paul.
Nelkin, D., & Pollak, M. (1981). *The atom besieged.* Cambridge, MA: MIT Press.
Rampell, C. (2013, July 8). Coveting not a corner office, but time at home. *New York Times*, p. A1.
Reider, R. (2009). *Dreaming the biosphere: The theater of all possibilities.* Albuquerque, NM: University of New Mexico Press.
Risen, J. (2013, July 8). Privacy group to ask Supreme Court to stop N.S.A.'s phone spying program. *New York Times*, p. A9.
Rohan, T. (2013, July 8). Beyond the finish line. *New York Times*, p. D1.

Sandberg, S., with N. Scovell. (2013). *Lean in: Women, work, and the will to lead.* New York, NY: Knopf.
Schumacher, E. F. (1973). *Small is beautiful: Economics as if people mattered.* New York, NY: Harper and Row.
Scott, J. C. (1998). *Seeing like a state: How certain schemes to improve the human condition have failed.* New Haven, CT: Yale University Press.
Sen, S. (1974). *A richer harvest: New horizons for developing countries.* New Delhi, India: Tata McGraw-Hill.
Strauss, L. L. (1954, September 17). Speech to the National Association of Science Writers, New York City, September 16. *New York Times*, p. 5.
Weber, M. (2004 [1918]). The vocation of science. In S. Whimster (Ed.), *The Essential Weber: A Reader* (pp. 270–287). London, England: Routledge.
Winner, L. (1986). *The whale and the reactor: A search for limits in an age of high technology.* Chicago, IL: University of Chicago Press.
World Commission on Environment and Development (WCED). (1987). *Our common future.* Oxford, England: Oxford University Press.

Heidegger on Techno-Posthumanism

Revolt against Finitude, or Doing What Comes "Naturally"?

Michael E. Zimmerman

Man is God only to the extent that he transcends the naturality and finitude of his spirit and elevates himself to God.
Hegel, Vorlesungen über die Philosophie der Geschichte

"Dead are all gods: now we want the Overman to live…" Thus spoke Zarathustra."
Nietzsche, Thus Spoke Zarathustra

Only a god can save us now.
Heidegger, interview with Der Spiegel

"So, does God exist? Well, I would say, not yet."
Kurzweil, Transcendent Man

Transcendent Man, a Film

The film begins with a voice-over in which a man describes a disturbing recurrent dream. He is alone in a house with many empty rooms, suggested by blurry light and dark images. This is what it means to be dead, concludes the speaker. This realization, so he tells us, gives rise to such a "profoundly sad, lonely feeling, that I can't bear it, so I go back to thinking about how I'm not going to die." Advancing technology, so the speaker hopes, will let him live far longer than ever before possible, possibly long enough to upload his consciousness into an advanced computer that will make him virtually immortal. By midcentury, he tells us, humans will merge with artificial intelligence that will be billions of times smarter than ordinary humans. A crucial outcome of this event, which he and some others call the Singularity, will be making the entire universe—including all its dumb matter—conscious, perhaps billions of years from now. Later in the film, the same man enters a warehouse room packed with his late father's memorabilia, music scores, recordings, diaries, financial records, photographs, and many other personal items. Using this collection, the man intends to recreate a version of his father when it becomes technically possible to do so.

The film under discussion is *Transcendent Man: The Life and Ideas of Ray Kurzweil* (Ptolemy, 2009).[1] Here and in Kurzweil's bestselling books, such as *The Singularity Is Near* (2006), he envisions for humankind and its artificial progeny powers and possibilities that were traditionally accorded only to a deity: virtual immortality, omniscience, mastery over nature, capacity to infuse everything with profoundly interconnected intelligence, and even the power to bring the dead back to life.[2]

Kurzweil's controversial views have become influential in part because he is such a brilliant inventor, targeting his products for that point in the future when they will not only be wanted and needed but also supported by the technological infrastructure. According to Kurzweil, science and technology are turning out findings—especially in the areas of artificial intelligence (AI), robotics, nanotechnology, and robotics—at an exponential rate, which he calls "the law of accelerating returns" (Kurzweil, 2006, p. 3). If his projections are right, and his track record is impressive, these discoveries and their applications will dramatically transform humankind and human culture. In coming decades, humans will become ever more proficient cyborgs, retrofitting and redesigning themselves with a cascade of scientific, medical, and technological breakthroughs. In December 2012, Kurzweil became a director of engineering at Google, which has the computing power needed to advance his major ambition: to create AI, which will help to make human immortality possible.

Enhancing the human is the central goal of *transhumanism* or transitional humanism. In the not-too-distant future, so we are told, transhumans will merge with super AI, which may evoke from ordinary humans the awe formerly associated with encountering the gods. Developing super AI is the goal of *techno-posthumanism*, which is promoted by Kurzweil and a growing number of others. This form of

---

1 The final epigraph above comes from the final scene in *Transcendent Man* (Ptolemy, 2009).

2 Perhaps the most visible transhumanist promoting life extension is Aubrey de Grey, cofounder and chief science officer of Sens Research Foundation, whose slogan is "Reimagining Aging," http://www.sens.org/about/leadership/executive-team. See de Grey, 2008.

posthumanism points to something very different from the philosophical-literary posthumanism developed by thinkers such as Jacques Derrida and Michel Foucault.[3]

In recent years, I have regularly taught a course focused on philosophy and science fiction to explore some of the issues taken up in this essay. Many contemporary philosophers have engaged in discussing films such as *The Matrix,* not only because such films push to the limit issues about subjectivity, morality, and the status of "reality" but also because they make visible a range of possible future worlds, both utopian and dystopian. Technoscience is an exceptionally powerful way of comprehending and deploying the forces of nature. While science fiction has been influenced by philosophical disputes about different genealogies of modernity and about different meanings of the "posthuman," science fiction also invites contemporary philosophers, scientists, and engineers to engage far more seriously than before the issues raised by a dizzying variety of futures, for the posthuman as well as for the human (see Geraci, 2010).

Critics dismiss transhumanism as merely offering a technologically transformed and positive version of the eugenics employed by Nazi Germany, whereas a number of engineers and scientists sympathetically engage transhumanism's vision of the coming posthuman age.[4] Despite legitimate concerns, the creation of advanced cyborgs and super AI may be inevitable. Surely, then, it is in our interest not only to locate, analyze, and evaluate the transformative ideas that dream such beings into existence but also to inform ourselves about the possible behavior and intentions of such beings. The possibility that super AI poses an existential risk to humanity's future can no longer be consigned to the realm of science fiction, but instead must now be taken seriously[5] (see Hawking et al., 2014, April 19).

---

3 Curiously, however, as Jean-Pierre Dupuy (2009, 2011) has argued, at the same moment when posthumanism was eliminating human subjectivity and replacing it with the structuring activity of language and social organization, proponents of cybernetics (which made techno-posthumanism possible) were also eliminating subjectivity and replacing it with the universal activity of computation. Cybernetics redefined the human being as a kind of complex computer and purged from "meaning" all traces of subjectivity. Whereas Heidegger conceived of cybernetics as the pinnacle of humanism, now understood as in the service of the "will to power," much of 20th-century French philosophy used cybernetics and its social science analogues to undermine humanism, ending up with the notion of subjectless cognition.

4 There is a significant literature on trans- and techno-posthumanism. Critiques include Fukuyama, 2003; Joy, 2000; Habermas, 2003; and Sandel, 2009. Optimistic assessments include Drexler, 2013; Naam, 2005; and Stock, 2002.

5 The Machine Intelligence Research Institute (MIRI) has been formed to examine and to bring to public awareness important aspects of AI, including its potentially destructive implications for the human future. See http://intelligence.org.

Kurzweil claims that super AI will emerge in the year 2045. Even if that deadline is not met, rapid technical progress will continue unless the global economy melts down or unless humankind is annihilated by a cosmic accident, including a collision with a large-enough meteor. In the meantime, techno-posthumanists plan to invent super AI that will escape from planet Earth altogether and seek its destiny in the stars. In this manner, self-conscious beings—which may be rare even in our vast universe—will survive the possible destruction of the biosphere. Humans will become or create the masters and possessors of nature, so techno-posthumanists aver, even though the attempt to do so may pose significant risks for humankind.

A central motivation for transhumanism and techno-posthumanism is to avoid death. While the ancients sought the mythical tree of life, early modern alchemists hoped to attain the elixir of life by transmutation of the elements and corresponding transformation of the soul. Many transhumanists, including Kurzweil and Aubrey de Grey, believe that this dream can finally be realized, given rapid technoscientific innovation (De Grey, 2008). As we drill down to the very substructures of matter, energy, and life, the prospect grows that humans can (a) construct life, (b) extend life, perhaps indefinitely; and (c) create a new, noncarbon-based form of life, endowed with AI that will bring about the *Singularity.* This term names that rupture in human and terrestrial history that would occur when and if AI becomes self-conscious and autonomous. At that moment, so we hear, AI will begin to redesign itself so such that, within a relatively short time, AI will be *billions of times* more intelligent than human beings. Something of the human will somehow be included in such AI, but ultimately it will leave us far behind. At least, so goes the optimistic story of many techno-posthumanists (Kurzweil, 2006).

The prospect of realizing such visions motivates many high-tech gurus. Some of the profits from our slick digital gadgets finance various aspects of research on AI. In 2012, Google named as a director of engineering Kurzweil, who has always regarded creating AI as his highest ambition. Other high-tech leaders, such as virtual-reality pioneer Jaron Lanier, explore the possible motives for creating new forms of autonomous and highly intelligent life. In a *New York Times* op-ed (2010) Lanier wrote,

> [C]omputer scientists are human, and are as terrified by the human condition as anyone else. We, the technical elite, seek some way of thinking that gives us an answer to death, for instance. This helps explain the allure of a place like the Singularity University. The influential Silicon Valley institution preaches a story that goes like this: one day in the not-so-distant future, the Internet will suddenly coalesce into a super-intelligent A.I., infinitely smarter than any of us individually and all of us combined; it will become alive in the blink of an eye, and take over the world before humans even realize what's happening. ….

> Yes, this sounds like many different science fiction movies. Yes, it sounds nutty when stated so bluntly. *But these are ideas with tremendous currency in Silicon Valley; these are guiding principles, not just amusements, for many of the most influential technologists.* (Lanier, 2010, my emphasis)

Heidegger on Techno-Posthumanism

How can Heidegger's thought shed light on the astonishing vision proposed by techno-posthumanism? Insofar as humankind is increasingly understood from the vantage point of technoscience—that is, as highly complex matter-energy that can at first be emulated and then dramatically enhanced—Heidegger would say that techno-posthumanism is the latest and perhaps most dangerous phase in the era of techno-industrial nihilism. In this era, all beings—including human beings—reveal themselves primarily as raw material for the purpose of enhancing power *for its own sake*, not for the sake of some identifiable *human* end. Although techno-posthumanists would disagree with Heidegger's view of modern technology (*Technik*) as the culmination of Western nihilism, they would agree with his supposition that super AI would continually enhance itself, becoming ever-more powerful, perhaps as an end in itself. Super AI would be, in effect, the ultimate ontical embodiment of what Heidegger—drawing on Nietzsche—calls the Will to Will.[6]

From Heidegger's viewpoint, techno-posthumanism is a new chapter in modernity's revolt against finitude, its concomitant desire to become God-like, and its effort to make the rationality of the "rational animal" the very ground of beings. Heidegger maintains, however, that what is essential to humankind is not rationality but rather "the clearing" (*die Lichtung*) that allows beings to manifest themselves—and thus "to be"—in their intelligibility. Arguably, the clearing is Heidegger's main philosophical topic (Sheehan, 2014). Synonyms for the "clearing" include "the world" (*die Welt*), "the nothing" (*das Nichts*), and "the appropriation" (*das Ereignis*). Although nowhere offering an account of how this clearing occurred in the first place—indeed, it would seem impossible to do so—Heidegger's attitude

---

6 Heidegger's term *Technik* is perhaps best translated as "technique" or "technics," rather than as "technology," although the latter term is usually used in translation. *Technik* refers to the know-how that lets us make and produce ever-more complex things in many different domains. Technology, on the other hand, refers to our study or investigation of *Technik*, just as biology refers to our investigation of life (*bios*) and ecology refers to our study of earth's household (*oikos*). Heidegger's study of the origins of nature of modern *Technik* constitutes an important instance of technology, in the way just defined.

toward it is one of profound reverence and appreciation. When he talks about techno-industrial nihilism that culminates the history of metaphysics, he means that modern humankind is oblivious to the nothing, *das Nichts*, or the clearing. Thus oblivious, humankind fails to appreciate its radical dependence on the clearing, which allows access to the Being of beings and, thus, to our own mode of Being.

According to Heidegger, humankind is "thrown" into this clearing, appropriated by it, as the site needed for humans to encounter beings *as* beings, that is, as things that *are*. The clearing is not a possession of humankind; instead, humankind exists in its service. In his famous "Letter on Humanism," published shortly after the end of World War II, Heidegger contrasted his view with humanism, which has evolved to the point that humans regard themselves as potential masters over the whole of beings:

> The essence of man, however, consists in his being more than merely human, if this is represented s "being a rational creature." More must not be understood here additively, as if the traditional definition of man were indeed to remain basic, only elaborated.... The "more" means: more originally and therefore more essentially in terms of his essence. But here something enigmatic manifests itself: man is in thrownness. This means that man, as the ek-sisting counter-throw of Being, is more than *animal rationale* precisely to the extent that he is less bound up with man conceived from subjectivity. Man is not the lord of beings. Man is the shepherd of Being.... Man is the neighbor of Being. (Heidegger, 1977a, 221 [Heidegger, 1967b, 172])

The clearing makes possible human finitude, mortality, and receptivity, thereby allowing us to be affected in ways that lets things *matter* to us. The human mode of Being is *care* (*Sorge*), without which sheer rationality is without meaning and purpose. In Heidegger's view, power-seeking and death-denying modernity threatens to close off the clearing, which has already become so constricted that today beings—including human beings—can show up primarily as flexible raw material. Like anything else, humans can now be ordered and exploited by cybernetics, which seems to be late-Heidegger's operational term for "enframing" (*Gestell*), a key term for his view of modern technology (see Heidegger, 1977b [Heidegger, 1954a]).

Human *Dasein*, according to Heidegger, is doubly finite. Not only are we dependent on beings (food, clothing, shelter) for survival, but also we are dependent for our very mode of being on the clearing that owns *us*. Human *Dasein* exists as—or perhaps better—*within* the temporal-historical clearing needed for the self-showing, that is, the Being of beings (*Sein des Seienden*) to occur. (In what follows, I capitalize "Being" to distinguish it from "beings.") The clearing, instantiated through humankind, allows us to encounter beings—through perception, understanding, and utilization—in a host of different ways. Human *praxis* plays a crucial role in disclosing things, but such disclosive activity always takes place *within* the clearing into which (and as which) humans have always already been thrown, and thus over which they exercise no control. Beings depend on the humanized clearing as the site in which to *reveal themselves* and also *to be revealed* or *brought out of concealment* by human language and *techne* (technical know-how). No subjective idealist, Heidegger insisted that natural beings manifest themselves as *always already there*, prior to our encountering, perceiving, and interpreting them. We do not *create* the natural phenomena that encounter us but depend on them for our continued existence.

What Heidegger might have called the *ontological* Singularity refers to the astonishing fact that beings appear *as* beings, that there is something rather than nothing, and that this something manifests within the clearing constituting human existence. The thinker and the poet may notice this extraordinary event, but most people never do.[7] In any possible experience, what show up are beings in their Being, that is, their self-manifesting. Our capacity for understanding the Being of beings allows us access to them. What allows for the self-showing is named the clearing, but this always goes unnoticed. Metaphysics studies the Being of beings, but not *das Nichts* that allows for beings to show up and thus "to be." Overlooking the clearing is not any human failing; instead, Heidegger says, the clearing *conceals itself.* According to Heidegger, this self-concealment—which initiated the 2,500-year-long metaphysical tradition—culminates in techno-industrial nihilism.

Plato initiated the metaphysical tradition by defining Being as *eidos*, the permanently present (eternal) form that reason alone can discern. Missing was any reference to the clearing within which such presencing (*eidos*) could occur. For metaphysics, Being names the *foundation and origin* of beings, that which allows them to endure, that which forms and structures them so that they persist. Gradually, humankind increasingly came to define itself as the rational animal and conceived of rationality as the ground or foundation of beings. Not surprisingly, beings increasingly revealed themselves first as objects for the rational subject and eventually as raw material over which technical mastery could be achieved. The consummation (*Vollendung)* of Western metaphysics is achieved in Nietzsche's doctrine of the Will to Power. "Consummation," so Heidegger writes, "means the *unimpeded development* of all the essential powers of beings, powers that have been reserved for a long time, to what they demand as a whole" (Heidegger, 1987, 7 [Heidegger, 1989a, 7], my emphasis.) Properly understood, according to Heidegger, the essence of Will is always to transcend its current stage of power in order to attain a higher stage. Power means the drive to attain something more, as in nonstop

7 Although the relation between Heidegger's *das Nichts* and Buddhism's *sunyata* (emptiness) is intriguing, it is also fraught with uncertainty.

progress, the ever-more elaborate techno-industrial ordering and organization of beings. Rightly understood, the Will to Power means the Will to Will.

Although humankind is not in control of the self-concealment of the clearing on which human beings depend, Heidegger sometimes talks as if humankind revolts against finite receptivity (the clearing) by asserting that the rational human animal is the ground (Being) of beings and, thus, the master and the possessor of nature. All this sounds to Heidegger like ontological blasphemy. While modern metaphysical humanism promises and increasingly delivers material well-being for many, humanism arises from the self-concealment of the clearing that allows us to encounter beings *as* beings in the first place. The quest for infinite power, the Will to Will, increasingly forecloses the finite receptivity that makes us specifically human. Indeed, in the technological era, humankind itself is no longer a rational subject standing over a totality of objects, but instead has become *the most important raw material*, useful for enhancing the endless growth of power for its own sake. Hence, the era of techno-industrial nihilism leads not only to world wars and nuclear weapons but also to the exponential growth of AI and related high-tech fields. Heidegger suggests that several centuries will probably unfold in accordance with "machination," the reducing of all beings to fungible raw material, as demanded by what Nietzsche called the Will to Power.

The metaphysics of the Will to Power is discernible in Kurzweil's prediction that the first thing an artificially intelligent computer will do is to redesign itself, so that it can become far more intelligent than *all human beings collectively*. If Heidegger is right, humankind—a crucial factor in the techno-industrial system—constitutes both the current preservation condition and also the enhancement condition needed to bring forth the next stage of the Will to Power: techno-posthumans. After the Singularity, AI itself will become the preservation condition needed for producing the *next* and more powerful stage of AI, that is, super AI. Everything becomes a means to another end, which, in turn, becomes yet another means. There will be only the bad infinity of endless striving, unless super AI develops ends of a sort that we cannot yet fathom—unless AI remains finite in a way that allows things to *matter* to it. I will have more to say about this later.

According to Heidegger, even though humans may think themselves to be in charge of technoscience, in fact, we are *servants* of the technological juggernaut. Technology is no longer a means to human ends, but rather an end in itself. In 1968, he wrote:

> [T]he conspicuous successes of the inexorable development of technicity continue to give the appearance that the human being is the master of technicity. In truth, however, he is the servant of the power that thoroughly dominates all technical production. The power of the challenging placing [*das Gestell*] shapes the human being

> into the mortal who is claimed, placed, and in this sense, used by this power and for it. The prevailing power in the presence of what-is-present needs the human being. (Heidegger, 2010, 218 [Heidegger, 2000, 627])

Heidegger holds out the possibility that technological nihilism might be overcome by "another beginning," given that the *current* disclosure of beings as raw material brings to a culmination the first beginning initiated by the ancient Greeks. Thomas Sheehan forcefully contends, however, that another beginning is ruled out by Heidegger's own conception of the interplay of being (presencing) and clearing (absencing) that allows human access to beings in the first place. For an entity "to be" means for it to reveal itself in its intelligibility within the clearing opened up through human beings. In apprehending this intelligibility, human *Dasein lets beings be*. If Sheehan is right, human *Dasein* is inevitably fascinated with and drawn toward beings. We are "fallen" into the world of things.[8] In seeking mastery over them, we are just *doing what comes naturally*, as it were. Indeed, beings seemingly invite us to investigate, appreciate, and also exploit them for purposes of our own. Even perception (*Wahr-nehmung*) is form of *grasping* that arises and cooperates with the showing up (Being) of beings. Grasping is a way of bringing to a stand the overwhelming surge of beings. Bringing beings to a stand, by understanding, perceiving, and manipulating them, allows beings to be what they are. Heidegger writes,

> [B]eings as such open themselves only to such a perceiving. This is what Parmenides' saying means: *To gar auto noein estin te kai einai*. "Perceiving and Being are the same." To be the same means to belong together in essence; beings are not in being *as beings*, that is, as present, without perceiving. But neither can perceiving take hold where there are no beings, where Being does not have the possibility of coming into the open. (Heidegger, 1987, 48 [Heidegger, 1989a, 93–94])[9]

The techno-industrial era's "industrial ontology" (Wilber, 1995), to which corresponds what Herbert Marcuse called "one-dimensional man" (Marcuse, 1964), is a highly efficient but also a highly restricted ways of "perceiving" beings.[10] If

---

8 For an insightful account of Heidegger's philosophical appropriation of Luther's concept of the fall, see McGrath, 2005.

9 I was reminded of this passage by Suhail Malik in his commendable essay "Nihilism and Life: Cosmobiology and Ontopoiesis in Heidegger's *Nietzsche*." Suhail, 2000, 91.

10 The term "one-dimensional" is from Herbert Marcuse's influential work *One-Dimensional Man* (1964). Marcuse wrote his dissertation under Heidegger's direction. The term "industrial ontology" is from Ken Wilber, *A Brief History of Everything* (1995). Wilber's work constitutes a major developmental counternarrative to Heidegger's account of Western history.

techno-posthumanists assume that we are making strides toward creating God-like beings, Heidegger would caution that these "strides" are, in fact, *making us* into what Michel Foucault called the biopower needed to carry out the destiny of the techno-industrial mode of disclosing beings. In coming decades, so Heidegger surmised, the Will to Power will allow and even demand that humans generate what today is depicted as autonomous, super AI. As Richard Polt has made clear, Heidegger's late reflections on cybernetics were prescient. He sensed the "draw" (*Zug*) of the hypertechnological future, even if details about it were lacking. Ever more incredible innovations, arising at an ever-increasing tempo, will make possible undreamt of powers. We may well create God-like beings, and we will seem like gods to ourselves, insofar as we created such beings. Despite the quote at the beginning of this essay, however, even Kurzweil agrees that humans cannot become what the Bible understands by "God."

Preparing the Humans Needed to Take Command of the Planet

Heidegger's view of modern technology was profoundly influenced in the early 1930s by his encounter with Ernst Jünger's writings, which were informed by an innovative (even if philosophically limited) interpretation of Nietzsche (Heidegger, 2004; Zimmerman, 1990). In the 1920s, most notably in *Being and Time* (*Sein und Zeit*), Heidegger indicated that beings primarily reveal themselves as ready-to-hand (*zuhanden*), that is, as always already involved with human productive activity undertaken with tools, whether in the workshop or in the field. The south wind, for example, reveals itself as oncoming rain needed for crops. Already, some say, Heidegger viewed nature primarily in terms of its use-value for humankind. After reading Jünger, Heidegger would conclude that modern technology discloses everything—including humankind itself—as raw material for enhancing power.

Informed by his heralded front-experience in World War I and by his study of Nietzsche's work, Jünger argued that technologically enhanced warfare heralded the rise of a new human type (*Typus*), stamped by the *Gestalt* of the worker-soldier. This *Gestalt* is the latest historical configuration of what Nietzsche called the Will to Power (Jünger, 2013 [1932]). Jünger's riveting literary accounts of the melding of cold steel and hot flesh, that is, the technological transformation of the human organism, struck Heidegger like a thunderbolt. Much of his subsequent philosophical research involved interpreting the essence of modern technology as the disclosure of all beings as resources for the Will to Will. Heidegger writes,

> The basic form of appearance in which the Will to Will arranges and calculates itself in the unhistorical element of the world of completed metaphysics can be stringently called "technology." This name includes all the areas of beings that equip the whole of beings: objectified nature, the business of culture, manufactured politics, and the gloss of ideals overlying everything. Thus "technology" does not signify here the separate areas of the production and equipment of machines. (Heidegger, 1972, 93 [Heidegger, 1954b, 72])

The technological-industrial, cybernetic era results from what Heidegger calls "the self-release of being into machination. This release takes man into unconditional service. It is by no means a decline and something 'negative' in any kind of sense" (Heidegger, 1972, 103 [Heidegger, 1954b, 82–83]). In many other places, however, Heidegger speaks negatively about what has become of humankind in the technological era. For instance, in *Contributions to Philosophy*, written in the mid-1930s, he says that modern humans have become "hexed" and "bewitched" by beings:

> We are used to calling the era of "civilization" the one that has *dispelled* all bewitchery, and this dispelling seems more probable—indeed, uniquely—connected to complete unquestionableness. *Yet ... [it] is just the reverse.* We merely need to know where the bewitchery comes from, namely, from the unbridled dominance of machination. When machination attains ultimate dominance, when it pervades everything, then there are no more circumstances whereby the bewitchery can be sensed explicitly and resisted. The hex cast by technology and by its constantly self-surpassing progress is only *one* sign of this bewitchery that directs everything toward calculation, utility, breeding, manageability, and regulation. (Heidegger, 2012, 98 [Heidegger, 1989b, 124])

In other words, what Max Weber famously called the "disenchantment" of the world is merely a transition to a mode of bewitchment far more comprehensive than any sorcerer or alchemist could have imagined. Heidegger's mention of bewitchment may be a nod to the role played by the *magus* not only in esoteric traditions like hermeticism but also in the rise of modern science and in the work of philosophers like Hegel, and by extension Marx. As Nietzsche once put it,

> *Preludes of science*—Do you really believe that the sciences would ever have originated and grown if the way had not been prepared by magicians, alchemists, astrologers, and witches whose promises and pretensions first had to create a thirst, a hunger, a taste for *hidden* and *forbidden* powers?[11] (Nietzsche, 1974, §300, 240)

---

11 There is a considerable literature on the role played by esoteric traditions in the origin and the development of modernity. See Fleming, 2013; Magee, 2008; Monod, 2013; and Zimmerman 2009.

Heidegger anticipated that Nietzsche's discourse would figure prominently in transhumanism. Transhumanists often cite Zarathustra's provocative claim that the human is a mere bridge between the ape and the Overman. Zarathustra's discourse offers intellectual panache to the ambitious projects of transhumanists, many of whom are atheistic humanists (Zimmerman, 2011). Although only infrequently referring to Zarathustra and ignoring altogether the fact that Zarathustra emphasized the importance of "dying at the right time" rather than yearning for immortality, Kurzweil takes seriously Zarathustra's pronouncement about the death of God. Comprehending the importance of *a new highest goal* for humankind, Kurzweil proposes that humankind aim to generate God-like progeny, the emergence of which will anoint us retrospectively as creators akin to the biblical God (see Zimmerman, 2008). The Singularity parallels in some ways, without thereby being identical with, what Christians regard as the Millennium (Geraci, 2010).

Interpreting Nietzsche's thought as ushering in techno-industrial nihilism, Heidegger describes the Overman as *what humankind must become* in order to serve machination, the techno-industrial mode of Being. The Overman, then, is not a great individual, but rather the transformed humankind necessary to *take command of the planet*:

> [N]ot just any kind of humanity is suited to bring about unconditional nihilism in a historical manner. Hence a struggle is even necessary about the decision as to which kind of humanity is capable of the unconditional completion of nihilism. (Heidegger, 1972, 103 [Heidegger, 1954b, 83])

During World War II, in an era that certainly seemed to be governed by the *Gestalt* of the worker, political regimes arose to enhance industrial output in the mode of "total mobilization." Hitler, Stalin, and (Heidegger would add) Roosevelt were fearless leaders evoked by the Will to Power. Infamous examples of efforts to generate the new kind of humanity needed for dominion over the earth are found in the "new Socialist Man" and the "new National Socialist Man," versions of what Jünger had in mind by "the worker" (*der Arbeiter*) (Rosenthal, 2002; Cheng, 2008). During World War II, Heidegger wrote: "Since man is the most important raw material, one can reckon with the fact that some day factories will be built for the artificial breeding of human material, based on present-day chemical research" (Heidegger, 1972, 106 [Heidegger, 1954b, 87]). Despite its devastation, World War II was only the prelude to ever-more titanic efforts to control and exploit beings, including human beings.

In the postwar era, Heidegger concluded that what Norbert Wiener called cybernetics was the step that would lead beyond the industrial era to what we now call the "age of information," which will further the endless process of planetary (and perhaps cosmic) exploitation. Humans are already subjected to the same kinds of control that cybernetics imposes on all other things and systems. The rise of cybernetics indicates, however, that the human era is coming to an end. Transhumanism aims to enhance the human organism, but eventually even enhanced transitional humans will be eclipsed by super AI, which will be far more efficient and ever-more powerful.

The enormous investment in weapons research during the Cold War continues in the United States today, under the leadership of organizations such as the Defense Advanced Research Projects Agency (DARPA). DARPA recently announced that it "has created a division that merges biology, engineering, and computer science to advance technologies for national security" (Malykhina, 2014, April 2). Working closely with private corporations in the fields of biotechnology, nanotechnology, robotics, and AI, DARPA recognizes that the organic human platform is far too limited (finite) to take advantage of the military power possibilities now coming into view.

Naturalizing Heidegger: Could the Clearing Ever Occur Again, This Time in Super AI?

So far, we have seen that Heidegger would regard the rise of super AI as the latest phase in techno-industrial nihilism, which arises due to the self-concealment of the clearing that allows humans to encounter beings. In what follows, I pose and offer a tentative answer to this question: What if super AI itself, rather than signifying the closing off of the original clearing, were associated with the emergence of a new clearing, one consistent with its own possibilities? Before turning to this question, we need to discuss how people like Kurzweil understand AI.

First, Kurzweil does not employ the vocabulary of "clearing" or "openness for Being." Instead, he speaks first of intelligence and second of consciousness. In some sense, even our smart phones are already "intelligent," in that they can exhibit computational activity that far exceeds what any individual person can do. By AI, however, most people—including Kurzweil—means not merely computational intelligence, but *self-conscious* intelligence. No one knows whether AI, if it ever occurs, would be conscious in any way that we recognize. Kurzweil, who admits that consciousness remains unexplained, posits that it is correlated with the enormously complex, hierarchical structure of the brain.[12] According to him, humans

12 Kurzweil (2006) offers a sophisticated discussion of consciousness and problems in defining it. See also Kurzweil (2012).

can understand what and how beings are because (a) aspects of them are accessible to human sensation and perception and (b) rational analysis can interpret and utilize beings that are accessible in that way. There is no discourse here about any sort of "clearing." Nevertheless, we may ask whether there could (unexpectedly) arise within AI a version of openness commensurate with AI.

In lectures from 1929 to 1930, Heidegger acknowledged that animals live within their own worlds, which grant very limited access to beings. Humankind, however, exists in a richer if still finite world, one that allows much greater access to the Being of beings (Heidegger,1995 [Heidegger, 1983]). Although at one time exploring the proximity between the human and the animal, Heidegger later claimed that an *abyss* separates the human from the animal (Heidegger, 1977a, 206 [Heidegger, 1967a, 157]). One motive for this shift was his conclusion that Aristotle's conception of the human as the rational animal eventually led to Nietzsche's metaphysics of the Will to Power. Anticipating what would become "social Darwinism," Nietzsche spoke approvingly of improving humankind by breeding. Another motive was that conceiving of humans as clever animals forecloses appreciation of their finite openness granted by the clearing.

How did the clearing arise in the first place? Bernard Stiegler's three-volume work *Technics and Time* attempts to answer this question, by arguing that humankind and technics are always already intertwined (Stiegler, 1998, 2009, 2010). Reversing Heidegger's interpretation, Stiegler maintains that *technics*, not temporality, forms the horizon (thus, the clearing) of human possibility. Indeed, because technics creates temporality, historical *transcendence* always depends on *empirical* (technical) circumstances. Stiegler recalls Friedrich Engels's insistence that the opposable thumb allowed humans to manipulate their environment in a way crucial for the formation of intelligence. Unlike Heidegger, who discounted the possibility that sciences such as archeology and anthropology could provide insight into human origins—that is, the origins of the clearing—Stiegler relies on scientific research into how tool use and the human arose together. Even he concedes, however, that this arising together constitutes an aporia. Did the human bring about tools, or did the tools bring about the human? Notably, as have other thinkers in the past when confronted with such an aporia, Stiegler appeals to a myth, that of Epimetheus, to help shed light on the technics–human relationship.

Heidegger resisted exploring archeological/anthropological contributions to understanding human origins in part because he lived during an era of positivism, behaviorism, and eliminative materialism, none of which were friendly to discourse about subjectivity and consciousness, much less to talk about the clearing. Since then, however, three important conceptual developments have occurred. The first two are Big Bang cosmology and quantum physics. The former has made possible cosmic grand narratives. One such narrative, associated with the Anthropic Principle, states that life (including self-conscious life) could not have evolved without the exquisite "fine-tuning" of crucial natural constants (Barrow and Tipler, 1988). Moreover, many quantum physicists emphasize that consciousness is inextricably involved in the quantum activity constitutive of physical phenomena (Rosenblum and Kuttner, 2011). In view of such conceptual developments, some people argue that the emergence of human consciousness is not only a possible but also perhaps a *necessary* feature of the universe. Humankind is required so that the universe can reveal itself to itself in its various ways (Davies, 2008; McIntosh, 2012; Swimme and Tucker, 2011). Proponents of techno-posthumanism would argue that the emergence of self-conscious life has *cosmic significance* not merely because of what has already occurred (the rise of self-conscious life) but also because of *what is still to come.*

These new, science-derived narratives often dovetail with the *third* above-mentioned development, namely, the reassertion of consciousness as an important aspect of philosophy of mind. It may seem obvious that—in addition to psychology and neurophysiology—philosophy of mind requires both a description (phenomenology) of first-person experience as well as an account of the conditions necessary for the possibility of such experience. Such was not the case until fairly recently, however, when behaviorism's predominance in social science ebbed. An important intervention was David J. Chalmers's 1996 book *The Conscious Mind* (see also Chalmers, 2010). In exploring what he calls "the hard problem" of consciousness, Chalmers argues that consciousness is both a *correlate* of complex material configurations and also a basic cosmic *constituent*, along with space, time, matter, and energy. According to this viewpoint, subjective consciousness is a late development of what begins as the barest of *affectivity*, the capacity of beings in some sense to *take into account* their environment. The capacity for taking-into-account can be so meager that it may go "all the way down," perhaps to the atomic level and even below, as suggested by Alfred North Whitehead's panpsychism. Chalmers proposes that this affective capacity eventually evolved into sensation, then into gradations of consciousness, and finally into human self-consciousness.

If we were willing to interpret primal affectivity as the seed that would eventually flower as human finitude, we could perhaps depict humankind as a remarkable development *within* cosmic evolutionary history, rather than as something *extra*cosmic. Defending this view is the burden of David Storey's book *Naturalizing Heidegger* (2015). Without adopting something like Storey's sophisticated naturalization, as Hans Jonas (1952, 1958) and Susan Taubes (1954) have argued, Heidegger promotes a kind of Gnostic acosmism.

Asserting in *Being and Time* that Being refers to "utter transcendence" (*Transzendenz schlechthin*), Heidegger intends Being to be understood as "wholly other" to beings (Heidegger, 1962, 62 [Heidegger, 1967a, 38]). Gradually, Heidegger made clear that the wholly other refers not so much to Being as to *the clearing*. His discourse about the wholly other is reminiscent of negative theology, according to which one cannot describe the utterly transcendent God, but must end by saying *neti neti*, "Not this, not that," as in the Upanishads.[13] Heidegger was influenced by Karl Barth's arguably neo-Gnostic theology, according to which God so transcends the world that we can say nothing positive about God without depicting God in terms suitable only for creatures. Insofar as humans exist within and perhaps even as the clearing/nothingness required for beings to manifest themselves, the human "essence" transcends the world as well. (The foregoing comparisons with negative theology are not meant to suggest that Heidegger's "clearing" refers to the biblical God, but rather to note certain nonaccidental affinities between Heidegger's ontology and negative theology.)

Despite Heidegger's reservations about evolutionary theory and cosmology, let us explore for a moment the possibility that the finite human clearing (affectivity, receptivity, temporality) was somehow associated with a complex set of empirical conditions that made possible a leap in the hierarchical structure of nature. Another such leap was made by organic life, which can take into account beings pertinent to an organism's survival and development. Could an analogous (even if very different) mode of finite receptivity or clearing somehow arise by virtue of the empirical conditions associated with super AI? Moreover, even if it were to arise, how would we ever *know* that it did? We are talking about a development, of course, that would far exceed anything measurable by the Turing Test.

Nevertheless, suppose that humans somehow invent AI that not only successfully demonstrates its self-consciousness (and, by inference, its finite receptivity) but also is vastly more intelligent and in many ways more capable than humans. How might a techno-posthumanist, who is informed about Heidegger's critique, assess such a development? Following Sheehan, she might say that our metaphysical destiny is to allow self-showing (Being) of beings to occur in ever-more profound and complete ways. Super AI, then, may be both an *inevitable* and *justifiable* development. Even if Heidegger is right that humans are "hexed" by modern technology, we will in any event pursue what such technology makes possible. Using, understanding, and manipulating beings are the only games in town for the vast majority of people, including the highly educated. What Heidegger calls "the self-release of being into machination" calls to mind Christ's *kenosis* (self-emptying), which Hegel interpreted as transcendent God pouring itself into creation as immanent *Geist*.

13 Richard Capobianco (personal communication) suggests that "More than this, more than that!" would better represent Heidegger's view.

Far from indicating a decline or something merely negative, then, the complete self-release of being into the machination of technoscience may, in fact, be the real "destiny of Being."

The technological Singularity associated with super AI may constitute what Nietzsche would call a redeeming work of art that gives a highest meaning to the efforts of countless scientists, engineers, investors, sci-fi writers, philosophers, and visionaries. The growing possibility of the Singularity will elicit ever-greater efforts on the part of those who want to help generate this next evolutionary leap. Such people want to enable such a leap, despite the risks and despite the fact that super AI will leave humankind behind, because super AI is the biggest possibility lying on the table.

Who will be ready, willing, and able to seize this opportunity? Are those currently committed to bringing forth super AI up to the challenge of doing so? Here we may recall Heidegger's contention that humankind must undergo a transformation if it is to correspond to what is required by the Will to Will. Some may regard transhumanism as such a transformation, but that remains to be seen.

A Note about Finitude

Finally, given that this essay began by discussing transhumanist aspirations to overcome mortality, let me offer a (tentative, speculative, and revisable) concluding word about finitude. As already noted, Heidegger once indicated that animals are endowed with limited modes of openness, even if world poor in comparison with humankind. Contemporary thinking invites us to interpret the human clearing as an evolutionary cosmic development, rather than as something that arose from nowhere. If another, posthuman mode of openness were to arise within super AI, however, it would not be because computer scientists would have intended or expected it, given that discourse about temporal openness is not widespread in the world of AI, despite the efforts of thinkers like Hubert Dreyfus (2007). Perhaps taking seriously the notion of affective, finite openness will be necessary, however, if AI researchers are ever to solve *the motivation problem.*

What is this problem? Let's assume that we have managed to produce AI that proceeds to redesign itself to be virtually immortal, enormously intelligent, and self-aware. Despite all these capacities, what would *motivate* such a super AI to do something now that could be put off for a century or even for a few millennia? In the face of millions of years, what's the hurry? Moreover, why would super AI want to do *anything at all* on its own at all, unless super AI were somehow to *matter* to itself? Arguably, for something to matter to Super AI would mean that it must somehow *care for itself* and perhaps for other beings as well. For super AI to be more than an awesome analytic engine, would its mode of Being not have to include something akin to *care*, as Heidegger defined humankind in *Being and Time*? If so, would super AI need to be *finite* and perhaps even *mortal*?

In his essay "The Thing," Heidegger underscored the relationship between mortality and the clearing by saying that "Death is the shrine of Nothing" (Heidegger, 1971, 178–179 [Heidegger, 1954c, 51). By "death," Heidegger means mortality, radical finitude, the capacity for death *as* death. This capacity belongs only to humans. The word "shrine" *(Schrein)* stems from a word meaning cabinet, chest, or reliquary and, in the Middle Ages, was also used to name the Ark (of the Covenant). Hence, mortality sustains and shelters the Nothing, that is, the clearing without which no being can be. To become who we really are, according to Heidegger, means to own up to the fact that we are always already owned by the clearing. To name such ownership, Heidegger uses the term *Ereignis*, sometimes translated "event of appropriation." Such owning up to one's being owned, which is called authenticity (*Eigentlichkeit*) in *Being and Time*, requires accepting one's mortality, finitude, and dependence. According to Heidegger, then, the goal of immortality posited by super AI amounts to forsaking what makes us human.

Finite AI would not prove satisfactory to those who crave immortality. Although Ray Kurzweil wants to transcend mortality, his goal in so doing is not to abandon his capacity to care for himself, for others, and for the whole of beings. Indeed, his stated goal is eschatological: to transform the mute universe into that which is conscious. If Heidegger is right, however, caring cannot be sustained with embracing mortality. Keeping this judgment in mind, I would agree to meld myself with a *non*finite, super AI only if there were a bail-out option available after the fact.

Choosing to bail out, however, would presuppose that I would still *care* about anything.

References

Barrow, J. D., & Tipler, F. J. (1998). *The anthropic cosmological principle*. New York, NY: Oxford University Press.

Chalmers, D. J. (1996). *The conscious mind: In search of a fundamental theory*. London, England: Oxford University Press.

Chalmers, D. J. (2010). *The character of consciousness*. Oxford, England: Oxford University Press.

Cheng, Y. (2008). *Creating the "new man": From Enlightenment ideals to socialist realities.* Honolulu, HI: University of Hawaii Press.

Davies, P. (2008). *The Goldilocks enigma: Why is the universe just right for life?* New York, NY: Mariner Books.

De Grey, A. (2008). *Ending aging: The rejuvenation breakthroughs that could reverse human aging in our lifetime.* New York, NY: St. Martin's Griffin.

Drexler, E. K. (2013). *Radical abundance: How a revolution in nanotechnology will change civilization.* New York, NY: PublicAffairs.

Dreyfus, H. (1988). Why Heideggerian AI failed and why fixing it would require making it more Heideggerian. *Philosophical Psychology, 20* (2), 247–268.

Dreyfus, H. (1992). *What computers still can't do.* New York, NY: MIT Press.

Dupuy, J.-P. (2000). *The mechanization of mind. On the origins of cognitive science* (translated by M. B. DeBevoise). New York, NY: MIT Press.

Dupuy, J. -P. (2011, September 1). H-: Cybernetics is an antihumanism: Advanced technologies and the rebellion against the human condition." *Metanexus.* Available at http://www.metanexus.net/essay/h-cybernetics-antihumanism-advanced-technologies-and-rebellion-against-human-condition.

Flanagan, O. (1993). *Consciousness reconsidered.* Cambridge, MA: Bradford Books/MIT Press.

Fleming, J. V. (2013). *The dark side of the Enlightenment: Wizards, alchemists, and spiritual seekers in the age of reason.* New York, NY: W.W. Norton and Company.

Fukuyama, F. (2003). *Our posthuman future: Consequences of the biotechnological revolution.* New York, NY: Picador.

Geraci, R. M. (2010). *Apocalyptic AI: Visions of heaven in robotics, artificial intelligence, and virtual reality.* New York, NY: Oxford University Press.

Habermas, J. (2003). *The future of human nature.* Cambridge, England: Polity Press.

Hawking, S., Tegmark, M., Russell, S., & Wilczek, F. (2014, April 19). Transcending complacency about superintelligent machines. *Huff Post. Science.* Available at http://www.huffingtonpost.com/stephen-hawking/artificial-intelligence_b_5174265.html.

Heidegger, M. (1954a). Die Frage nach der Technik. In *Vorträge und Aufsätze*, Teil I (5–36). Pfullingen, Germany: Günther Neske.

Heidegger, M. (1954b). Überwindung der Metaphysik. In *Vorträge und Aufsätze*, Teil I (63–91). Pfullingen, Germany: Günter Neske.

Heidegger, M. (1954c). Das Ding. In M. Heidegger, *Vorträge und Aufsätze,* Teil II (pp. 163–181). Pfullingen, Germany: Verlag Günther Neske.

Heidegger, M. (1962). *Being and time.* (J. Macquarrie & E. Robinson, Trans.). New York, NY: Harper & Row.

Heidegger, M. (1967a). *Sein und Zeit.* Tübingen, Germany: Max Niemeyer Verlag.

Heidegger, M. (1967b). Brief über den "Humanismus." In *Wegmarken* (pp. 145–194). Frankfurt am Main, Germany: Vittorio Klostermann.

Heidegger, M. (1971). The thing. In A. Hofstader (Trans.), *Poetry, language, thought* (pp. 178–179). New York, NY: Harper & Row.

Heidegger, M. (1972). Overcoming metaphysics. In J. Stambaugh (Trans.), *The end of philosophy* (pp. 84–110). New York: Harper & Row.

Heidegger, M. (1977a). Letter on humanism. (D. F. Krell, Trans.) In D. F. Krell (Ed.), *Basic writings* (pp. 213–265). New York, NY: Harper & Row.

Heidegger, M. (1977b). The question concerning technology. (W. Lovitt, Trans.). In D. F. Krell (Ed.), *Basic Writings* (pp. 307–342). New York, NY: Harper & Row.

Heidegger, M. (1983). *Die Grundbegriffe der Metaphysik. Welt—Endlichkeit—Einsamkeit.* In F.-W. von Herrmann (Ed.), *Gesamtausgabe*, vols. 29/30. Frankfurt am Main, Germany: Vittorio Klostermann.

Heidegger, M. (1987). *The will to power as knowledge.* In J. Stambaugh, D. F. Krell, & F. A. Capuzzi (Trans.), *Nietzsche* (pp. 100–158). San Francisco, CA: HarperSanFrancisco.

Heidegger, M. (1989a). *Nietzsches Lehre vom Willen zur Macht als Erkenntnis* (Summer semester 1939). In E. Hanser (Ed.), *Gesamtausgabe*, vol. 47. Frankfurt am Main, Germany: Vittoria Klostermann.

Heidegger, M. (1989b). *Beiträge zur Philosophie (Vom Ereignis).* (F.-W. von Hermann, Ed). Frankfurt am Main, Germany: Vittorio Klostermann.

Heidegger, M. (1995). *Fundamental problems of metaphysics: World, finitude, solitude* (W. McNeill & N. Walker, Trans). Bloomington, IN: Indiana University Press.

Heidegger, M. (2000). *Reden und andere Zeugnisse eines Lebensweges (1910–1976).* In H. Heidegger (Ed.), *Gesamtausgabe* (Vol. 16, 620–33). Frankfurt am Main, Germany: Vittorio Klostermann.

Heidegger, M. (2004). *Zu Ernst Jünger „Der Arbeiter."* In P. Trawny (Ed.), Gesamtausgabe, Vol. 90. Frankfurt am Main, Germany: Vittorio Klostermann.

Heidegger, M. (2010). On the question concerning the determination of the matter for thinking. (R. Capobianco & M. Göbel, Trans.). *Epoche, 14* (2), 213–233.

Heidegger, M. (2012). *Contributions to philosophy.* (R. Rojcewicz & D. Vallega-Neu, Trans.). Bloomington, IN: Indiana University Press.

Jonas, H. (1952). Gnosticism and modern nihilism. *Social Research, 19* (4), 430–452.

Jonas, H. (1958). "Gnosticism, nihilism, and existentialism." In *The Gnostic religion* (pp. 320–340). Boston, MA: Beacon Press.

Joy, B. (2000, April). "Why the future doesn't need us." *Wired, 8* (4). Available at http://archive.wired.com/wired/archive/8.04/joy.html.

Jünger, E. (2013 [1932]). *Der Arbeiter: Herrschaft und Gestalt.* Stuttgart, Germany: Klett-Cotta-Verlag.

Kurzweil, R. (2006). *The singularity is near: When humans transcend biology.* New York, NY: Penguin.

Kurzweil, R. (2012). *How to create a mind: The secret of human thought revealed.* New York, NY: Viking.

Lanier, J. (2010, August 9). "The first church of robotics." *The New York Times.*

Lazier, B. (2012). *God interrupted: Heresy and the European imagination between the world wars.* Princeton, NJ: Princeton University Press.

Magee, G. (2008). *Hegel and the hermetic tradition.* Ithaca, NY: Cornell University Press.

Malik, S. (2000). *Nihilism now! Monsters of energy.* (K. Ansell-Pearson & D. Morgan, Eds.) New York, NY: St. Martin's Press.

Malykhina, E. (2014, April 2). New DARPA office merges biology and technology. *Information Week, Government.* Available at http://www.informationweek.com/government/leadership/new-darpa-office-merges-biology-and-technology/d/d-id/1141668.

Marcuse, H. (1964). *One-dimensional man.* Boston, MA: Beacon Press.

McGrath, S. J. (2005). The facticity of being godforsaken: The young Heidegger's accommodation of Luther's theology of the cross. *American Catholic Philosophical Quarterly, 79* (2), 273–290.

McIntosh, S. (2012). *Evolution's purpose.* New York, NY: Select Books.

Monod, P. K. (2013). *Solomon's secret arts: The occult in the age of Enlightenment*. New Haven, CT: Yale University Press.
Naam, R. (2005). *More than human: Embracing the promise of biological enhancement*. New York, NY: Broadway Books.
Nietzsche, F. (1974). *The gay science*. (W. Kaufmann, Trans.). New York, NY: Vintage Press.
Polt, R. (2104). A Heideggerian critique of cyberbeing. In H. Pedersen & M. Altman (Eds.) *The horizons of authenticity* (pp. 179–199). Dordrecht, the Netherlands: Springer.
Ptolemy, R. B. (Director). (2009). *Transcendent man: The life and times of Ray Kurzweil* [Documentary]. United States: Ptolemaic Films.
Rosenblum, B., & Kuttner, F. (2011). *Quantum enigma: Physics encounters consciousness*. New York, NY: Oxford University Press, 2011.
Rosenthal, B. G. (2002). *New myth, new worlds: From Nietzsche to Stalinism*. University Park, PA: Pennsylvania State University Press.
Sandel, M. J. (2009). *The case against perfection: Ethics in the age of genetic engineering*. Cambridge, MA: Belknap Press.
Sheehan, T. (1998). Nihilism: Heidegger/Jünger/Aristotle.I In B. C. Hopkins (Ed.), *Phenomenology: Japanese and American perspectives* (pp. 273–316). Dordrecht, the Netherlands: Kluwer Academic Publishers.
Sheehan, T. (2002). Nihilism and its discontents. In F. Raffoul & D. Pettigrew (Eds.), *Heidegger and practical philosophy* (pp. 275–300). Albany, NY: State University of New York Press.
Sheehan, T. (2014). *Making sense of Heidegger*. Lanham, MD: Rowman & Littlefield.
Stiegler, B. (1998). *Technics and time, 1: The fault of Epimetheus*. (R. Beardsworth & G. Collins, Trans.). Stanford, CA: Stanford University Press.
Stiegler, B. (2009). *Technics and time, 2: Disorientation*. (S. Barker, Trans.). Stanford, CA: Stanford University Press.
Stiegler, B. (2010). *Technics and time, 3: Cinematic time and the question of malaise*. (S. Barker, Trans.). Stanford, CA: Stanford University Press.
Stock, G. (2002). *Redesigning humans: Our inevitable genetic future*. New York, NY: Houghton Mifflin.
Storey, D. (2015). *Naturalizing Heidegger: His confrontation with Nietzsche, his contributions to environmental philosophy*. Albany, NY: SUNY Press.
Swimme, B. T., & Tucker, M. E. (2011). *Journey of the universe*. New Haven, CT: Yale University Press.
Taubes, S. A. (1954). The Gnostic foundation of Heidegger's nihilism. *The Journal of Religion, 24* (3), 155–172.
Wilber, K. (1995). *A brief history of everything*. Boston, MA: Shambhala.
Zimmerman, M. E. (1990). *Heidegger's confrontation with modernity: Technology, politics, art*. Bloomington, IN: Indiana University Press.
Zimmerman, M. E. (2008). The singularity: A crucial phase in divine self-actualization?" *Cosmos and History: The Journal of Natural and Social Philosophy, 4* (1–2), 347–370.
Zimmerman, M. E. (2009). Religious motifs in technological posthumanism. *Western Humanities Review, 63* (3), 67–83.
Zimmerman, M. E. (2011). Last man or overman? Transhuman appropriations of a Nietzschean theme. *The Hedgehog Review: Critical Reflections on Contemporary Culture, 13* (2), 31–44.

Part II

Ethics and Politics of Envisioned Futures

Transhumanism Is Humanism, and Humanism Is Transhumanism

5

Micha Brumlik

1 Transhumanism and Philosophical Anthropology

To understand transhumanism, it is necessary first to clarify what exactly "humanism" is. If we understand the term to be more than just a "philanthropic" or normative-theoretical attitude (Baab, 2013) according to which—taught as early as the Greek sophists—"man is the measure of all things," we require a definition of this unique biological species. Such a definition can be accomplished only if we differentiate the human species from various species of nonhuman animals on earth as well, as distinguish between humans and other postulated kinds of presumably reasonable beings, such as ghosts, machines, angels, or gods and God. The serious attempt to theorize human nature began in earnest in what we commonly call the Renaissance when what is today is called "transhumanism" was postulated for the first time. To put it more precisely, already in the Renaissance one recognized that any kind of humanism would also be "transhumanism." This idea was stated in Giovanni Pico della Mirandola's "Oration on the Dignity of Man" (1486). There Pico depicts God, the Creator, speaking to the first human, his creature:

> Neither a fixed abode nor a form that is thine alone nor any function peculiar to thyself have we given thee, Adam, to the end that according to thy longing and according to thy judgment thou mayest and possess what abode, what form, and what functions thou thyself shalt desire. The nature of all other beings is limited and constrained within the bounds of laws prescribed by us. Thou, constrained by no limits, in accordance with thine own free will, in whose hand we have placed thee, shalt ordain for thyself the limits of thy nature. We have placed thee at the world's center that thou mayest from thence more easily observe whatever is in the world. We have made thee neither of heaven nor of earth, neither mortal nor immortal, so that with freedom of choice and with honor, as though the maker and molder of thyself, thou mayest fashion thyself in whatever shape thou shalt prefer. Thou shalt have the power to degenerate into the lower forms of life, which are brutish. Thou shalt have

> the power, out of thy soul's judgment, to be reborn into the higher forms, which are divine. (trans. In Cassirer, Kristeller, & Randall, 1948, pp. 224–225)

Pico's famous words indicate that, as early as the Renaissance, some thinkers recognized that the nature of the biological species *Homo sapiens* is undetermined because humans have the capacity and the freedom to be either like the beasts or like the angels. In this rather revolutionary understanding of being human, there is no fixed human nature; rather, human beings possess the innate capacity to make themselves through will and deliberate action (Kristeller, 1979).

This revolutionary insight was clearly stated and further developed by the intellectual tradition school of thought known as philosophical anthropology that flourished from the 1920s on in German-speaking countries (Witteriede, 2009; Delfgaauw, Holz, & Natua, 1985). In this tradition, philosophers, biologists and others aimed to combine the human sciences with philosophical reflection.

In his 1928 work *The Human Place in the Cosmos* (German 2005; English 2008), Max Scheler argued that humanity's special position was due to our spirit, which guaranteed open mindedness, self-awareness, and facticity. Scheler's ideas influenced Arnold Gehlen, Helmuth Plessner, and Ernst Cassirer, among other German intellectuals. Gehlen's main work, *Der Mensch: seine Natur und seine Stellung in der Welt* (1940), postulated that the human being, in comparison to nonhuman animals, is a "deficient being."[1] Because of the relative inferiority of bodily organs and instincts, humans must compensate by creating institutions and making tools that function as "artificial" extensions of their organs (Delitz, 2011). In this formulation, enhancement by means of technology is the human response to a natural limitation.

A different reflection on human nature was suggested by Helmuth Plessner, the son of a Jewish father who was forced to leave Germany in 1933. Plessner believed humans to be characterized by an "eccentric positionality"—by the fact that humans may and must "get out of themselves"—and that their existence is characterized by their relation to their somatic basis—by the reflective doubling of "the body I am" and "the body I have" (Plessner, 1975). In Plessner's philosophical biology, humans are environmentally interactive organisms that alternate between open and closed intentionality. Humans are different from other animals by virtue of the balance between their "inner" and "outer" self. When individuals transcend their outer self and realize their inner life, they are open to perceptions, experiences, and expressions that have a greater sociological and historical significance. Thus, the

---

1 Gehlen joined the National Socialist German Workers Party (NSDAP) in 1933. Although he published the book in 1940, it is not a racist text.

"eccentric positionality" of humans integrates their physical, mental, cultural, and social dimensions. As Jürgen Habermas wrote with regard to Plessner, a "person 'has' or 'possesses' her body only through 'being' this body in proceeding with her life. It is from this phenomenon of being a body and, at the same time, having a body (*Leibsein und Körperhaben*) that Helmut Plessner set out to describe and analyze the 'eccentric position' of man" (Habermas, 2003, p. 50).

Prior to Plessner, an integrative view of the human had already reached a very elaborate and systematic exposition in works by Ernst Cassirer that revolved around the insight of the human being as a "symbolic animal." Cassirer's *Philosophy of Symbolic Forms* (1923) was an original attempt to articulate a coherent philosophy of humankind that bridges scientific and nonscientific modes of thought. In his *An Essay on Man* (1944)—published while Cassirer was in exile in the United States—he succinctly reiterated his conviction that humans can become aware of themselves only by way of symbols:

> He has surrounded himself with linguistic forms, artistic images, mythical symbols or religious rites to such an extent that he is unable to see or recognize anything without this artificial medium getting between himself and reality. (Cassirer, 1996, p. 50)[2]

As symbolic animals, then, humans interpose systems of signs or systems of expression between themselves and the world, thereby bridging the gap between nature and culture as well as between thought, tools and artifacts.

Philosophical anthropology can be traced back not only to the Renaissance but also to German philosophers of the late 18th and 19th centuries. One influential reflection comes from Johann Gottfried Herder. About the nature of human beings, Herder avers in *Ideen zur Philosophie der Geschichte der Menschheit*,

> Man is the first freedman of creation; he stands upright. The scales of the good and the bad, of good and evil, of what is wrong and what is true, depend on him; he can research, he is supposed to choose. Just as nature has provided him with two free hands as tools and with a far-looking eye to guide his steps, just the same he has the power not only to set the weights and measures but also to himself act as a weight on the scales. (IV, 4, in German)

Herder's reflections on human mind, language, culture, and history were remarkably profound, containing the blueprint for modern anthropology and opened new lines of thought that viewed the human being as an animal that generates and is embed-

2 Unless a translator is specified in the references, all translations from German texts were done for this publication.

ded within culture. In this regard, Herder's approach was very different from the universalist anthropology developed by Kant. Although Herder is not commonly associated with German idealism as a philosophical movement, his ideas deeply influenced its leading exponents, including Hegel, Schelling, and Fichte (Pinkard, 2002) and also romanticism (Berlin & Hardy, 2013). The discipline of modern anthropology is also indebted to Herder since Franz Boas, who is considered the founder of modern anthropology, was deeply immersed in Herder's legacy.

These strands of reflection on being human that attended to embodiment and historicity were woven together in the philosophy of Friedrich Nietzsche whose notion of the "Superhuman" (*Übermensch*) has been widely taken up in contemporary transhumanist thought (for example, Sorgner, 2009). Nietzsche's doctrine of the Superhuman has been charged with fascism, not without good reason. In *Thus Spoke Zarathustra*, Nietzsche states,

> Man is a rope connecting animal and Superhuman—a rope across an abyss. … What is great with man is that he is a bridge and not a purpose: what can be loved with man is that he is a transition and a fall. (Nietzsche, 1988, p. 16, in German)

The debate about Nietzsche's concept of the Superhuman continues (Niemeyer, 2011): whereas some consider it the nucleus of a fascist idea of the systematic breeding of the human species to become something superior, others understand the Superhuman as a concept of education because the human being, as Kant had already stated, is nothing than what he becomes due to education. Whether human self-improvement can be accomplished by education or by genetic engineering is at the heart of the debate about transhumanism.

Nietzsche's devastating critique of the Enlightenment ideal of rationality had a complex reception in 20th-century thought, influencing a long list of philosophers, most notably Michel Foucault. Indeed, according to Ernst Behler, "Foucault's entire text can be seen as a re-enacting of Nietzsche in our time, an actualization of Nietzsche toward the end of the twentieth century" (1996, p. 317). In *The Order of Things* (1969), Foucault announced that soon "man will disappear like a face made of sand," by which he meant that the humanistic outlook inaugurated by the Renaissance is no longer philosophically credible. Foucault's sustained concern with becoming "what we are," while recognizing the inherent paradox of human becoming, tracked Nietzsche's imperative: "Be Yourself! All that you are now doing, thinking and desiring is not really yourself." Foucault's declaration of the "end of Man" was a critique of the philosophical concept of the human as a being endowed with a rational soul that functions as an internalized rule. Like Nietzsche's declaration of the "death of God," Foucault's announcement dismantles key metaphysical verities in the Western philosophical tradition. These ideas proved to be intellectual entry points for transhumanist and posthumanist ideas

This brief intellectual history enables us to make a preliminary remark about transhumanism. To think properly about transhumanism requires us to develop a philosophical anthropology that explains how embodied human beings develop their ideas about themselves in relation to their physical environment. Those belonging to the biological species *Homo sapiens* are exactly those creatures who have transformed and further developed their existing physical body. As symbolic animals, human beings cannot but understand these somatic developments to be symbolically significant. But precisely because human beings think symbolically, they must embed everything concerning their bodies, their physique, and their somatic existence in mythical, religious, artistic, ethical, or scientific and scholarly contexts of meaning—contexts that are also always characterized by claims to political authority and rulership. The uniqueness of human beings (Plessner's "eccentric positionality") entails that human attempts at enhanement by technology will necessarily have ethical and political significance.

2 Transhumanist Practices: Armament and Jewelry

Philosophical anthropology, however, is not only about the extension or strengthening of organs in the sense of improving performance or increasing the chances of survival but also about religious representation—such as body ornaments. Weapons and jewelry are typical examples of that what is considered an extension of organs or symbolic interpretation of the body. This becomes obvious, for example, in the current fiercely debated question about the ethical meaning of the use of "drones" in war. Is there a crucial difference between the spear or bow and arrows, long-range weapons already known in Stone Age, and missile-carrying drones today?

On the face of it, there is no difference between drones and slingshot that helped young David kill Goliath. After all, both the slingshot and drones are weapons, and weapons, we need to remind ourselves, are nothing but "extensions of organs." Whereas animals use their teeth, claws, and beaks to improve their situation in the struggle for survival, the animals belonging to the human species invented hand axes, clubs, and daggers. In the struggle for survival, it becomes obvious that the one who is capable of injuring or killing the enemy without close contact or by being faster than the opponent will have the upper hand. All weapons, whether they are slingshots, spears, bow and arrows, are mechanism of human survival. By the same token, the species that succeeds in reducing its own vulnerability by help of artificial fur, shields, and other protective devices will have the upper hand in the natural struggle for survival.

In short, from an anthropological perspective, weapons are nothing but organs perfected by distance and speed. But when we look at these weapons from ethical and political perspectives, we discover something else: the improvement of distance, speed, and armor leads to radically new ways of mutual killing, thus, to new, extended ways of rule and exploitation. Technological inventions like weaponry are never neutral; they always carry deep political implications. For example, without the invention of the machine gun, there would have been no colonialism; without tanks, no way to bring World War I to an end; without the nuclear bomb, no end to war in the Pacific as early as 1945.

However, not all weapons are of the same quality. The nuclear bomb presents a qualitative turning point in the history of human arms making: with it, the evolution of artificially extended organs has come to an end because quantity finally changed into quality. As a weapon of mass destruction, the nuclear bomb presents a reversal of purpose and means: because it can bring about the killing of all humankind, the nuclear bomb means death for all those who even keep such a weapon. We grant that drones cannot be compared to nuclear weapons, as they are hardly anything more than extremely modernized variants of slingshots or bow and arrows. Drones are highly efficient long-range weapons, lacking any chivalry (that is to say, any possibility that the fighter might endanger him- or herself) and thus rightly suspected of lowering the threshold to murderous violence.

Hegel analyzed the inherent connection between modernity and efficient long-range weapons, which, in his days was the needle rifle. In the *Grundlinien der Philosophie des Rechts*, he writes,

> The principle of the modern world-thought and the universal has given courage a higher form, because its display now seems to be more mechanical, the act not of this particular person, but of a number of a whole.... It is for this reason that thought had invented the gun, and the invention of this weapon, which has changed the purely personal form of bravery into a more abstract one, is no accident. (1911, # 329, in German).

The making of weapons, then, expresses the human capacity to extend the physical boundaries of the human body as much as it expresses human capacity for symbolic meaning.

If weapons illustrate the fateful ways of extending the human body, the making of jewelry (or fashion more broadly) is a less destructive form of human enhancement. The *Homo sapiens* species spread all over the world because of its ability to create for itself an artificially made "fur" that enabled it to survive harsh climates. The human species has gradually lost its natural protection against cold—fur—and developed alternative substitutes in the form of shelter and clothing. Being a "symbolic animal," as Cassirer has taught us, humans had to give this kind of substitute organ a certain shape. Clothing and shelter are, thus, forms of human enhancement that enabled humanity to evolve and dominate other species.

But what are we to make of jewelry and other forms of fashion that characterize human clothing? On the face of it, they do not contribute to survival and, therefore, should not be discussed in the same breath with weaponry. Again, if we take philosophical anthropology seriously, we may appreciate the deep significance of jewelry and fashion and other forms of human decoration. Several philosophers, anthropologists, and social theorists help us think about these modes of human behavior as forms of enhancement. For example, the philosopher and literary critic Hans-Georg Gadamer echoes Kant's *Critique of Judgement* when he reflects on architecture. Buildings point to "decoration" as a highly condensed way of representing objects and their meaning in space and time (Gadamer, 1972, p. 152). The sociologist Georg Simmel participated in this philosophical discourse by offering insightful reflections on fashion in his compilation of essays *Philosophische Kultur* (*Philosophical Culture*; 1998 [1911]). There he interpreted fashion as a phenomenon both of conscious individualization and community formation. Fashion, more than other forms of human self-expression, expresses predominant social inequality with particular clarity (Simmel, 1998, pp. 38–63).

If fashion so functions in highly developed class societies, what about jewelry and "fashion" in tribal societies that are devoid of class consciousness and class rule? In his travel report *Traurige Tropen* (*Sad Tropics*; 1970 [1955]), Claude Lévi-Strauss showed that, in tribal societies, jewelry is an indispensable element of world-presenting rituals (174). But rituals, as recently shown by Jürgen Habermas, are necessary modes of social communication and—occasionally—of solidarity that are prior even to spoken language (Habermas, 2012). The anthropological evidence lends further support to the insight that humans are symbolic animals who are engaged in social communication and ritualized behavior that go beyond mere physical survival through overcoming external threats and adapting to new environments. The anthropological meaning of jewelry, fashion, and decoration compels us seriously to consider the human as a "symbolic animal" precisely as Cassirer has instructed.

Homo sapiens is a species of *animal* whose existence is dependent on biological reproduction and whose ways of life have been maintained over the generations, as with other animal species. But since humans are symbolic animals, their ways of life reflect the capacity to express themselves in symbolic forms, giving rise to culture. So how does reproduction relate to culture in the case of the symbolic animal? When we take philosophical anthropology seriously, we can have a better framework to assess transhumanism.

3 Philosophical Anthropology and Intergenerational Relationships

The core of philosophical anthropology is the timeliness of human life. It focuses on phenomena such as plasticity, responsibility, or educatability, the capability to learn, and the capacity for expression and communication. However, on closer examination, these terms directly or indirectly refer to the phenomenon of development and, thus, to the temporality of human existence. The substrate of human biography is the human body; it predetermines stimulations for changes of the human individual by itself as well as by its significant others. The human body makes social interaction possible and, with it, the emergence of culture. But the human body itself is subject to changes, aging, and decay against and despite the human will. Previously, notwithstanding all historically and culturally relative differences, human embodiment was understood as being something different to—and more than—the human mind, and the cultivation of the "body I am," including the mind (as opposed to "the body I have") stood at the center of education and cultural development. With the dawning of its technological reproducibility, it is questionable, however, if the human body will retain the naïve spontaneity that up to now has enforced and made possible the organization of body and life.

If the traditional concept of a human being has come to an end, then the end of what we have called "education" and "pedagogy" might also be near. But before we strike up the final swansong of traditional human corporeality, let us once again reflect more theoretically about human corporeality sustained through reproduction. Taking procreation into account, human life is necessarily structured relationally. Both from an anthropological and a social point of view, the relational character is constituted by the duality of sex difference as a precondition for procreation. These two distinctions (that is, first, the distinction between the sexes and, second, the distinction between generations) constitute the basic nature of human development, highlighting the double nature of the human.

Luise Winterhager-Schmid enables us to appreciate that double nature:

> "[B]ig" are those who are capable of conceiving "little ones." That is something "little ones" are not yet capable of doing....The genealogical inter-generational relationship is previous to all other generational relationships. It is insolubly connected to man's physical nature of consisting of two sexes. The difference between the sexes is the

> precondition for generativity and generation. (Winterhager-Schmid, 2000, p. 26, in German)

The sexes, as well as the difference between generations, originate from and are caused by the process of evolution. For the time being at least, the natural-historical rootedness of humankind, which cannot do other than understand itself historically, is not limited to humans only but is shared by all warm-blooded mammals. The sexual and double-gendered reproduction of animal species happens by melting the genomes of two, sexually different, fertile and prolific, nonhermaphroditic partners. In this context, even a strictly biological view no longer considers sexuality the driving force of the replication of the genome but rather a method of repairing and varying individual genomes (Wickler and Seibt, 1998, p. 36).

In the course of evolution, the replication of the genome by two differently specialized, unisexual members of a species has pushed through first against parthenogenesis, then against hermaphroditic reproduction because, first, hermaphroditism allows for a greater variety of the genome and, second, the specialization in just producing oocytes, on the one hand, and sperms, on the other, provides the respective partners with greater replication opportunities than does hermaphroditism. From then on, producers of sperms (males) may support the spread of their genomes not only by sexual intercourse but also by fighting and excluding the offspring of others, without taking care for upbringing, whereas the producers of oocytes (females) can dedicate themselves to caring and upbringing and can leave the exclusion of alien offspring to the sperm producers. Against this background, sociobiology and population-genetics have identified two fundamentally different replication strategies: whereas under conditions of risky environments that make survival improbable the best solution seems to be the fast and frequent conception of offspring and rather lenient parental care, stable environments reward strategies of rare and slow conception and birth, as well as highly intensive parental care that require a high investment of power, time, money, love, and concern by the parents (Scheunpflug, 2001, p. 117).

Concerning its phylogenetic development, the offspring of the species *Homo sapiens*, being a previable and nursing animal, is oriented toward strategies of the latter kind. Thus, among all primates, *Homo sapiens* shows the longest pregnancies and the lowest fertility rates. In accordance with this is the evolutionary meaning of the prefertility period of life, which is also particularly long with this species and is here called "biological childhood." It allows for a long "apprenticeship" period to make the adult function successfully. This point is well stated by Athanasios Chasiotis:

> The variables of the social environment which must be learned, such as cultural norms, change over time, but this happens comparably slowly in comparison to man's

> lifespan. Thus the best evolutionary strategy is establishing "sensitive situations" which can be learned particularly easily in early childhood, thus making it easier to acquire certain ways of behavior. In this context the child's need of care is seen as a precondition by way of which man becomes capable of raising his offspring to being "better," more than average reproductively successful adults. To achieve this, for the first five years of life we must assume a sensitive period when the child learns how to adopt the reproduction strategies of its adult family members. (Chasiotis, 1999, p. 14, in German; cf. Blaffer Hrdy 2000, p. 550)

These findings accord not only with insights from psychoanalysis and the empirical findings of attachment research but also with the findings of development-oriented brain research (Eliot, 2001). In this sense, transcultural, comparative studies are also capable of identifying a clearly definable, biological childhood of the offspring of *Homo sapiens* that differs at best concerning its speed and kind but not concerning its sequence and developmental logic.

These evolutionary-psychological hints make ethical-moral and, thus, educational sense when we take into consideration the species's typical reproduction and child-rearing strategies, on the one hand, and its typical "eccentric positionality" (to use Plessner's terminology), on the other. The primary person of reference for the newborn child in the overwhelming majority of cases is still the mother. The human newborn behaves according to a strange mixture of preconscious selection strategies, moral imperatives, and partnership-related life scripts. Accordingly, when it comes to the riddle of the adoption of obviously weak children, the conclusion of anthropologist and evolutionary psychologist Sarah Blaffer Hrdy is that

> [i]n contrast to other animals, humans are capable of consciously making decisions which contradict their own interests. … Such a voluntary behavior, which contradicts all biological self-interest, shows features of true heroism, a moral kind of heroism, as George Eliot had in mind when distinguishing humans from Darwin's "ridiculous animals" and emphasizing that our actions predetermine us to the same extent as we decide about our actions. (2000, p. 253, in German)

Actions, however, develop in the course of societal evolution into institutions and systems of norms that, in the course of the individual child's socialization, determine or limit its self-conception and options for action. Thus, concerning the ethical dimension of the development of sociopedagogical theories, it is essential to assess how far a realistic idea of the evolutionarily pregiven cross-generational relationship is compatible to the imperatives of a universalist morality.

To summarize, among all primates, *Homo sapiens* shows the longest period of pregnancy and the lowest fertility rate. This fits with the evolutionary significance of that period of life before the maturity of the reproductive system, which is also particularly long with this species and is here called "biological childhood." The latter's meaning is to provide for a long period of apprenticeship to allow the adult to function successfully. Attending to procreation and intergenerational dependency reveals what is at stake in technological reproduction. Can this "apprenticeship" be cut short or perfected by way of the purposeful technological use of reproduction strategies starting with the genome?

4 Purposeful Replication of Humans, Transhumanist Reproduction: The Species-ethical Problem

Jürgen Habermas (2003) correctly understood that transhumanism raises problems of "species ethics." In the traditional terminology of philosophy, the term "ethics" means the doctrine of "good life," in the context of which it is, after all, impossible to decide which ways of achieving a good life deserve to be systematically normatively distinguished. In contrast to "ethics," however, all those normative systems that are called "morality" deal with the question of certain ways of behavior, or more exactly certain ways of acting or not acting, that is, duties, that can be systematically and clearly proven. Then again, we must distinguish "right" from "morality" and "ethics," the former's norms of acting or not acting usually being backed by the state's coercive means.

The nature of a species-ethical problem has become most obvious when Habermas (2003) discusses what he calls "liberal eugenics" (cf. Bernard, 2014). Habermas reflects on the question whether—and if so which—ethical and moral criteria are reasonable and necessary for the research, analysis, and technological processing of the human genome. Can we state serious reasons that people who want children should not "compose" their child by use of genetic methods, as if they are making, so to speak, a fine meal, and determine in advance not only the child's gender but also the color of its eyes, the shape and color of its hair, and even its intelligence and other mental traits? We may grant that, in the future, mothers will take the genome from qualified sperm banks, from males who, for whatever reasons, cannot find a female partner for procreation or that women who can no longer have children, due to illness or old age, will clone their own genome and have the thus purposefully constructed child be borne by a surrogate mother. Somatically speaking, such a child would be an exact copy of the cloning father or cloning mother, but the cloned child *will not become a perfect* 1:1 copy of the cloning parents due to differences in epigenetics and to the socialization processes that the child, once born, will be subject to.

Thus, as a species-ethical dilemma, the problem of "transhumanism" is this: to what degree is it a good thing for those belonging to the species *Homo sapiens* to intensify and enhance their pregiven capabilities? Human enhancement today is not just the intensification of certain physical capacities of specific organs such as eyes and ears but the fusion of the somatic-human substrate with machines. Is the human-machine interface good for *Homo sapiens*? From this species-ethical question, we must again distinguish the moral question whether such further developments of the somatic basis might possibly violate deontologically grounded interpersonal demands or duties; finally, it must be assessed if and, if at all, which legal regulations must be established in this field to protect existing rights.

This is the context in which Habermas raises the question of whether there are "postmetaphysical" answers to the question of a "good life": by first postulating that the abstract, universalist morality of human dignity and human rights "is itself sustained by a prior ethical self-understanding of the species, which is shared by all moral persons"(Habermas, 2003, p. 40). He must give an answer to the question "whether the instrumentalization of human nature changes the ethical self-understanding of the species in such a way that we may no longer see ourselves as ethically free and morally equal beings guided by norms and reasons (Habermas, 2003, pp. 40–41). And he argues that

> [w]hat hitherto was "given" as organic nature, and could at most be "bred," now shifts to the realm of artifacts and their production. To the degree that even the human organism is drawn into this sphere of intervention, Helmuth Plessner's phenomenological distinction between "being a body" and "having a body" becomes surprisingly current: the boundary between the nature that we "are" and the organic endowments we "give" to ourselves disappears. (Habermas, 2003, p. 12)

Referring to Helmut Plessner's philosophical anthropology, Habermas assumes that the knowledge of cloned humans, namely, those who have been "artificially" and "technologically" produced with all their installations would fundamentally change the way in which they understand themselves. This would result in a dedifferentiation of the distinction between the "grown" and the "made." Habermas warns us that this situation

> might usher in the vertiginous awareness that, as a consequence of a genetic intervention carried out before we were born, the subjective nature we experience as being something we cannot dispose over is actually the result of an instrumentalization of a part of our nature. The realization that our hereditary factors were, in a past before our past, subjected to programming, confronts us on an existential level, so to speak, with the expectation that we subordinate our being a body to our having a body. (Habermas, 2003, pp. 53–54)

On the basis of his assumption that individuals can perceive themselves to be one with themselves and their body only if the latter is experienced as "naturally grown" and not "technologically made," Habermas argues that also all kinds of a universalist morality postulating the uniqueness and dignity of a human individual must be based on the assumption that, concerning their actions, decisions, and their moral behavior (which again is based on "reasons" gained from insight), individuals must understand themselves as being exempt from human command on the physical level.

In light of Hannah Arendt's concept of "natality," then, Habermas is allowed to assume that actors may claim those performative self-attributions for themselves without which they would not be able to understand themselves as the initiators of their own actions and claims. Following Arendt, one may say that any human birth, "as a divide between nature and culture," marks a new beginning (Habermas, 2003, p. 59). Now, of course, we may ask if the awareness of individuals that they were "naturally" conceived and then purposefully raised from whichever parents does not also make them have the impression that they have been made as "characters." Is it really a fact that the awareness of having been "technologically made" affects the subjectively perceived awareness of freedom more than the awareness of, say, a miserable childhood? And what about, we may further ask, those early societies that necessarily did not have any genetic knowledge or even awareness that semen provokes a pregnancy? Indeed, we know of tribal societies where the conception of a child is said to result from a blow of the wind or a bath in water. And is it really true, as Habermas believes, that it is only our awareness that one's genetic beginning is not subject to control that guarantees a minimum of the awareness of freedom? Thus, Habermas, who supports a strict regulation of embryo research and preimplantation genetic diagnosis (PID), is able to speak out in favor of species-ethical self-restriction:

> Normative barriers in our dealings with embryos are the result of the point of view taken by a moral community of persons that fends off the pacemakers of a self-instrumentalization of the species in order to safeguard—let us say: out of concern for itself, but in the broader perspective of the ethics of the species—its communicatively structured form of life. (Habermas, 2003, pp. 71–72)

Is, therefore, the awareness of coincidental conception and of the contingency of our having been developed the guarantee for our awareness of freedom and also of moral autonomy? At the end of his considerations, Habermas speaks out in favor of staying away from cloning, PID, and the like because the knowledge of the coincidental melding of semen and ovum allows for self-awareness in the sense of contingency and, thus, freedom. However, we must then ask if, in societies that do not have this knowledge, humans are not individuals insofar as they do not have any awareness of freedom. What are we supposed to think of societies where a woman's pregnancy is believed to be the result of a blow of the wind, a bath in water, or a ghost? (Blaffer Hrdy, 2000, p. 290).

No matter how convincing Habermas's species-ethical objections against cloning and PID look, shifting the whole burden of proof onto individuals' awareness of the contingency of their creation is much less convincing. Instead of following this path of species-ethical argumentation, in what follows, I assess by way of a thought experiment if there are limits to such a consequently thought-out kind of transhumanism and, if so, which limits.

5 Extreme Transhumanism: Cyborgs and Other Rational Beings

In their scandalous bestseller *Empire* (2000), the authors Michael Hardt and Toni Negri refer not only to the "death of mankind," as it has been announced by Michel Foucault, but also to the "Cyborg Manifesto" of Donna Haraway, a radical-feminist philosopher of science. With this, the fantasies of post- and transhumanists, or extropians, which previously had been sneered at, leave the realm of nutcases and sectarians and knock at the gates of serious-discourse communities:

> We certainly do need to change our bodies and ourselves, and in perhaps a much more radical way than the cyberpunk authors imagine.... We have to arrive at constituting a coherent political artifice, an artificial becoming in the sense that the humanists spoke of a homohomo produced by art and knowledge. (Hardt & Negri, 2000, p. 216)

Hardt and Negri articulate what, as mentioned above, was already thought in detail by Friedrich Nietzsche in his considerations on the Superhuman: "Man is a rope," he says in *Zarathustra*, "spanning between animal and Superhuman—a rope across an abyss." Nietzsche completes this consideration by stating that a human being is worthy to be loved precisely because he or she is a transition and not a decline. Whoever believes all this to be science fiction may be reminded of the serious debate between Peter Sloterdijk and Habermas (Sloterdijk, 2008) in which Habermas later—in his discussion of "liberal eugenics"—attempted nothing less than an ethics of the human species, even of human nature, an attempt to keep it in its traditional form. In the future, and much earlier than many believe, political thought will also have to face this challenge. In the following, I will sketch the possible consequences of transhumanism for political thought.

In the Western tradition, human and political self-awareness are the same. On the one hand, politics are always politics by humans and for humans; on the other hand; political life is the highest way of life humans can achieve. Technology must bring about the end of this connection, thus changing the Western concept of the human being. Humans are, Aristotle says in the *Nichomachean Ethics* (1162a), political beings by nature. Although antiquity drew a clear distinction between nature first order and nature second order, we must assume that here Aristotle referred to the human nature first order, that is, the genetically pregiven order, not to that human nature that is the result of education and habit, talents and aspirations. Furthermore, we must assume that here the term *polis* actually suggests the city ruled by the will of its citizens and not—as suggested by the Latin translation *animal sociale*—just any arbitrary community of humans. However, the citizens of the polis are always living beings (in Latin, *animalia*), and, even as organic beings, they have more in common with other animals than with a mere circulatory system.

In the course of its history, this political being with the capability to manipulate nature instrumentally has not only succeeded in overpowering nature but also in technologically shaping and reshaping the nature of its own species. Now, to use Heidegger's formulation (1954), the human physique is made subject to the rule of the framework. Under the grasp of technology, physique and polis, previously united within the human being, are separated. This separation of physique and polis as a result of technology illustrates the reality of humans who neither can nor are permitted to be citizens—the problem of slaves and so-called subhumans is well known from history, but this entails also the possibility of citizens no longer belonging to the human species. Nonetheless, Aristotle believed it possible to consider some members of the species, namely, slaves, as organic tools (in Greek, *andrapoda*), viewing them as mere human chattels, and he found it extremely difficult to explain, for instance, the possibility of friendship among citizens and slaves. The paradoxes of slavery, which extend to the anything but paradoxical forced labor camps of National Socialism, give evidence to this.

Today, by a radical reversal of this discriminating practice, human rights—not citizenship rights—are granted even to members of higher primate species (Cavalieri & Singer, 1994). For example, the utilitarian philosopher Peter Singer (1984) offers justification for this by referring to the modern age and its egalitarian intuition that full rights can be granted only to individuals, but then to all individuals. However, according to Singer, individuals are beings who feel pain and lust, who have intentionality and the capacity to desire a continued future existence. On the basis of these criteria, a wide group of animals would have to be granted basic rights, while a number of members of the human species, such as babies below the age of two and comatose patients, would not enjoy these rights.

The question of whether there can be beings who are neither primates nor hydrocarbon-based and who are not reproduced by sexual intercourse and natural birth but by machines is not one that can be answered philosophically *a priori*; it has to be answered empirically. Independent of this, we can make the empirical observation that, in the course of its history and possibly even shortly after its evolutionary inception, the human species has departed from its hydrocarbon primate nature and, by way of the principle of extension and improvement of the body's in-built capacities, has incorporated an ever larger element of machinery. Increasingly, humans are becoming—and have always been—cyborgs.

While glasses, skis, roller skates, or hearing aids do not provoke for science fiction-like nightmares, we must insist—in view of the logic of development—that the species *Homo sapiens* will perfect the capability of extension by way of science and technology, thus shifting the already existing machine aspects of human life from its organic basis to something else. Thus, in a certain sense, members of this species have always, since emerging from the realm of animals, been cyborgs: already the first hand axe paved the way toward mechanization. Up to this point, however, artificial organs (for example, clothing and shelter, as discussed above) have made their way from the surface of the body into the environment. The disturbing feature of recent technological developments is that now technologies do not necessarily move outward from the surface of the body into the environment but also inward, toward the interior of the body. The path from this already existing transhumanism toward some future posthumanism (Braidotti, 2014) leads from prostheses of inner organs via functional equivalents of the sensory organs to machine substitutes of nerves and parts of, or perhaps eventually the entirety of, the human brain.

This in principle imaginable and foreseeable development runs in the direction of creating more and more hybrids that progressively transition from a human organism to beings with an exclusively machine-based personality, even if only a weak one. Thus, the question of speciesism emerges. By "speciesism," Peter Singer, who coined the term in his fight for animal rights (Singer, 1996), especially the rights of great apes, means a kind of moral particularism that arbitrarily grants rights and claims only to those beings/persons belonging to the species of *Homo sapiens* and are not ready to withdraw these rights from members of the human species who have not yet developed a personality. Robert Spaemann objects to this, for example, saying that the idea that some humans could withdraw rights from other humans according to arbitrarily chosen criteria would violate equality criteria within the species (Spaemann, 1984). Humans must privilege other humans—all members of the species—to a certain degree simply because the way of life of this species is intersubjective: as has been shown with the above-mentioned question about the intergenerational constitution of the species of *Homo sapiens*, humans are by their very nature necessarily dependent on other humans. This is also and especially true if one considers the species to be the only biological species that, unlike all other nonhuman animals, is capable of developing reason—a kind of reason due to which they enjoy inalienable dignity. Nevertheless, it may be imagined that human rights can be granted to nonhuman individuals belonging to the family of primates without violating the claims of members of the human species. But, if the criterion for granting rights is personality, no matter how weak, what is right for individuals belonging to the family of primates must *a fortiori* be right for all kinds of conscience and intentionality, even if it is machine-made. Moral universalism requires the acceptance of apes, machines, and all imaginable kinds of intelligent entities who should have legal standing by virtue of their intelligence (Donaldson & Kymlicka, 2013).

Thus, here is the problem of isomorphy and isonomy, already known from the ancient polis yet increased. Whereas the history of the emancipation of humankind has been and still is a contradictory history of ever-increasing inclusion of members of one's own species with different gender, of different age, different means, different color, and so forth, now it seems as if the emancipation of allomorphic individuals—individuals of a different kind—is imminent.

To judge the possibility of the emancipation of allomorphs, at first we must answer the question of whether the principles not only of a community, a *corpus sociale*, but also of a legal fellowship, a *polis*, is possible based only on the single criterion of weak personality. Kant's carefully chosen term "rational being" seems to suggest something like this. But what exactly do gorillas, humans, and machines endowed with intentionality have in common? With respect to the animal side, Alastair MacIntyre (2001) has tried to delineate the reproduction, care, and upbringing programs of mammals—including dolphins—as that most affect-based communality that engenders common interests.

Which interests, however, will machine-based kinds of conscience develop, given that they are lacking these affective ties because they have completely different reproduction mechanisms? The characters of science fiction, from Stanley Kubrick's Hal computer in *2001*, the replicants in Ridley Scott's *Blade Runner*, and the artificial boy in Steven Spielberg's *A. I. Artificial Intelligence*, have either shown a desire for continuation, for ties, even for love, or have displayed a coldness that is deadly for humans. A polis of man and machine—an also politically exercised program of a transition from transhumanism to posthumanism—is currently unimaginable because the speciesism criticized by Singer can at best be overcome in the realm of mammals and because there it can be implemented only through proxy via the advocacy of those humans who will speak on behalf of those beings who cannot speak for themselves. The reason for this is that, for conceptual reasons, it is as impossible to answer the question of what it is like to be a machine, as it is to answer to the question of what it is like to be a bat (Nagel, 1974).

Notions of personality that are not based on the primate program refer to extremely restricted forms of human life, such as people paralyzed from the neck down. Some in this condition wish to die, whereas others are prepared to assume the burden of existing as human-machine hybrids. It is readily clear that this kind of conundrum does not apply to forms of machine consciousness. After all, beings who have never been able to move would not perceive a lack of movement and lack of sensory contact as a deficit. Should we, thus, imagine personally gifted machines as completely paralyzed autistics? Perhaps, but why should we grant machines that have no experience of what is plausible—and significant—for us humans the power to codecide our interests?

6 Conclusion

I have argued that the notion of transhumanism as political inclusion of intelligent machines reveals the following. First, it shows that the debate on so-called transhumanism is senseless insofar as classical philosophical anthropology has already dealt with it in showing that what is commonly understood by "transhumanism" has always been part of the biological species of *Homo sapiens*. Second, I have shown that one possible ethical problem for transhumanism is, at best, that the creation of humans as such might no longer be left to the slightest coincidence but may instead be carefully planned. Third, I have demonstrated that a kind of transhumanism that dares the theoretical transition to posthumanism is either trivial or empty. It becomes trivial if it postulates that intelligent machines, irrespective of their outer appearance, have emotional desires, needs, and abilities comparable to the biological species of *Homo sapiens*, for "if you double the causes, you double the effects." However, it is empty because, until this has been proven, there is nothing to say about it: it is an empty idea.

But this raises another species-ethical question: Is it desirable and good to contribute to the construction of such machines and thus to an evolutionary leap whose consequences are unpredictable?

References

Baab, F. (2013). *Was ist Humanismus? Geschichte des Begriffes, Gegenkonzepte, säkulare Humanismen heute*. Regensburg, Germany: Pustet.

Behler, E. (1996). Nietzsche in the twentieth century." In Bernd Magnus & Kathleen M. Higgins (Eds.), *The Cambridge companion to Nietzsche* (pp. 281–322). Cambridge, England: Cambridge University Press.

Beiser, F. C. (2002). *German idealism: The struggle against subjectivism, 1781–1801*. Cambridge, MA: Harvard University Press.

Berlin, I., & Hardy, H. (2013). *The roots of romanticism*. Princeton, NJ: Princeton University Press.

Bernard, A. (2014). *Kinder Machen: Neue Reproduktionstechnologien und die Ordnung der Familie. Samenspender, Leihmütter, Künstliche Befruchtung*. Frankfurt am Main, Germany: S. Fischer.

Blaffer Hrdy., S. (2000). *Mutter Natur: Die weibliche Seite der Evolution*. Berlin, Germany: Berlin Verlag.

Bloch, E. (2000 [1923]). *The spirit of utopia*. Stanford, CA: Stanford University Press.

Braidotti, R. (2014). *Posthumanismus: Leben jenseits des Menschen*. (T. Laugstien, Trans.) Frankfurt am Main, Germany: Campus.

Cassirer, E. (1996). *Versuch über den Menschen. Einführung in eine Philosophie der Kultur*. Hamburg, Germany: Felix Meiner.

Cavalieri, P., & Singer, P. (1994). *Menschenrechte für die Großen Menschenaffen: Das Great Ape Projekt*. (H.-J. Koskull, Trans.). Munich, Germany: Wilhelm Goldmann. In English: *The great ape project: Equality beyond humanity*. London, England: Fourth Estate, 1993.

Chasiotis, A. (1999). *Kindheit und Lebenslauf: Untersuchungen zur evolutionären Psychologie der Lebensspanne*. Bern, Switzerland: Hans Huber.

Delfgaauw, B., Holz, H. H., & Nauta, L. (Eds.) (1985). *Philosophische Rede vom Menschen: Studien zur Anthropologie Helmuth Plessners*. Frankfurt am Main, Germany: Peter Lang International Academic Publishers.

Delitz, H. (2011). *Arnold Gehlen*. Konstanz, Germany: UVK.

Donaldson, S., & Kymlicka, W. (2013). *Zoopolis: Eine politische Ethik der Tierrechte*. (J. Schulte, Trans.). Berlin, Germany: Suhrkamp.

Eliot, L. (2001). *Was geht da drinnen vor? Die Gehirnentwicklung in den ersten fünf Lebensjahren*. Berlin, Germany: Berlin Verlag.

Foucault, M. (1970). *The order of things: An archaeology of the human sciences*. New York, NY: Pantheon Books.

Gadamer, H-G. (1972). *Wahrheit und Methode* (3rd ed.). Tübingen, Germany: Mohr.

Gehlen, A. (1940). *Der Mensch: seine Natur und seine Stellung in der Welt*. Berlin, Germany: Junker and Dünnhaupt. English: McMillan, C., & Pillemer, K. (Trans.). (1988). *Man, his nature and place in the world*. New York, NY: Columbia University Press.

Habermas, J. (2003). *The future of human nature*. Cambridge: Polity (orig. in German: (2001). *Die Zukunft der menschlichen Natur: Auf dem Weg zu einer liberalen Eugenik?* Frankfurt am Main, Germany: Suhrkamp).

Habermas, J. (2012). Eine Hypothese zum gattungsgeschichtlichen Sinn des Ritus. In J. Habermas, *Nachmetaphysisches Denken* II (pp. 77–95). Berlin, Germany: Suhrkamp.

Hardt, M., & Negri, A. (2000). *Empire*. Cambridge, MA: Harvard University Press.

Hegel, G. W. F. (1911). *Grundlinien der Philosophie des Rechts*. Leipzig, Germany: F. Meiner.

Heidegger, M. (1954). Die Frage nach der Technik. In *Vorträge und Aufsätze*, Teil I (pp. 5–36). Pfullingen, Germany: Günther Neske.

Herder, J.G. (1965). *Ideen zur Philosophie der Geschichte der Menschheit.* Berlin, Germany: Aufbau-Verlag.

Kant, I. (1964). Vorlesungen über Pädagogik. In W. Weischedel (Ed.), *Schriften zur Anthropologie Geschichtsphilosophie Politik und Padagogik.* Frankfurt, Germany: Insel-Verlag. In English: G. Zöller & R. Louden (Eds.) *Anthropology, History and Education* (pp. 434–484). Cambridge, England: Cambridge University Press, 2007.

Kristeller, P. O. (1979). The dignity of man. In M. Mooney (Ed.), *Renaissance thought and its sources* (pp. 168–181). New York, NY: Columbia University Press.

Levin, Y. (2008). *Imagining the future: Science and American democracy.* New York, NY: Encounter Books.

Levi-Strauss, C. (1970 [1955]). *Traurige Tropen.* Frankfurt am Main, Germany: Suhrkamp.

MacIntyre, A. (2001). *Die Anerkennung der Abhängigkeit: Über menschliche Tugenden.* (C. Goldmann, Trans.). Hamburg, Germany: Rotbuch.

Nagel, T. (1974). What is it like to be a bat? *Philosophical Review, 83* (4), 435–450.

Niemeyer, C. (2011). *Nietzsche verstehen: Eine Gebrauchsanweisung.* Darmstadt, Germany: Lambert Schneider.

Nietzsche, F. (1988). *Also sprach Zarathustra.* Munich, Germany: dtv.

Pico della Mirandola, G. (1988). *Über die Würde des Menschen.* (A. Buck, Ed.). Zürich, Switzerland: Manesse. English: (1948). Oration on the dignity of man. In E. Cassirer, P. O. Kristeller, & J. H. Randall Jr. (Eds.), *The Renaissance philosophy of man* (223–255). Chicago, IL: University of Chicago Press.

Pinkard, T. (2002). *German philosophy, 1760-1860: The legacy of idealism.* Cambridge, England: Cambridge University Press.

Plessner, H. (1975). *Die Stufen des Organischen und der Mensch.* Berlin, Germany: de Gruyter.

Scheler, M. (2005). *Die Stellung des Menschen im Kosmos.* Bonn, Germany: Bouvier.

Scheunpflug, A. (2001). *Biologische Grundlagen des Lernens.* Berlin, Germany: Cornelsen Scriptor.

Simmel, G. (1998). *Philosophische Kultur: Über das Abenteuer, die Geschlechter und die Krise der Moderne.* Berlin, Germany: Wagenbach.

Singer, P. (1984). *Praktische Ethik.* Stuttgart, Germany: Reclam. In English: *Practical ethics.* New York, NY: Cambridge University Press, 1979.

Singer, P. (1996). *Animal Liberation: Die Befreiung der Tiere.* (C. Schorcht, Trans.). Reinbek, Germany: Rowohlt. In English: *Animal liberation.* New York, NY: Harper, 1975.

Sloterdijk, P. (2008). *Regeln für den Menschenpark: Ein Antwortschreiben zu Heideggers Brief über den Humanismus.* Frankfurt am Main, Germany: Suhrkamp.

Sorgner, Stefan L. (2009). Nietzsche, the overhuman, and transhumanism. *Journal of Evolution and Technology, 20* (1), 29–42.

Spaemann, R. (1984). Tierschutz und Menschenwürde." In Ursula M. Händel (Ed.). *Tierschutz—Testfall unserer Menschlichkeit* (pp. 71–81). Frankfurt am Main, Germany: Fischer Taschenbuchverlag.

Wickler, W., & Seibt, U. (1998). *Männlich—Weiblich: Ein Naturgesetz und seine Folgen.* Heidelberg, Germany: Spektrum.

Wietteriede, H. (2009). *Eine Einführung in die Philosophische Anthropologie: Max Scheler, Helmuth Plessner, Arnold Gehlen.* Frankfurt am Main, Germany: Peter Lang International Acaemic Publishers.

Winterhager-Schmid, L. (2000). "'Groß' und 'klein'—Bedeutung der Erfahrung mit Generationendifferenz im Prozeß des Heranwachsens." In L. Winterhager-Schmid (Ed.), *Erfahrung mit Generationendifferenz* (pp. 15–37). Weinheim, Germany: Deutscher Studien-Verlag.

Three Transhumanist Types of (Post)Human Perfection 6

Stefan Lorenz Sorgner

Human perfection is a notion that is closely connected to the tradition of virtue ethics, which was the dominant ethical theory in ancient and medieval philosophy. It does not play such a central role in utilitarian or deontological approaches to ethics, however. Human perfection in ancient and medieval traditions is also closely connected to the notion of *eudaimonia*, which means possessing a good *daimon* whereby daimons used to be the gods responsible for human fate. Consequently, *eudaimonia* can also be translated as possessing a good fate. Nowadays, it is common to identify *eudaimonia* with the good life. The answers given by virtue ethicists concerning the question of the good life have been manifold. Plato stresses that, once you cognitively grasp the good, then you have to possess the virtues associated with it—that is, the four cardinal virtues—and live in accord with them. Aristotle, on the other hand, disagrees, referring to *akrasia*, the weakness of will, and stresses the relevance of external goods and bodily goods for the good life, in addition to both the ethical and intellectual virtues. Stoics uphold the unity of virtues, explaining why, if you have one virtue, then you have to possess all of them and that being virtuous is necessary and contingent for living a good life. The Christian philosophical tradition introduced the importance of the three Christian virtues—love, hope and faith—whereby love is seen as the most important of the three. This tradition is still very strong in contemporary ethical debates. Vattimo (1997) claims that the ethics of Jesus as put forward in the New Testament actually corresponds with a postmodern pluralist approach of ethics, as long as one acts on the basis of love. It is his reinterpretation of Augustine's love-and-do-what-you-want principle. Sandel (2007), on the other hand, draws upon the concept of unconditional love, to argue that it is vicious to use enhancement technologies, both with respect to oneself as well as with respect to one's children, claiming that only depraved parents employ gene technologies at their children.

The central connection in all these approaches is the one between human perfection and leading a good life. If you possess temperance, courage, wisdom, and justice, then you lead a good life (Plato). If you are prudent, then you possess *phronesis*, and you know how to apply correctly the various ethical and intellectual virtues, which also help you acquire and keep bodily and external goods; in turn, all of these capacities help you lead a good life (Aristotle). If you possess the virtue of love, then you lead a good live (Augustine). In the contemporary debate, Vattimo and Sandel reinterpret the Christian virtue ethical tradition and apply it to contemporary challenges.

Some transhumanists also put forward suggestions concerning the good life and the human perfections necessary for realizing it. Nick Bostrom refers to Giovanni Pico della Mirandola as an ancestor of transhumanism because Pico also held that "that Man does not have a ready-made form and is responsible for shaping himself" (2005, p. 2). Referring to Pico as a leading representative of Renaissance humanism, Bostrom asserts that this philosophical approach, like transhumanism, "created the ideal of the well-rounded person, one who is highly developed scientifically, morally, culturally, and spiritually" (ibid.). However, these comments are false in many ways. Indeed, Pico della Mirandola does not belong to the Renaissance humanist tradition. Renaissance humanism is best represented by Giannozzo Manetti, who stressed and glorified the bodily aspects of human existence while still interpreting the world from a Christian perspective. Pico della Mirandola, rather, belongs to Renaissance Neoplatonism. Even though, he acknowledges that human beings are free to form themselves the way they want to, Pico also stressed that human perfection consists in the *unio mystica*, which one can promote mainly by purifying oneself from any earthly attachment. Human perfection for Pico consists in the process of purification of any earthly element, so that the likelihood gets increased for the state of mystical union with the one, the *hen*, or, in other words, with the Christian God, as his approach to Neoplatonism was a Christian one. This process gets promoted by means of intellectual virtues, according to Pico. A well-rounded personality was definitely not Pico's ideal of perfection (see Sorgner, 2010, pp. 65–81).

Still, one might argue that, even though Bostrom's historical inspiration for his ideal of human perfection is based on a misunderstanding, it does not follow that the ideal he is suggesting is false. However, I will show that this *is* also the case. In this article, I will present his concept of human perfection and another one by Julian Savulescu, which, although slightly weaker than Bostrom's conception, can still count as detailed concept of human perfections, and analyse the plausibility of both. I am referring to Savulescu's concept in this context because it is complex, interesting, and closely connected to transhumanist thinking, even though he does not regard himself as a transhumanist. Due to there being an immediate connection between the concepts of human perfection and the concept of the good life, I will be using both words in this context. A description of human perfection states the conditions necessary for living a good life. If you ask for a good life, then you get an analysis of human perfection. Furthermore, I am distinguishing between strong and weak concepts of the good, whereby I am drawing upon Martha Nussbaum's use of these distinctions (1999, p. 28). A strong concept of the good presents attitudes and elements of a good life that is concerned with all aspects of life, which is not the case for a weak concept of the good. If the attitudes and elements of a good life are described in a complex manner, then the concept can be described as a detailed one. If this is not the case, the concept in question is a vague one. A traditional Catholic concept of the good is concerned with all aspects of life; hence, it is a strong concept of the good. In addition, it also presents very specific conditions for living a good life and, thus, is also a detailed concept of the good and can, therefore, be referred to as strong and detailed concept of the good. Nussbaum's concept of the good, on the other hand, is concerned with all aspects of life but merely provides some general advice concerning these various issues in question (1999). Hence, her account can be referred to as a strong but vague concept of the good life, which, indeed, is her own way of referring to her position. A philosophical approach that merely stresses the relevance of a specific attitude for the good life has to be classified as a weak concept of the good. Depending on whether this specific element is described in a complex manner or not, the approach is a weak and detailed one or a weak and vague one. Savulescu's account of the good represents a weak but detailed concept of the good because it is concerned only with some selected elements of human existence, but he describes them in a detailed manner. Bostrom's account has to be referred to as a strong concept of the good that is also detailed in many respects. Both concepts can be identified with mainstream transhumanism. I will show that both of their concepts are implausible, and I will argue for an alternative concept that can be classified as a weak and vague concept of the good. In addition, this alternative concept not only belongs to the transhumanist tradition but is also strongly influenced by Nietzschean and posthumanist thinking. Thereby, one can refer to this approach as "metahumanist" because it lies beyond a dualist understanding of humanism but between trans- and posthumanism. The prefix *meta* covers both aspects of this understanding because it means both "beyond" and "between." Thereby, metahumanism, which belongs to the neo-Nietzschean and neo-Heraclitean traditions, represents an inclusive alternative to the other two philosophical approaches (Del Val & Sorgner, 2011). Some further clarifications concerning the use of the concepts trans- and posthumanism might be helpful in this context.

Transhumanists affirm technologies in order to increase the likelihood of the coming about of posthumans. The concept of the "posthuman" is an open one. It can be that the posthuman refers to a silicon-based entity in cyberspace. The posthuman can also stand for a carbon-based entity that no longer belongs to the human species or one that still belongs to the human species but has at least one capacity that goes beyond the capacities currently living human beings possess. A strong version of transhumanism affirms that there is a moral but not a legal obligation to use specific enhancement techniques. The bioliberal[1] Julian Savulescu, who is not a transhumanist but closely associated with this movement, represents this perspective well (Savulescu, 2001; Savulescu & Kahane, 2009). Another strong version of transhumanism holds that enhancement technologies necessarily promote good lives for all human beings that this insight ought to be legally relevant, that is, governments should finance research concerning the technologies in question. Aubrey de Grey can be seen as representative of this tradition (De Grey, 2007, pp. 335–339) A weak version of transhumanism, on the other hand, might affirm that enhancement techniques increase the likelihood of many people leading good lives without this insight implying necessary legal or moral duties, even though it does have consequences concerning moral and legal rights (Sorgner, 2013a, 2013b).

Posthumanists, on the other hand, are concerned with nondualist thinking and acting. A strong version of posthumanism assumes that human beings are merely gradually different from other natural beings and that this insight ought to be legally obligatory. Peter Singer (2002, 2011a) can be seen as defender of such a view. A weak version of posthumanism might also affirm that human beings are merely gradually different from other natural beings and that this insight ought to be a legally legitimate perspective, that is, it should not be legally impossible to think and act accordingly (Sorgner 2013b).

By integrating and further developing central concepts of both contemporary approaches, a metahumanist approach was developed. It is founded on the insight that both movements do not differ from each other as strongly as many of their proponents assume; indeed, central premises of both approaches are commonly shared. By weakening central claims of both approaches, potentially violent implicit demands also get reduced. Thereby, the option of human beings leading good, flourishing lives according to their own idiosyncratic standards gets promoted.

---

1 Bioliberals stress the right of freedom concerning legal regulations of bioethical questions. Bioconservatives, on the other hand, assume the priority of a common account of the good with respect to the question of rights, which leads to their being restrictive concerning legal regulations of bioethical questions.

Having clarified some elements of the intellectual background concerning the question of human perfection, the core lines of arguments can be briefly described. In this article, I explicate three concepts of the good life common in the transhumanist discourse and then argue in favor of one of the concepts: (a) the Renaissance Ideal as represented by Bostrom (2001); (2) a commonsense account of the good associated with Savulescu (2001; Savulescu & Kahane, 2009); and (3) a radically pluralist conception of the good inspired by Friedrich Nietzsche, which I regard as most compelling (Sorgner, 2010). I will outline the various concepts of good and analyze their intellectual plausibility. The notion of plausibility is closely connected to three elements: Judgments are plausible, if they are (1) affirmed by larger interest groups, (2) supported by the latest scientific insights, or (3) related to the norm of negative freedom (Sorgner, 2013c).

1 The Renaissance Human Ideal

The Renaissance ideal can be found in various authors and often gets associated with transhumanism within public debates. Upon closer consideration, however, it is rather difficult to find thinkers who uphold this position over a long period of time. Bostrom argues for this ideal in a paper from 2001, but he has subsequently altered and weakened his position on this question (for example, 2009). Still, in his essay "Transhumanist Values" (2001), he writes the following: "Transhumanism imports from secular humanism the ideal of the fully-developed and well-rounded personality. We can't all be renaissance geniuses, but we can strive to constantly refine ourselves and to broaden our intellectual horizons" (Bostrom, 2001). Two things become clear: Bostrom sees transhumanism in the Enlightenment secular-humanist tradition, and he claims that, from this tradition, transhumanism is supposed to have taken over the concept of the "fully-developed and well-rounded personality," which then was developed further by integrating it into a transhumanist framework. It implies that all human beings wish to become "Renaissance geniuses" (2001), even though it is also clear that, for most humans, this is not a feasible option. Still, by being aware of this wish and acknowledging its existence and relevance, Bostrom holds that we as human beings can nevertheless try to "strive to constantly refine ourselves and to broaden our intellectual horizons" (2001).

Even though, Bostrom upholds such a strong concept of the good, there is no doubt that (most) transhumanists (including him) argue on the basis of a version of liberalism concerning political issues. The varieties of liberal attitudes among transhumanists are very wide, and one can both find libertarian thinkers, like

Max More, and more social-democratic ones, like James Hughes (2014). It is also the case that there does not have to be a conflict between holding a strong concept of the good life while acknowledging the relevance of a political liberalism that allows people to act according to a great variety of attitudes. The same tension can be found in the position of Savulescu, who also upholds a strong account of the good, which even implies rigid moral duties, but at the same time argues in favor of a rather libertarian political system (2001; and with Kahane, 2009)

Interestingly enough, this type of tension can also be found in Nietzsche. On the one hand, he stresses that the physiologies of all human beings differ radically from each other and that this has an enormous influence on what is important for living a good life. On the other hand, he stresses the relevance of the classic ideal, which he associates with the great strength; according to Nietzsche, all human beings are attempting to reach a great strength, due to their being and being driven by the will to power (Sorgner, 2007, pp. 39–65). The following three passages from his *Nachlass*, his writings not published by himself, provide some hints concerning the relevance and meaning of the classical ideal in Nietzsche's philosophy, even though it must be acknowledged that his understanding of the classical does not refer to the concept of the good life only but also applies to various other realms, like the realms of cultural, artistic and literary creations. In *The Will to Power* (1968), he writes:

> The highest type: the classical ideal—as the expression of the well-constitutedness of all the chief instincts. Therein the highest style: the grand style. Expression of the "will to power" itself. The instinct that is most feared dares to acknowledge itself. (p. 341)

> The classical style is essentially a representation of this calm, simplification, abbreviation, concentration—the highest feeling of power is concentrated in the classical type. (p. 799)

> Classical taste: this means will to simplification, strengthening, to visible happiness, to the terrible, the courage of psychological nakedness.... (p. 868)

In the cases of both Nietzsche and of some transhumanists, what constitutes the Renaissance genius, or the classical ideal, is a "fully-developed and well-rounded personality" that attempts "to constantly refine" itself (Bostrom, 2001); this can be further associated with the following qualities: "calm, simplification, abbreviation, concentration—the highest feeling of power" (Nietzsche, 1968, p. 799), an "expression of the well-constitutedness of all the chief instincts" (ibid., p. 341) and a "will to simplification, strengthening, to visible happiness" (ibid., p. 868). Translating these various aspects in a more accessible language, the ideal is characterized by bodily beauty, strength, and health; intellectual and cognitive excellence; enormous philosophical wisdom and knowledge; and musical and artistic sensibilities and capacities, which coexist in a balanced and integrated personality and come along with a beautiful, strong, and healthy body. In other words, the Renaissance genius can be identified as an Einstein with the body of a Schwarzenegger, the face of a Johnny Depp, and the wisdom of an Aristotle. At least the male version of the ideal can be characterized thus. A female Renaissance genius might be a Marie Curie with the body of a Heidi Klum, the face of a Kate Moss, and the wisdom of a Hildegard of Bingen.

The implication of such a concept of the good is that it would promote the likelihood of living a good life for all human beings, if someone continuously tries to enhance her own capacities with respect to the various aspects of her existence. By means of sport, one's bodily beauty and strength could be trained. By means of music making, one's capacity of making and appreciating music would be promoted. By means of memorizing poems, one's memory would be improved. This list could be continued with respect to all the other aspects mentioned before, which are associated with the Renaissance genius. However, it is not only by means of drawing upon traditional technologies with respect to the aforementioned qualities that would increase the likelihood of someone's leading a good life, but also by drawing upon emerging technologies the likelihood of approaching the Renaissance ideal can be promoted. Morphological enhancement (like plastic surgery), pharmacological enhancement (like the use of Ritalin), genetic enhancement both by modification and by selection, and cyborg enhancement (like the use of deep-brain stimulation) could all be appropriate means for constantly refining oneself, so that one gets closer to the ideal of turning into a Renaissance genius.

2 A Commonsense Account of the Human Good

A seemingly weaker account than the first one is the commonsense account of the good life presented by Savulescu. However, it is also a detailed account of the good for several reasons. First, he firmly believes that there is such a thing as the best life: "A common objection to PB [the principle of procreative beneficence] is that there is no such thing as a better or best life. It is hard to defend such a claim" (Savulescu & Kahane, 2009, p. 278).[2] Furthermore, it is clear to him that we regularly rank

2 The principle of procreative beneficence was developed by Julian Savulescu to stress the moral, but not the legal, obligation to select the child after in-vitro fertilization and preimplantation genetic diagnosis that has the highest likelihood of living a good life because such acts bring about the best moral consequences. In a recent article, I explained

the goodness of lives on the basis of the right criterion: "There are plenty of cases where we can rank the goodness of lives. We do so in numerous moral decisions in everyday life" (ibid., p. 279). He even stresses that the concept of the good life is based on an incontrovertible commonsense consensus that we are already supposed to apply in many everyday situations:

> But although there is this philosophical disagreement, there is considerable consensus about the particular traits or states that make life better or worse, a consensus that would rule out many procreative choices as grossly unreasonable.... PB doesn't rely on some special and controversial conception of well-being. All it asks us is to apply in our procreative decisions the same concepts we already employ in everyday situations. (Savulescu & Kahane, 2009, p. 279)

Though Savulescu asserts that we are aware of the appropriate criteria of a commonsense concept of the good life, it turns out that his concept is not as commonsensical and straightforward as he suggests. In an earlier essay, he maintains that it is good to have good memory and a strong intelligence (2001, p. 420), which he still maintains in the later paper, where he stresses that a high intelligence is helpful for a good life:

> If parents could increase the prospects of future children's lives by selecting children who are far more intelligent, emphatic or healthier than existing people, then PB instructs parents to select such future children. (Savulescu & Kahane, 2009, p. 290)

In the later paper, having a strong memory and being able to concentrate well and understand other people's feelings also seem central for Savulescu:

> How can the capacity to remember things better, concentrate longer, be less depressed, or better understand other people's feelings have the effect that one will be less likely to achieve the good life? (Savulescu & Kahane, 2009, p. 284)

Then, he also stresses that it is bad to have a disposition for depression (ibid., p. 281) or a disability (ibid., 286), whereby he alters his definition of what counts as a disability, moving from a functionalist understanding toward one that considers the cultural context, which is probably the most important alteration concerning his account of the good between 2001 and 2009.

Even though he presents quite a detailed analysis of several aspects that are important for human lives, Savulescu has not worked out all the implications of his theory of the good. Does he mean that a human entity is always better off with a

---

why this principle ought to be seen as an immoral rather than as a moral one (Sorgner, 2014).

higher intelligence, a stronger memory, or more intensive capacity to concentrate? He does not make any judgment by means of which he makes clear that this is not what he means. Yet I am doubtful whether it is actually the case that any strengthening of one's memory is necessarily always in one's own interest. For example, while it is good to have good memory (and I definitely wish to have a better memory), I am also grateful for the capacity to forget things selectively. If I permanently remembered all the bad things that other people have done to me, then I would most probably decide to move away from civilization and live like a hermit on the top of a mountain. Hence, I would be hesitant to claim with Savulescu that having a better memory necessarily is always a good thing and that we have a moral reason for always choosing the fertilized egg after in-vitro fertilization and preimplantation genetic diagnosis with a better memory. Likewise, it can be asked whether it is always in the interest of a child to have a higher intelligence. If the child is the only one with such a high intelligence, while all the people around him are fools, then I would be hesitant to claim that having such a high intelligence necessarily is in the best interest of the child, as this capacity might have the consequence that the child will be excluded from social life. Hence, it seems to me that social setting is of central relevance for the qualities necessary for leading a good life.

Savulescu seems to have realized and acknowledged the impact of this line of thought, especially with respect to the question of disability, which can and ought to get transferred also to other domains of a good life. In his 2001 article, Savulescu upholds the following position: "[I]t is bad that blind and deaf children are born when sighted and hearing children could have been born in their place" (p. 423). In his coauthored 2009 paper, he develops his views concerning disabilities further (p. 284) According to Savulescu and Kahane, "disability is a context and person-relative concept. What may make it harder to lead a good life in one circumstance may make it easier in another" (Savulescu & Kahane, 2009, p. 286) As a consequence of Savulescu's altered approach concerning disability, he and Kahane stress that "on our account of disability, people do have reasons not to have a future child who is likely to be disabled if they have the option of choosing another who is expected to have less or no disability, although whether it would be wrong to do so would depend on the overall balance of moral reasons" (2009, p. 286) They even reach the following conclusion given their new approach:

> If a case can be made that deafness is not a disability in our sense—if it can be shown that deafness does not reduce well-being, or at least that in a given context deafness is not expected to be a disability—then PB would not give any moral reason not to select deafness. (2009, p. 289)

These final remarks concerning disability are quite interesting and have some plausibility. However, it is not clear which remarks actually represent Savulescu's theory of the good life, which serves as the justification for the principle of procreative beneficence. Savulescu's writings are replete with incompatible statements concerning the good life, as if he has a theory of the good for every difficult question that might be posed by the critics. On the one hand, he upholds a perfectionist theory of the good life that can be identified with being more intelligent, healthier, having a stronger memory, and so on. On the other, he refers to what we regard as good life, which can be named as commonsense approach to the good life. Yet I am pretty certain that, if large groups of Western citizens were asked what a good life is, it would not be a perfectionist theory of the good that would be upheld. My assumption is supported by the fact that many mothers in the United States who ordered sperm were interested in sperm from good-looking and athletic Ivy League students rather than in the sperm of Nobel Prize winners who are the very best intellectuals or scholars in their field (Caplan, 2012, p. 156). Unfortunately, the question whether they would have preferred good-looking, athletic Noble Prize winners to good-looking, athletic Ivy League students cannot be answered on the basis of the given data. Still, these data could provide some evidence in favor of the Renaissance genius as a widespread ideal of the good. Nevertheless, Savulescu might reply that this argument does not go against his theory. Intelligence still plays a central role concerning the choices of these American mothers, as they desire to have an at least fairly intelligent Ivy League student as sperm donor. Important for noticing the incompatibility of his various remarks concerning the good life is that in a further statement he points out that PB is neutral to classical philosophical theories of the good life (Savulescu & Kahane, 2009, p. 279) However, it could be argued that PB can be combined with various theories of the good, as it is logically separated from theories of the good. He combines it with his commonsense approach due to its being the appropriate universally valid concept. In any case, his commonsense approach (with Kahane, 2009) focuses primarily on the following qualities as necessary constituents of a good life:

1. Not being disabled whereby disability is seen as context dependent quality;
2. No disposition for mental illnesses;
3. Having good health;
4. Good capacities for communication, memory, and empathy;
5. High intelligence.

Points 1 to 3 can be summarized under the category "health" and 4 and 5 under the category "cognitive capacities." Hence, the commonsense approach differs from the Renaissance genius concept insofar as the Renaissance ideal includes the possession of strong bodily capacities like being forceful, beautiful, and athletic, attributes that are not demanded by the commonsense approach to the good. From my perspective, it makes sense to exclude these bodily capacities from the list of necessary qualities for living a good life, for several intuitively plausible reasons. For instance, the French singer and actor Serge Gainsbourg seems to have had quite a fulfilled life without having had exceptional bodily capacities (at least from the perspective of a classic or Renaissance ideal). I will provide some philosophical reasons for this position in the next section.

3 A Radical Pluralistic Account of the Human Good

Many people might find either the Renaissance ideal or the commonsense approach not so bad and possibly even plausible. However, in the previous section, I have already mentioned some reasons that a strengthening of the cognitive capacities might not always be in everyone's best interest. The remaining issue in question concerning the commonsense suggestion put forward by Savulescu is that of health and whether health can actually count as an all-purpose good or a good that necessarily promotes the good life for all human beings. The argument I will be dealing with in this context will also reveal why further philosophical doubts can be raised concerning the Renaissance ideal suggestion. Even Jürgen Habermas seems to accept health as an all-purpose good and, consequently, allows genetic interventions to promote health (2001, pp. 48, 92). Yet, thinkers like Habermas and Savulescu do not sufficiently recognize the philosophical demand of what it means to claim that a judgment is universally valid such that applies to all entities that belong to the human species at all times independent of their personal and cultural circumstances and psychophysiological demands. These psychophysiological demands determine what is required for someone to live a good life, and these demands differ radically from person to person and from time to time because they change during the various stages of life. I am talking about psychophysiological demands in order to stress the intimate connection between faculties that have long been seen as separated from one another under the distinction between immaterial soul and material body. This is a highly implausible account of our human constitution. A nondualist, relational account of human ontology is a far more promising approach. Each psychophysiology consists of specific affects, drives, and wishes, whose fulfillments can be identified with the good life of the human entity in question.

Given these psychophysiological demands, human choices that differ from the ones made by the majority of people should not be taken lightly. It is far more likely that the choices of the majority of people follow the model of Martin Heidegger's "man." The majority of people act according to the rules "one" is expected to act on. Social demands and cultural constructs influence their acts more strongly than their own psychophysiological demands, which actually correspond to their needs for living a good life. Only by listening to and acting in accord with someone's psychophysiological demands does a person become authentic. Consequently, the following acts can be understood as being based on authentic wishes. This does not have to be the case in all instances, but it can be the case: (1) Person A wishing to die; (2) Person B desiring to have her healthy leg removed; (3) Person C wishing to eat parts of himself; (4) Person D not wanting to be cured from her manic depression; (5) Person E regarding his deafness as an advantage but not as a disablement. The list of potential examples could easily be continued. If the Renaissance genius account or the commonsense account of the good were universally valid, these wishes could not be accepted as authentic ones, but would have to be seen as expressions of an ill mind.

I do not think that this has to be the case. By claiming that these wishes represent insane states of the mind, these persons are treated paternalistically and violently: their wishes are not recognized as their own, and others claim that they know better than oneself what is in one's own interest. I regard such a way of treating people as highly problematic because the otherness of someone else's wishes does not get appreciated appropriately. On the one hand, there are culturally dominant paradigms of leading a good life; on the other hand, there are the needs of one's own psychophysiology that do not necessarily correspond to these general demands. A pregnant woman who wishes to have sex with men other than the father of her child, a student who enjoys sexual intercourse with several people at the same time, and a young girl who is longing for erotic encounters with a woman thirty years her senior represent three examples in question. All of these desires do not get approval from culturally dominant paradigms of a good life; but there are people with such desires, and it is aggressive, violent, and paternalistic to approach them by claiming that they do not understand themselves in an appropriate manner because these types of acts do not correspond with the concept of the good life, which the culture in questions regards as true. Initially, it was difficult for me, too, to imagine that it can be that a deaf person is not disabled but merely different. However, by recognizing the wide range of preferences, choices, tastes, and cultures in all parts of the world at various times, I came to realize how important it is to recognize that a different human being might regard different capacities as important and different shapes as attractive.

In the fields of artistic appreciation, sexual preferences, and preferred tastes of food and drinks, it becomes particularly clear that an individual preference differs from time to time and from person to person. It is even the case that the individual set of preferences is usually not being matched by anyone else's set of preferences. Person A might hate spinach, white bread, beer, and honey, while person B cannot eat spinach, white bread, beer, and sugar, and C might reject consuming cucumbers, white bread, wine, and honey. Any attempt to give a universally valid nonformal account of someone's preferences seems to be bound to fail because there always seems to be someone who can convincingly claim that her authentic wish does not correspond to the concept in question. To acknowledge, recognize, and accept such a position means to affirm a radical pluralist account of the good. Given such an account of the good implies that human perfection consists in acting and thinking in correspondence with one's own psychophysiological demands. Hence, a nonformal account of human perfection is not an option, if this concept of the good is accepted as most plausible.

By referring to cases of different tastes and sexual preferences when arguing in favor of this account of goodness, I am not claiming that the good life can be reduced to questions of taste. Yet these examples provide particularly clear evidence of how difficult it can be to reach a common ground concerning individual preferences, wishes, desires, affects, drives, and goals. Health might be a strong case for an all-purpose good. However, there have been many human beings who regarded their diseases as blessings because they enabled them to develop capacities or have experiences they would not have had without the diseases in question and that for them were central for living fulfilled lives, for example, because it enabled them permanently to work at themselves for developing capacities, which they would not have had otherwise.

I am not claiming either that it is always morally and legally acceptable to act in accord with one's own authentic wish. If you get sexual gratification only by raping someone, then your concept of the good is necessarily in conflict with what is legally acceptable in most countries today, and it is good that this is the case because, in this example, your own concept of the good is (most probably) in conflict with the concept of the good of the person you wish to rape.

In addition to the philosophical reasons mentioned in favor of a radically pluralist account of the good, there is a further, separate aspect that needs to be tackled in this context. It is independent of the question which account of the good life represents the most plausible one. In the following passages, I tackle the question whether it is actually appropriate to refer to the radically pluralist account of the good as a transhumanist one. The radically pluralist concept of the good could be criticized as not being a proper transhumanist concept of the good since transhumanism aims for an increased likelihood of the coming about of the posthuman who could be either a further developed carbon-based (genetically enhanced) or a further developed silicon-based entity (an uploaded mind), and it seems clear that the likelihood of human survival and further development is increased if specific capacities like intelligence and cognitive capacities are promoted. However, this argument is flawed, as there is no necessary conflict between this concept of the good and the potential to increase the likelihood of the coming about of the posthuman. Some of Charles Darwin's basic principles must not be forgotten. The survival of the fittest does not imply that the strongest, tallest, or blondest human being survives. The fittest is always the one who is adapted best to the environmental and cultural circumstances. It cannot be excluded that there will be a situation in the future in which there will be so much loud noise that only the deaf will be able to bear the noise level and survive. It seems to be impossible for us to know which capacities will be the ones that will guarantee our own survival and our living well in the future. We can only decide from our past experiences on which capacities we regard as being in our best interest. I am not denying either that the Renaissance genius and the commonsense concept have a far-reaching appeal in this context. It is also the case that psychological and other empirical studies confirm that these concepts in question seem to be shared and affirmed by a great percentage of human beings (compare Bostrom, 2009, pp. 113–116). I am merely stressing that it is inappropriate for both concepts to claim universal validity. Furthermore, it is not guaranteed that the capacities affirmed by the initial two concepts of the good will be the ones that promote our fitness and our being best adapted to future environmental and cultural circumstances. Hence, there is the need to argue for the importance of the radical pluralist concept of the good life that is closely connected to and founded in the plurality of all of our psychophysiological demands, needs, and desires. Besides the philosophical reasons mentioned here in favor of the radical pluralist account, it could also be in the interest of a society to acknowledge the relevance and plausibility of a radically pluralist account: "Don't put all eggs into one basket." As it is uncertain which qualities will promote our future survival, it is likely that an affirmation of plurality will increase the chances of human survival. Hence, both philosophical and social reasons support the radically pluralist position, for which I argue, and they also provide reasons for classifying this account as a transhumanist one.

Conclusion

The transhumanist discourse necessarily requires reflection on the meaning of perfection and brings to mind Nietzsche's imagined figure of the overhuman.[3] The transhumanist discourse, however, is ambiguous because it employs various meanings of perfection; indeed, when one examines the transhumanist discourse closely, as this essay seeks to do, one can easily see that transhumanist philosophers hold different and even incompatible notions of the good. The above analysis shows that the transhumanist discourse employs at least three distinct concepts of the good. Whereas the first two represent traditional varieties of transhumanism, the third one, in favor of which I argue, might not represent mainstream transhumanism, but it is at least compatible with the central premises of transhumanism, as I explained at the end of part three of this article.

By stressing the problems related to the claim of putting forward a universally valid nonformal concept of the good, several reasons underline the relevance of a radically pluralist account of the good. Thereby, I am not denying that, for instance, an increase of the health span is relevant for a great percentage of human beings, which, again, ought to provide governments with a reason for promoting research in this field. At the same time without contradicting myself, I am stressing that even an increased health span does not necessarily belong to the concept of the good of all human beings. It is an enormous cultural achievement to recognize and allow the radical plurality of the good. By trying to move beyond a nonformal account of the good, individuals get treated in a violent and paternalistic manner. Luckily, we have already moved far beyond cultural situations in which religious or aristocratic leaders have the right to decide on the concept of the good according to which their subordinates have to live. This does not mean that we are already living in a cultural and social situation in which the need to criticize the legal and political structures no longer exists. A violent and morally problematic type of paternalism is still widespread and can be found in cultures all over the world. Once I realized what an enormous achievement it has been to be able to live freely in accord with your own needs, desires, instincts, and wishes and to be able to express your worries, criticisms, and opinions freely, I also became conscious of promoting this achievement of making people aware of what has been gained as a consequence of the processes of enlightenment that have taken place. Again, I cannot claim that citizens of enlightened countries who do not have to fear the paternalistic consequences of nonenlightened legal systems are living in paradise. There are

---

3 See "Was Nietzsche a Transhumanist?" Institute for Ethics and Emerging Technologies, available at http://ieet.org/index.php/IEET/more/pellissier20120423.

many highly problematic issues still to be found even in such countries. However, by becoming aware of the relevance of a radically pluralist concept of the good, it also gets easier to detect potentially problematic paternalistic structures, so that it becomes possible to share one's insights with others (Sorgner, 2010, pp. 239–266).

By realizing, acknowledging, and living in accordance with the insight of a radical plurality of the good, violence against individuals gets reduced, and human flourishing gets promoted. The above reflections merely provide an initial hint concerning the relevance of this concept. However, the question concerning the good and human perfection is of central relevance for all other philosophical, ethical, cultural, and artistic fields, topics, and domains and ought to be analyzed and considered appropriately.

References

Bostrom, N. (2001). Transhumanist values. Version of April 18, 2001. Available at http://www.nickbostrom.com/tra/values.html.

Bostrom, N. (2005). A history of transhumanist thought. *Journal of Evolution and Technology, 14* (1), 1–30.

Bostrom, N. (2009). Why I want to be a posthuman when I grow up. In B. Gordijn and R. Chadwick (Eds.), *Medical Enhancement and Posthumanity* (pp. 107–136). New York, NY: Springer.

Caplan, R. (2012). *Selfish reasons to have more kids: Why being a great parent is less work and more fun than you think.* New York, NY: Basic Books.

De Grey, A. (2007). *Ending aging: The rejuvenation breakthroughs that could reverse aging in our lifetime.* New York, NY: St. Martin's Griffin.

Del Val, J., & Sorgner, S. L. (2011). A metahumanist manifesto. *The Agonist, 4* (2), 1–4.

Habermas, J. (2001). *Die Zukunft der menschlichen Natur. Auf dem Weg zu einer liberalen Eugenik?* Frankfurt am Main, Germany: Suhrkamp.

Hughes, J. (2014). Politics. In R. Ranisch & S. L. Sorgner (Eds.), *Post- and transhumanism: An introduction.* Frankfurt am Main, Germany: Peter Lang.

Nietzsche, F. (1968). *The will to power.* (W. Kaufmann & R. J. Hollingdale, Trans.). London, England: Weidenfeld and Nicholson.

Nussbaum, Martha C. (1999). *Gerechtigkeit oder das gute Leben.* (I. Utz, Trans.). Frankfurt am Main, Germany: Suhrkamp Verlag.

Sandel, M. (2007). *The Case against perfection: Ethics in the age of genetic engineering.* Cambridge, MA: Harvard University Press.

Savulescu, J. (2001). Procreative beneficence: Why we should select the best children. *Bioethics, 15* (5–6), 413–426.

Savulescu, J., & Kahane, G. (2009). The moral obligation to create children with the best chance of the best life. *Bioethics, 23* (5), 274–290.

Singer, P. (2002). *Animal liberation.* New York, NY: Harper.

Singer, P. (2011a). *Practical ethics* (3rd ed.). Cambridge, England: Cambridge University Press.
Singer, P. (2011b). Vorwort [Foreword]. (S. L. Sorgner, Trans.). In T.-L. Eissa & S. L. Sorgner (Eds.), *Geschichte der Bioethik: Eine Einführung* (pp. 13–15). Paderborn, Germany: Mentis.
Sorgner, S. L. (2007). *Metaphysics without truth: On the importance of consistency within Nietzsche's philosophy* (2nd rev. ed.). Milwaukee, WI: University of Marquette Press.
Sorgner, S. L. (2010). *Menschenwürde nach Nietzsche. Die Geschichte eines Begriffs*. Darmstadt, Germany: WBG.
Sorgner, S. L. (2013a): Evolution, education, and genetic enhancement. In S. L. Sorgner & B.-R. Jovanovic (Eds.), *Evolution and the future: Anthropology, ethics, religion* (pp. 85–100). Frankfurt am Main, Germany: Peter Lang.
Sorgner, S. L. (2013b). Human dignity 2.0: Beyond a rigid version of anthropocentrism. *Trans-Humanities, 6* (1), 135–159.
Sorgner, S. L. (2013c): Paternalistic cultures versus nihilistic cultures. *European Journal of Science and Theology, 9* (1), 55–60.
Sorgner, S. L. (2014). Is there a "moral obligation to create children with the best chance of the best life"? *Humana Mente. Journal of Philosophical Studies, 26*, 199–212.
Vattimo, G (1997).*Glauben–Philosophieren*. (C. Schultz, Trans.). Stuttgart: Reclam. Original Italian: *Credere di Credere* (Garzanti Editore, 1996). English: *Belief* (L. D'Isanto and D. Webb, Trans.) (Stanford University Press, 1999).

Reduced Heritage

How Transhumanism Secularizes and Desecularizes Religious Visions

7

Thorsten Moos

Transhumanism and Religion

The relationship between transhumanism and religion has been noted by several scholars.[1] Hava Tirosh-Samuelson conceptualizes transhumanism as a "secularist faith" (Tirosh-Samuelson, 2012a, p. 710) with a double movement: it takes up religious topics in an immanentist framework and, in return, loads technology with religious significance. Ronald Cole-Turner (2012) links transhumanist ideas especially to the worldview of American evangelical Christians. No doubt: There is *something religious* in transhumanism, even though transhumanism presents itself as a secular discourse. But to what extent and why can "religion" be a useful hermeneutical framework to understand what transhumanism is about?

Obviously, the diagnosis of transhumanism as "religious" serves different types of arguments. Transhumanists themselves have described their aims in religious categories, motives, and narratives[2] and pointed to religious ancestors of transhumanist thinking. In his early essay from 1990 (revised 1996), Max More explicates his "Extropian philosophy," which he understands as "the most developed form of transhumanism," as a functional equivalent to religion. Disregarding all differences, this philosophy equals religion in being "deeply value-laden," a "meaning-fostering system" for human life and agency, with a major concern to "abolish the greatest evil: death" and a corresponding "brilliant vision for the future" (More, 1996; see also Bostrom, 2003). Transhumanism, as this argument puts it, legitimately replaces religion: it fulfills the core function of religion to give meaning to the lives of humans, and it adopts some religious content, especially

1 For an overview, see the contributions in Tirosh-Samuelson & Mossmann, 2012.

2 2. More, 1996 [1990], 2013; Bostrom, 2003; FAQ 3.0, n.d.; for the early transhumanists in the 1920s, see Coenen, 2008, 2009; Tirosh-Samuelson, 2012b.

the quest to overcome mortality. But it does so in a different way: it substitutes reason for dogmatism, science for faith, progress for stagnation (More, 1996), and active "paradise-engineering" for the passive hope for divine intervention (Bostrom, 2003). This self-ascribed transhumanist secularization of religion in function and content has also been diagnosed by scholars of religious studies who are critical of transhumanism. They judge differently: transhumanism is not the legitimate successor of religion but an idolatry that illegitimately expropriates religion without being able to substitute it.[3] Moreover, observers from the social sciences use religious parallels to understand transhumanism. In many cases, their understandings of this parallel are connected with normative statements as well. The authors mark the exaggerated claims of this technology-fixated "redemption religion" (Schummer, 2009, p. 64) or "technocratic eschatology" (Coenen, 2009, p. 153), or they denote its lack of reflective potential (Saage, 2009, p. 133). The diagnosis of a social phenomenon as "religious" or as "secularized" is always value-laden, as Hans Blumenberg has convincingly argued: to understand the modern world and its phenomena as secularized means to conceptualize them as dependent on, and perhaps even culturally indebted to, religion ("Kulturschuld," Blumenberg, 1988, p. 33). By using the term "secularization," questions of legitimacy are negotiated, in either an obvious or a hidden manner. Thus, it is important to be careful while integrating transhumanism in the modern history of religion—and to clearly show aims and context: for the sake of what argument?

The following contribution is not apologetic but systematic. Being a theologian, I do not try to defend religious beliefs against transhumanist offenses, and I do not refer to transhumanism as an illegitimate successor of religion, expropriating functions and material that are *essentially* religious. Instead, I want to analyze three aspects of the relation between transhumanism and religion. First, in its production of future visions, transhumanism adopts religious motives, images, and strategies in a specific, sometimes surprisingly direct manner. Second, transhumanism has merit because it stresses some important points that sometimes tend to be neglected in religious and theological discourses. Third, transhumanism adopts only *some* religious strategies and dissociates itself from others (such as God-talk, for example). By thus baking the cake with only half the recipe, so to speak, transhumanism loses the specific rational potential that traditional world religions have developed in dealing and coping with visions, wishes, and hopes. The religiously inspired *vision management* of transhumanism (the way visions are produced, communicated, and criticized) is insufficient—not from a believer's point of view but from a rational standpoint (which is, by the way, an opposition that I do

3 See Coenen, 2008; Tirosh-Samuelson, 2012a, p. 718, and the literature cited there.

not support[4]). So I do not talk about belief but about reason: about the rationality immersed in religion and its symbols that still might positively contribute to the history of reason (Habermas, 2005).[5]

In doing so, I presuppose that transhumanism is not a *mere* instrument to generate funding (Schummer, 2009, pp. 67–78; Coenen, 2009, p. 136) but also a discourse about the role of technology for human self-understanding, about ethics of technology, and about priorities in current and upcoming research. So it is worthwhile discussing the truth claims and, above all, the intellectual tools of transhumanism. Are these tools adequate for the subjects under discussion? This essay answers the question in the negative. My concern is not that transhumanism could endanger traditional religions, but rather that it could be a step backward in the history of "applied reason" (Bostrom, 2003).

By talking about transhumanism, I refer to authors such as Max More and Nick Bostrom who not only call themselves transhumanists but also claim representativeness for the transhumanist movement. Especially, I look at programmatic texts by Max More (1996 [1990] and 2013) and Nick Bostrom (2003; FAQ 3.0, n.d.). Those texts claim to give an overview on transhumanism in general.[6]

Vision Management

Religions have always dealt with visions, especially with imaginations of the future. In religious communities, visions are produced, circulated, and handed over to subsequent generations. But visions are also reinterpreted, weakened, defused, and kept in check. Religion is not only the business of production and proliferation but also of containment of future visions that claim ultimate relevance for our present lives. Transhumanist visions inherit religious visions in a specific manner.

The following essay analyzes three main types of religious visions of the future, articulated by Jewish-Christian traditions as I, a Protestant theologian, understand them. The future is imagined (1) as the *time of perfection*, (2) as *apocalypse*, consisting of a detailed sequence of states to come, or (3) as *parousia in delay*, as the expected

4 For the rationality of belief systems, see, for example, Barth, 2004.

5 When speaking about "religion," I refer to traditional religions, especially to the Jewish-Christian tradition. For the argument developed here, I do not have to presuppose a specific concept of religion, might it be functional or material or something in between.

6 For Bostrom as a representative of transhumanism, see Münch, 2012, p. 287.

return of the redeemer that has not (yet) happened.[7] Each of these types provides a specific structure of mediation between the present human condition and status of society, on the one hand, and ideals of a better life, on the other hand. For each type, the symbolic material will be analyzed, before the potential anthropological and social implications of the imaginary will be discussed.

In a second step, transhumanism is related to these religious materials and the underlying experiences that are reflected in them. It will be shown that the visionary structure of transhumanism is nearly as complicated as the Christian one, as far as visionary anthropology and time diagnosis are concerned. But then, there is a blind spot: religious visions deal with visionary ideals while admitting that the human condition is—and for systematic reasons will always be—marked by deep ambiguities. Transhumanist visions do not. As will be shown below, they answer the problems of finitude and ambiguity by means of what might be called *technological denial.*

The Golden Age and the Sense of Possibility

And God shall wipe away all tears from their eyes; and there shall be no more death, neither sorrow, nor crying, neither shall there be any more pain: for the former things are passed away. (Rev 21:4)

Posthuman beings would no longer suffer from disease, aging, and inevitable death. ... Posthumans would also have much greater cognitive capabilities, and more refined emotions (more joy, less anger, or whatever changes each individual prefers). (More, 2013, p. 4)

First, there is a strong strand in Jewish-Christian tradition where the future is imagined as the *time of perfection*. Everything that restricts us *now* will be overcome *then*. All the miserable aspects of finitude—illness, death, but also ignorance, moral and social deprivation—will be turned into their opposite: in the kingdom of God, cumulating in the *visio beatifica*, the eternal happiness in God's presence.[8]

Let's do some "vision assessment" (Grunwald, 2007): what is the truth claim and what is the social impact of this kind of visions? It is, basically, an anthropological

7 Parousia is strictly a Christian (not a Jewish-Christian) concept.

8 For the history of utopian thinking, see Schölderle, 2012; for Jewish utopianism, see Fischer, 1988, and, for Christian utopianism, Seligman, 1988.

one. Visions like these allow, first of all, communication about what it means to be human: to be a finite, imperfect being, to live in improper conditions, to long for overcoming one's finitude, at least for improving one's condition. Thus, the double characteristic of human beings is expressed in this vision: awareness of finitude and eagerness to transcend it. According to the philosopher Peter Sloterdijk, humans of all eras and cultural spaces are clamped in "vertical tensions" (Sloterdijk, 2009, p. 27, my translation) that make them strive for improving their condition. Vision production up to transhumanism in our day might be seen among the most powerful articulations of this "upward" intention.

It is important to note that not only the future can take the role of the time of perfection. The past can do as well: the Golden Age or a paradise is a state of perfection before the course of history started (see Schölderle, 2012, p. 51). Future or past: generally the perfect ages in this sense are not part of history but beyond it. Time is not the link between imperfection and perfection, but the symbol for the distance that is between us and the perfect state.

This corresponds to the fact that this concept of a perfect state could not only be attributed to a certain period of *time*—might it be in the past or in the future—but also to a certain area in *space*: to a place where—somewhere, not here—all the restrictions that apply to us do not hold.[9] This is the original idea of utopia: a place that is a nonplace, disconnected from the space that we inhabit. We might long for it, but we cannot reach it. The biblical paradise is a utopia in this sense: a garden "eastward" (Gen. 2:8), the ideal conditions of which are, according to biblical myth, blocked by cherubim—thus, inaccessible to us now.[10] As we know from Karl Mannheim, utopias like this do not only serve as means of anthropological communication but allow us to think about a world that is completely different from ours (Mannheim, 1986a and b; Dickel, 2009). Utopias sharpen a "sense of possibility" (Musil, 1970, p. 16). In this sense, they are an important resource not only of anthropological but also of ethical thinking.

Does all this also apply to transhumanism? Indeed, some transhumanists emphasize that they do not claim any state of perfection, any utopia or paradise (More, 2013; Bostrom, 2003). Not a final state, but continuous progress is the guiding vision that they want to establish. The reason is simple: every fixed state—no

9 For the concept of chronotopia, where the utopian imaginary is connected with a notion of accelerated time that "makes things happen," see, for example, Armitage & Roberts, 2002.

10 I follow the functionalist definition of "utopia" in the tradition of the sociology of knowledge (Mannheim, 1986b). Others exclude religious and mythological narratives from utopianism by definition (Saage, 2009); I will refer to utopia in this sense as *secular utopianism*. See Dickel 2009.

matter how desirable it may be—is still finite in the sense that one can still think of overcoming it. Nothing is so good that you could not wish to leave it behind. So, according to Max More (1996), only the idea of continuous progress and growth can provide our lives with unsurpassable meaning. Nevertheless, it is obvious that transhumanists *do* work with visions of concrete desirable states: curing all diseases, becoming immortal, solving world hunger, reviving bodies of the deceased. Although every state can be thought of as improvable, transhumanist visions get their power precisely from the formulation of utopian states.

In this light, transhumanism can be read as a utopian discourse on human finitude, in function and some content equivalent to religious visions of the Golden Age, the kingdom of God, the heavenly Jerusalem, and such.[11] And, as I see it, it is in some respect a good one, allowing us to think of illness and, above all, of death as a *scandal*. Boldly opposing philosophical traditions that try to integrate death into life, transhumanists condemn the attempts to reconcile the living with their forthcoming death as "deathist philosophies" (Bostrom, 2003, p. 37). For transhumanists, death is a scandal because it is the negation of life; and there is nothing glorious about illness. Transhumanists, especially those involved in radical life-extension projects (see, for example, De Grey, 2007), continue to insist on that basic anthropological truth.

Indeed, especially in modern philosophy, there have been powerful attempts to cope with the sheer fact of death (for the following, see Hügli, 1998). For Hegel, it was the step from particularity to universality; for Schopenhauer, the longed-for redemption from being an individual. For Georg Simmel, death is the aspect that gives life a form. Max Scheler understands our certitude to die as a necessary implication of our concept of life: life is a whole, so it must have an end. For Søren Kierkegaard and the existentialists following him, being aware of one's own death is the key to real, serious, proper living. And for the postmodern philosopher Jacques Derrida, death is present inside live, being its principle of organization. Among bioethicists, this view has been strongly upheld by Leon Kass (2001, 2003).

By contrast, there have always been philosophers who were able to acknowledge the scandal of death. Jean-Paul Sartre and Albert Camus thought of death as the absolute negation of live. Herbert Marcuse and Theodor W. Adorno protested against the acceptance of death as a way of collaboration with a repressive society. Adorno

11 11. Nevertheless, the ontological status of transhumanist utopias differs from the ontological status of religious eschatologies. Even if the kingdom of God might, as a future utopia, eventually be reached, it cannot be thought of as being overcome afterward. Its location in time is in the future or present, but never in the past. A transhumanist utopia is, on the contrary, just one in a sequence of future states. It is part of history, not the end of it. This has important consequences, some of which are explored below.

stated that "without the conception of, yes, life without chains, freed from death, the thought of the utopia could not be thought of at all" (cited in Hügli, 1998, col. 1237, my translation).

In the Christian theological tradition, there has always been a tension between the conception of death as a curse resulting from sin, on the one hand, and death as a part of God's good creation on the other. There have been many attempts to solve this tension, the approach of Friedrich Schleiermacher being the most prominent among them. For Schleiermacher, death is not the consequence of sin but belongs to human nature. But it is sin that makes death an evil for us. Nevertheless, according to Eberhard Jüngel, this tension cannot be solved. Death from a theological point of view has to be understood as a curse first, as an essentially negative moment. Death is an "anomaly" (Tilly, 2012, p. 14). After this is said, the question of a possible acceptance of death and dying in the light of salvation can be asked (Jüngel, 1971; 2005, p. 440). This has ethical consequences: especially in bioethical contexts, theologians have sometimes presented themselves as advocates of "natural dying" (Koch, 2005, p. 443), putting medical attempts to prolong live under suspicion. But in a time of developed medicine, the talk of "natural dying" is not convincing. Moreover, any reference to the nature of the human being in bioethical contexts is highly problematic—in its metaphysical as well as in its naturalistic variants.[12] With the transhumanists, death has to be conceptualized basically not as an element of human nature in the normative sense of the word. Death, and with it illness and suffering, are not acceptable; they are legitimate objects to fight against.[13] Theology must not sacralize the "natural" in the sense of limited, restricted, suffering lives (Körtner, 2005). It is important to emphasize that this protest against death and illness has got nothing to do with any Christian or non-Christian hostility against the body and the physical (against Coenen, 2009, p. 149; 2008, p. 56; see also Münch, 2012). Especially if one takes the conception of embodiment seriously, death as the extinguisher of body, mind, and spirit altogether appears unacceptable.[14]

Let me specify this point. The argument against the acceptance of death, illness, and suffering, restrictions and limitations, is only legitimate if *one's own death* is

---

12 See Clausen, 2006, and the discussion of naturalness in the context of bioethics in Birnbacher, 2006.

13 Interestingly, a significant change has taken place between the transhumanist visions of the 1920s and of today. In the 1920s, the overcoming of death was often thought of as dissolution of the individual. On the contrary, the transhumanists of our day are too individualistic to support that view. Instead, they state the overcoming of death by every individual in itself (Coenen, 2009, pp. 147, 153).

14 This has to be distinguished from the question of whether mere existence is already a good in itself (see Watson, 2009).

regarded. Turning it into a general statement is highly problematic since it might lead to degradation and devaluation of life with disabilities and illnesses. I might long for overcoming my own restrictions because they make me suffer. But I am not allowed to judge the lives of others. The ethicist and scientist Gregor Wolbring indicated the dangers of general visions of human enhancement as far as the social status of humans with disabilities is concerned. Despite its emphasis on individual choice, transhumanism advocates a general ideal of "able-ism" (Wolbring, 2007; see also Bülow, 2012) that, in fact, would socially exclude the choice to accept one's condition—here is where the blind spots of transhumanism begin to emerge. Nevertheless, there is some important rational potential connected with transhumanist vision production. It is a tool of anthropological deliberation and may bring a sense of possibility, resulting from the utopian aspect of overcoming death and suffering.

Apocalyptics and Striving for Realization

For this we say unto you by the word of the Lord, that we which are alive and remain unto the coming of the Lord shall not prevent them which are asleep. (1 Thess. 4:15)

Almost all of those who do think that there will be a singularity believe it will happen in this century, and many think it is likely to happen within several decades. (FAQ 3.0, n.d.)

For many deceivers are entered into the world, who confess not that Jesus Christ is come in the flesh. This is a deceiver and an antichrist. (2 John 7)

The urgency of the need to replace religions with other types of meaning-fostering systems is all the more evident when we think of the inherent irrationalism of religion and its entropic retardation of progress (More, 1996).

Religions do not simply establish and hand down visions of a future serving as a fruitful contrast to how the world is. They also deal with the issue of realization. In this context, the state of perfection is a *future* state in a strong sense. The ideal is consequently temporalized: It *will* happen—maybe under certain conditions. This issue of realization opens up new questions: the question of agents: who is responsible to make it happen? the question of instruments: which means and tools will get us there? the question of time: how long will it take until we get there?

Jewish as well as Christian traditions provide two basic models of how the coming state of salvation will be reached (Erlemann, 1995, p. 393). The *evolutionary model* sees a continuous progressive realization of the hoped-for state in the course of history. Divine power and human actions will lead to improvement. Signs in history show things are getting better. This is—*mutatis mutandis*—the model of the Hebrew prophets of salvation (Is. 40–55), of the progressive notion of history in early Christianity, of the reclamation of inner-worldly hope in 17th-century Pietism (Johann Jakob Spener; see Breul, 2012), but also, in a secularized version, of the enlightened concept of progress, or reimported into theology again, of the idea of a gradual establishment of the reign of God due to cultural progress in 19th-century Protestant theology (Richard Rothe).

And, then, there is a second model, which is prominently represented in apocalyptics: *revolution* instead of continuous progress. In apocalyptic writings, things get more complicated. The future is imagined not only as the perfect state but as a detailed sequence of states that will follow each other until the perfect state is reached. A typical sequence includes times of hardship, then a reign of peace on earth, lasting 1,000 years, followed by a disastrous fight between the powers of good and evil. Finally, the latter will be annihilated, before the state of perfection will arise. Each stage is marked again by a set of signs that can be read. You see them if you know: signs—political or natural catastrophes—tell where we stand and what is to do now. Apocalyptics, be it in its Jewish, Christian, or secular variants (Tilly, 2012, pp. 124ff.), has always been a reaction to the experience that the ideal state of society and of human condition could not be imagined as an end of the road. The current era is so negative in its state and tendency that a continuous process from now to then cannot be conceived of. Thus, not development, only revolution can get us from here to there. Apocalyptics is an answer to the question of mediation between reality and the ideal under difficult conditions.

Both the evolutionary and revolutionary models try to imagine a road from imperfection to perfection. Time serves no longer just as a symbol for the distance but as a viable way from now to then. We will get there, in the course of history. This vision is not for anthropology's sake, but for ethics and for the philosophy of history. Its aims are time diagnosis (that is, "where are we?")[15] and coordination of actions (that is, "what are we to do?"). And it aims to convey a clear message: there will be a good end, even if current events do not seem to point in that direction. Our actions, but also catastrophes and divine intervention, will transport us from our present to the desired future.

---

15 For a historical perspective on time diagnosis, see Meumann, 2012.

It is easy to see that transhumanism employs both models of realization. On the one hand, continuous technological progress will get us to the state we desire. On the other, the ideal seems to be so ambitious that evolution might be too slow compared with the urgency in which the ideal is desired. For example, 150,000 people dying each day—in a normal, not even a catastrophic scenario—is an intolerable loss of lives that will lead to a "sense of urgency" (Bostrom, 2003). Thus, any retardation of technological progress in the field of life extension is not acceptable. Some transhumanists are convinced that one cannot rely just on evolutionary progress. Their answer is *the singularity* (Vinge, 1993; Kurzweil, 2005; Eden et al., 2013).[16] Not our restricted brains but a "smarter-than-human intelligence" (FAQ 3.0, n.d.) will from some future moment on guide the development. If we manage to set up such a superior technological intelligence, progress will speed up and finally get an infinite slope: evolution becomes revolution. This idea of the singularity especially has been diagnosed as religious or pseudo-religious (Bringsjord, Bringsjord, & Bello, 2013; Schummer, 2009, p. 63).

But the parallel goes deeper. Apocalyptics always goes along with a strong dualism—in multiple respects. First, there is a social dualism, a dualism of *agents*. The "children of light" and the "children of darkness" stand against each other, and everyone has to know which side to choose. On the one side are fighters for the good powers; on the other are the ones who hinder or at least slow down the victory of the good.[17] In Christian apocalypticism, antichrist figures appear as part of the apocalyptic struggle. For Max More, ironically enough, religion itself takes the role of the antichrist. It is the prominent power of "entropy," advocating stagnation, whereas the children of light—in More's term, the "extropy" party (1996 [1990])—try to overcome this resistance. In a similar sense, Steve Fuller distinguishes between "up-wingers" and "down-wingers." It is not the old dichotomy of political life—the opposition of "right-wing" and "left-wing" —that will dominate future controversies. Instead, it is the antagonism between progress-oriented optimists on the one hand, combining libertarian Rightists and technocratic Leftists, and earth-bound premodernists, on the other, where the conservative Right and the communitarian Left join (Fuller, 2013, October 24).[18]

16 As Jacob Taubes (1991) notes, "The 'soon' belongs to the nature of apocalyptic belief" (p. 32, my translation).

17 For the context of this concept in Jewish sectarian apocalypticism, see Tilly, 2012, pp. 66–67.

18 For this dualism of agents in early 19th-century transhumanism, see Coenen, 2010, p. 73.

Along with this strong social dualism comes a dualism of *actions*, an ethical dualism. Apocalyptical movements were always connected with a strong, sometimes ascetic, bulk of moral imperatives. Apocalyptical visions call to action. *Now* is the time of decision: the sentence of the expected Last Judgment will depend on whether you stuck to the community of the righteous and their rules or whether you have failed to obey (Tilly, 2012, pp. 44–46). This ethical dualism is also at the core of transhumanism: "If we do the right things, a wonderful posthuman future with limitless opportunities for growth and flourishing may lie ahead" (Bostrom, 2003, p. 54). And then, all that techno-optimism comes up, and the instructions oscillate between sales promotion texts ("Sign up for cryonics"), mother's exhortations ("Live healthily and avoid unnecessary risks"), and options for legislation and a research agenda.[19]

Finally, there is a third dualism of apocalyptical movements: the dualism of the *worlds*. Since a continuous road from here to there cannot be imagined, the forthcoming world of plenty cannot be imagined as having the same ontological characteristics as the present world (Tilly, 2012, pp. 13–16). Some transhumanists vigorously oppose the argument that their thinking is dualistic (More, 2013, p. 7). In fact, transhuman ideas rely heavily on dualistic assumptions. Take the idea of "uploading": when the neuronal structure of the brain can be perfectly simulated on computers or other information technological systems, then we can survive on another physical basis. This basis is still physical, so it is not a Cartesian substance dualism but rather computer functionalism (Münch, 2012). Nevertheless, it is presupposed that our "wetware" existence now and our future existence uploaded on a computer are *the same*. This identity assumption implies that mind and brain, hardware and software, spirit and body, inner representations and outer objects can be separated from each other. This dualistic approach is accompanied by a strong hostility against the body and the physical in transhumanism (Coenen, 2010, p. 71; Münch, 2012), even if transhumanists explicitly state the opposite (More, 2013, p. 15). More precisely, in transhumanism we see the concurrence of (naturalistic) monism and (computer functionalism-based) dualism, a phenomenon that Jacob Taubes has in general described as the metaphysical core of apocalyptics (1991, pp. 34–35).

Especially in the adoption of apocalyptical motives, transhumanism inherits some of the reflexive potential of religion: the question of agents and ethics, the problem of mediation between reality and the ideal. But it also loses reflective potential. By deliberately excluding the divine agent in the evolution or the apocalyptic revolution, respectively, the questions of power, powerlessness, and control cannot

19 Bostrom, 2003, p. 52. See also pp. 34–35, where he cites the advertisement text of the "Cryonics Institute."

be directly addressed. With the symbol "God," religions have a strong tool to reflect on the distribution of power and especially on the finitude of human agency. The careless naïveté of the transhumanists' concept of control, especially of self-control, can be seen in the vision of Johan Nebbeling: "We will become managers of the mixing console of our brains" (cited in Coenen, 2009, p. 149). Who is that "I" that will be sitting at the mixing console of my brain? It must be a kind of homunculus, a little agent within the brain that is not identical with the brain, a replication of me inside me. This is the infinite regress of the problem of self-control, a *reductio ad absurdum*. By imagining ourselves as perfect subjects of self-control, we become nothing but objects of control; subjectivity vanishes. There is much to say about this age-old philosophical problem. Here, it only can be stated that the concept of God has always been a powerful means within religious thinking to articulate this problem and to reflect on possible solutions.[20] Hence, the problem of transhumanist apocalyptics without God is not so much its antireligious orientation: it is the lack of rationality that comes along with it.

Parusia's Delay and Reflective Vision Management

And when he was demanded of the Pharisees, when the kingdom of God should come, he answered them and said, The kingdom of God cometh not with observation: Neither shall they say, Lo here! or, lo there! for, behold, the kingdom of God is within you. And he said unto the disciples, The days will come, when ye shall desire to see one of the days of the Son of man, and ye shall not see it. And they shall say to you, See here; or, see there: go not after them, nor follow them. (Luke 17:20–23)

Sign up for cryonics. (FAQ 3.0, n.d.)

Third and last, there is one weak side of the revolutionary expectations within apocalyptics: historical falsification (Erlemann, 1995, pp. 393–394). What if the revolution does not happen? For the early Christians, the time of perfection to come soon was the parusia, the return of Christ. But the lesson was painful: parusia was delayed. Christ has not come. The expected and hoped-for future remained where

20 Take, for example, the famous theological debate between Martin Luther and Erasmus of Rotterdam on the question of the free will, or Schleiermacher's concept of the feeling of ultimate dependence (*schlechthinnige Abhängigkeit*) being the core of religion.

it was—in the future. The disappointment of hopes for immediate perfectionist fulfillment opened up a space of history (in the proper sense of the word), a new time span that is characterized by a deep ambiguity: finitude on the one hand and the expectation of the good-to-come on the other. Parusia's delay opens up *life on probation*.

Times of visionary disappointment were among the most dynamic eras in history: examples are the early Christian missionary movement and Pietism around the year 1700. Visions of perfection were not given up but transformed into orientations for living in present ambiguity. Thus, religious hope remained alive, but it abandoned the naïve fantasies of a coming land of plenty where we will live happily forever, and it was transformed to the idea of a fulfillment in God that is beyond imagination. A separation took place: the trusting endeavor for improvement on the one hand (which is ours) was separated from the idea of completion and perfection on the other (which is God's). By this, religious visions achieved their deepest impact on modern culture. I will call this aspect *differentiation* between improvement and perfection.

So, a third model of mediation between reality and the ideal arose: the *revelatory* model. All that is needed for perfection—which means all that is needed to solve the problem of our finitude—has already happened. It has to be discovered (Erlemann, 1995, p. 394). The only thing to expect is the hidden to become obvious, the new reality to enter our lives. There is a broad theological discussion about how this can be conceived of; for our argument, it is only important to notice that the religious differentiation between perfection and improvement, salvation and cure, God's termination of history and our actions within history again opens up an important reflexive potential. I will elaborate this by stressing two aspects. First, this differentation bears in itself a profound element of *secularization*. Conduct of life, societal institutions, politics, ethics, technology: human agency in all its contexts can be understood as secular, as unburdened from the demand to bring about salvation or perfection. It belongs, as Dietrich Bonhoeffer termed it, not to the last but to the next-to-last. This secularization on religious grounds implies a second element of reflexive potential. To say it pithily, it allows one to free oneself from the content of one's own hopes. It allows one to ask the question "Will that which I hope for be a solution for problems that I have today?" For example, pleasure surely is worth striving for. But a state of eternal pleasure is nothing I can want to achieve in history. The *visio beatifica*, the eternal pleasure in God's presence, is a visionary narrative about the throne room of God, not a guiding orientation for human actions in history. In the latter case, this ideal immediately changes into a dystopia (Coenen, 2009, p. 140), such as the totalitarian consumerist vision in

Aldous Huxley's *Brave New World* (1932).[21] The reason for this dystopian turnaround is the extensive elimination process that would be needed to get every source of displeasure out of the way. Inner antagonisms, outer conflicts, any confrontation with something or someone beyond my control—including myself!—would have to be eliminated. What would remain is an empty self—the perfect mystic or the perfect, unresisting consumer.

In the light of the above question, it is easy to see that the problem of finite human existence is not solved by an infinite lifespan. The ideal is not a solution for the actual problem. The problem is that human life will never be perfect and complete. It is full of ambiguities, broken expectations, corrupt relationships, dilemmas, failure, and guilt. Living longer, as the transhumanist proponents of radical life extension envision, will not solve these problems of human existence. For example, if I cause a car accident in my twenties and carelessly kill somebody, then an extended lifespan will give me the opportunity to experience that guilt longer, but it will not make the experience of forgiveness more accessible. And, important to say, this is not a psychological problem. Even if I could suppress the *emotion* of guilt by means of psychotropic drugs, guilt as a moral fact will remain. Yes, death is a scandal, and it is worthwhile fighting against it. But the abolition of death will not solve the problems of human finitude.

This can be elaborated by means of the religion philosophy of the theologian Ulrich Barth (2003). According to Barth, it is the fundamental act of religious self-interpretation of the human subject to realize that, in its finitude, fragmentarity, limitation in time, and contingency, it can claim a meaning ("Sinn") and purpose ("Bestimmung") that go beyond all that relativity. But then, the religious subject becomes aware of the fact that its actual, limited life, on the one hand, and its claim for ultimate meaning, on the other, fall apart and cannot be brought together under the conditions of human existence. Even an unlimited life will be lived under conditions of time (see also Kass, 2001, p. 22); and even complete self-control will not be able to overcome the contingency and particularity of human living. Religious symbols such as the distinction between human and divine agency (and, thus, between improvement and perfection) allow us to become aware of and to communicate these insights.

This aspect of differentiation between improvement and perfection is the blind spot in transhumanism. Transhumanists share the early Christians' expectation that the state of perfection is near. And they as well are afraid of parousia's delay: What if I die before technology is able to prolong my life as much as I desire? Concern is great. But the answer transhumanists give us is not that we must face

21 For the totalitarian character of apocalyptic politics, see Gray, 2009.

the ambiguities of life on this bank of the river that—yet—separates us from the land of plenty. Rather, the transhumanist answer is the *technological denial of ambiguity*: cryonics. Get yourself deep-frozen, switch to stand-by. They will surely defrost you as soon as they will have developed the technologies that you now have to miss. In the fridge, it is easy to skip the problem of disappointment that comes with every vivid vision of perfection. Correspondingly, other technological visions of transhumanism can be understood as technological denial of ambiguity: so, the technologies of emotional self-control by means of psychotropic drugs or the technology of overcoming the contingencies and limitations of bodily existence by "uploading" one's mind on a computer. The first loses the self it aims to control, while, in the second case, even the "uploaded" mind will face contingencies and limitations—now in the computer code it "lives" by.

To sum up, the distinction between perfection and improvement is an important accomplishment of religious thought, one of the religious antidotes to cope with the visionary power traditional Western religions come along with. It allows us to articulate visions that express deep longings and hopes originating from the conditions of human existence; and it allows us to distinguish between these possibly overwhelming visions and political programs or individual goals. Thus, in this religious figure, we find a profoundly antitotalitarian and humane rationality. Nonetheless, it also bears a danger. Late 19th-century liberal Protestantism especially used this differentiation to transfer all religious hopes and all content of religious visions to inside the subject. It is all about personality, subjectivity, the internal human. As a consequence, the religious concern about the outer world was lost. Religious cosmology was suspended, handed over to the natural sciences (Moos, 2012). Hope could only be articulated for single subjects or for human society. The reign of the physical, nature, the human body were excluded from the process of salvation. It is exactly this gap that naïve crypto-religious eschatologies such as transhumanism try to fill—at the expense of the concept of a rich inner life, by substituting the empty consumer-self for it.

Conclusion: The Lack of Reflexivity

Indeed, the transhumanist narrative of technology-driven process secularizes traditional religious beliefs. Transhumanism inherits religious vision production in content and form. And it is a respectable heir to traditional religion, insofar as it opposes the glorification of and the cheap reconciliation with sickness and death. But the share of its heritage is only partial: transhumanism adopts all that religious overdrive that can be connected with visions of a coming state of perfection, but it does not inherit the reflexive dams that religions developed to cope with that visionary overdrive. Transhumanism did not inherit the strong secularizing force of Christianity that allows it to keep in check the overwhelming visions of a land of plenty in all their naïveté. Transhumanism desecularizes these traditions and preaches salvation from our finitude.

Hence, transhumanism can be understood as an "eschatological mutation of a naturalistic world view" (Coenen, 2010, p. 86, *my translation*). By the structural similarities to religious eschatologies, transhumanism participates in the power of visionary thinking. By taking only half of it—and by not even adopting the reflexive potential of secular utopian thinking that also separates perfection (utopia as a nonplace) and improvement (realization) (Dierse, 2001; Saage, 2009, p. 133; Nida-Rümelin & Kufeld, 2011)—transhumanism lacks the rationality that is needed to cope with visions and ideals. Secular utopianism and religions have, each in its own way, developed strategies to keep in check their own visionary power. Transhumanism has not. It is quite ironic that this halfhearted criticism of religion is responsible for the lack of rationality and reason within transhumanist thinking. Thus, I argue for a deliberate use of the secularizing power of religions—especially their capacity to differentiate between salvation and improvement, leading to a critical distance to the content of one's own hopes—to find a realistic, effective, and humane (!) way to deal with modern technologies and the clouds of visions and countervisions surrounding them.

References

Armitage, J., & Roberts, J. (2002). Chronotopia. In J. Armitage and J. Roberts (Eds.), *Living with cyberspace: Technology and society in the 21st century* (pp. 43–56). New York, NY: Continuum.

Barth, U. (2003). Was ist Religion? In U. Barth, *Religion in der Moderne* (pp. 3–27). Tübingen, Germany: Mohr Siebeck.

Barth, U. (2004). *Aufgeklärter Protestantismus*. Tübingen, Germany: Mohr Siebeck.

Birnbacher, D. (2006). *Natürlichkeit*. Berlin, Germany: de Gruyter.

Blumenberg, H. (1988). *Die Legitimität der Neuzeit*. Frankfurt am Main, Germany, Suhrkamp.

Bostrom, N. (2003) *The transhumanist FAQ. A general introduction. Version 2.1.* Available at: http://www.transhumanism.org/resources/FAQv21.pdf.

Breul, W. (2012). "Hoffnung besserer Zeiten." Der Wandel der "Endzeit" im lutherischen Pietismus um 1700. In A. Landwehr (Ed.), *Frühe neue Zeiten. Zeitwissen zwischen Reformation und Revolution* (pp. 261–282).Bielefeld, Germany: Transcript.

Bringsjord, S., Bringsjord, A., & Bello, P. (2013). Belief in the singularity is fideistic. In A. H. Eden, J. H. Moor, J. H. Sorakher, & E. Steinhart (Eds.), *Singularity hypotheses: A scientific and philosophical assessment* (pp. 395–409). Heidelberg, Germany: Springer.

Bülow, M. H. (2012). Das gute alte Hirn. Wie die Sorgen um eine alternde Gesellschaft und die Ideen zum kognitiven Enhancement in den Neurowissenschaften interagieren. In M. Eilers, K. Grüber, & C. Rehmann-Sutter (Eds.), *Verbesserte Körper—gutes Leben? Bioethik, Enhancement und die Disability Studies* (pp. 269–285). Frankfurt am Main, Germany: Peter Lang.

Cady, L. E. (2012). Religion and the technowonderland of transhumanism. In H. Tirosh-Samuelson & K. L. Mossman (Eds.), *Building better humans? Refocusing the debate on transhumanism* (pp. 83–104). Frankfurt am Main, Germany: Peter Lang.

Clausen, J. (2006). Die "Natur des Menschen": Geworden und gemacht. Anthropologisch-ethische Überlegungen zum Enhancement. *Zeitschrift für medizinische Ethik. Wissenschaft, Kultur, Religion*, 52, 391–401.

Coenen, C. (2008). Verbesserung des Menschen durch konvergierende Technologien? Christliche und posthumanistische Stimmen in einer aktuellen Technikdebatte. In H. Böhm and K. Ott (Eds.), *Bioethik:Menschliche Identität in Grenzbereichen* (pp. 41–123). Leipzig, Germany: Evangelische Verlagsanstalt.

Coenen, C. (2009). Transhumanismus und Utopie. Ein Abgrenzungsversuch aus aktuellem Anlass. In R. Steltemeier, S. Dickel, S. Gaycken, & T. Knobloch (Eds.), *Neue Utopien. Zum Wandel eines Genres* (pp. 135-168). Heidelberg, Germany: Manutius.

Coenen, C. (2010). Zum mythischen Kontext der Debatte über Human Enhancement. In C. Coenen, S. Gammel, R. Heil, & A. Woyke (Eds.), *Die Debatte über "Human Enhancement. Historische, philosophische und ethische Aspekte der technologischen Verbesserung des Menschen* (pp. 63–90). Bielefeld, Germany: Transcript.

Coenen, C., Gammel, S., Heil, R., & Woyke, A. (Eds.). (2010). *Die Debatte über "Human Enhancement". Historische, philosophische und ethische Aspekte der technologischen Verbesserung des Menschen.* Science studies. Bielefeld, Germany: Transcript.

Cole-Turner, R. (2012). The singularity and the rapture: Transhumanist and popular Christian views of the future. Zygon, 47 (4), 777–796.

De Grey, A. (2007). *Ending aging: The rejuvenation breakthroughs that could reverse human aging in our lifetime.* New York, NY: St. Martin's Press.

Dickel, S. (2009). Utopische Positionierungen. Vorüberlegungen zu einer Wissensoziologie des Utopischen und deren Anwendungsmöglichkeiten am Beispiel populärwissenschaftlicher Enhancement-Visionen. In R. Steltemeier, S. Dickel, S. Gaycken, & T. Knobloch (Eds.), *Neue Utopien. Zum Wandel eines Genres* (pp. 169–200). Heidelberg, Germany: Manutius.

Dickel, S. (2011). *Enhancement-Utopien. Soziologische Analysen zur Konstruktion des Neuen Menschen.* Baden-Baden, Germany: Nomos.

Dierse, U. (2001). Utopie. In J. Ritter, K. Gründer, & G. Gabriel (Eds.), *Historisches Handbuch der Philosophie.* Band 11: U–V (pp. 510–526). Darmstadt, Germany: Wissenschaftliche Buchgesellschaft.

Eden, A. H., Moor, J. H., et al. (Eds.). (2013). *Singularity hypotheses: A scientific and philosphical assessment.* Heidelberg, Germany: Springer.

Eilers, M., Grüber, K., & Rehmann-Sutter, C. (Eds.). (2012). *Verbesserte Körper—gutes Leben? Bioethik, Enhancement und die Disability Studies. Praktische Philosophie kontrovers.* Frankfurt am Main, Germany: Peter Lang.

Erlemann, K. (1995). *Naherwartung und Parusieverzögerung im Neuen Testament. Ein Beitrag zur Frage religiöser Zeiterfahrung.* Tübingen, Germany: Francke.
FAQ 3.0. (n.d.) *Transhumanist FAQ 3.0.* Available at http://humanityplus.org/philosophy/transhumanist-faq/.
Fischer, S. (1988). Jewish salvational visions, utopias, and attitudes towards the halacha. *International Journal of Comparative Sociology*, 39 (1–2), 62–75.
Franssen, T. (2012). Prometheus steigt herab: Beeinträchtigung oder Enhancement? In M. Eilers, K. Grüber, & C. Rehmann-Sutter (Eds.), *Verbesserte Körper—gutes Leben? Bioethik, Enhancement und die Disability Studies* (pp. 205–227). Frankfurt am Main, Germany: Peter Lang.
Fuchs, R.-P. (2012). Gegen die Apokalypse? Zukunftsdiskurse im Dreißigjährigen Krieg. In A. Landwehr (Ed.), *Frühe neue Zeiten. Zeitwissen zwischen Reformation und Revolution* (pp. 237–259). Bielefeld, Germany: Transcript.
Fuller, S. (2013, October 24). *Ninety-degree revolution.* Aeon. Available at http://aeon.co/magazine/living-together/right-and-left-are-fading-the-future-is-black-and-green/.
Geraci, R. M. (2008). Apocalyptic AI: Religion and the promise of artificial intelligence. *Journal of American Academy of Religion*, 76 (1),138–166.
Gray, J. (2009). *Politik der Apokalypse. Wie Religion die Welt in die Krise stürzt.* Stuttgart, Germany: Klett-Cotta.
Grunwald, A. (2007). Grenzmanagement. Kontingenzsteigerung und neuer Orientierungsbedarf am Beispiel der technischen Verbesserung des Menschen. *Lifis Online.* Available at http://www.leibniz-institut.de/de/lifis-online.
Habermas, J. (2005). Religion in der Öffentlichkeit. In J. Habermas, *Zwischen Naturalismus und Religion. Philosophische Aufsätze* (119–154). Frankfurt am Main, Germany: Suhrkamp.
Hansell, G. R., & Grassie, W. (Eds.) (2011). *H+/: Transhumanism and its critics.* Philadelphia, PA: Metanexus.
Hügli, A. (1998). Tod. In J. Ritter & K. Gründer (Eds.), *Historisches Wörterbuch der Philosophie.* Band 10: St–T (pp. 1227–1242). Darmstadt, Germany: Wissenschaftliche Buchgesellschaft.
Jüngel, E. (1971). *Tod.* Stuttgart, Germany: Kreuz.
Jüngel, E. (2005). Tod VII. Dogmengeschichtlich und dogmatisch. In H. D. Betz, D. S. Browning, B. Janowski, & E. Jüngel (Eds.), *Religion in Geschichte und Gegenwart. Handwörterbuch für Theologie und Religionswissenschaft.* Band 8 T–Z (pp. 439-441). Tübingen, Germany: Mohr Siebeck.
Kass, L. (2001). L'Chaim and its limits: Why not immortality? *First Things*, 113, 17–24.
Kass, L. (2003). Ageless bodies, happy souls: Biotechnologies and the pursuit of perfection. *The New Atlantis*, 1, 9–28. Available at http://www.thenewatlantis.com.
Kipke, R. (2011). *Besser werden. Eine ethische Untersuchung zu Selbstformung und Neuro-Enhancement.* Paderborn, Germany: Mentis.
Koch, T. (2005). Tod VIII. Ethisch. In H. D. Betz, D. S. Browning, B. Janowski, & E. Jüngel (Eds.), *Religion in Geschichte und Gegenwart. Handwörterbuch für Theologie und Religionswissenschaft.* Band 8 T–Z (pp. 441-443). Tübingen, Germany: Mohr Siebeck.
Körtner, U. H. J. (2005). *Lasset uns Menschen machen. Christliche Anthropologie im biotechnologischen Zeitalter.* München, Germany: Beck.
Kurzweil, R. (2005). *The singularity is near: When humans transcend biology.* New York, NY: Viking.
Landwehr, A. (Ed.). (2012). *Frühe neue Zeiten. Zeitwissen zwischen Reformation und Revolution. Mainzer historische Kulturwissenschaften.* Bielefeld, Germany: Transcript.

Mannheim, K. (1986a). Das utopische Bewußtsein. In A. Neusüß (Ed.), *Utopie. Begriff und Phänomen des Utopischen* (pp. 265–285). Frankfurt/Main, Germany: Campus-Verlag.
Mannheim, K. (1986b). Utopie. In A. Neusüß (Ed.), *Utopie. Begriff und Phänomen des Utopischen* (pp. 113–120). Frankfurt/Main, Germany: Campus-Verlag.
Meumann, M. (2012). Der Zeitgeist vor dem Zeitgeist. Genius saeculi als historiographisches, mnemonisches und gegenwartsdiagnostisches Konzept im 17. und 18. Jahrhundert. In A. Landwehr (Ed.), *Frühe neue Zeiten. Zeitwissen zwischen Reformation und Revolution* (pp. 283–317). Bielefeld, Germany: Transcript.
Moos, T. (2005). *Staatszweck und Staatsaufgaben in den protestantischen Ethiken des 19. Jahrhunderts.* Münster, Germany: Lit.
Moos, T. (2012). Die kosmologische Verlegenheit moderner Religion. Zur praktischen Notwendigkeit einer theologischen Kosmologie. In F. Vogelsang, H. Meisinger, & T. Moos (Eds.), *Gibt es eine Ordnung des Universums? Der Kosmos zwischen Messung, Anschauung und religiöser Deutung. Festschrift für Professor Dr. Jürgen Hübner* (pp. 201–229). Bonn, Germany: Evangelische Akademie im Rheinland.
More, M. (1996 [1990]). *Transhumanism: Towards a futurist philosophy.* http://de.scribd.com/doc/257580713/Transhumanism-Toward-a-Futurist-Philosophy#scribd.
More, M. (2013). The philosophy of transhumanism. In M. More and N. Vita-More (Eds.), *The transhumanist reader: Classical and contemporary essays on the science, thechnology, and philosophy of the human future* (pp. 3–17). West Susssex, England: John Wiley & Sons.
Münch, N. (2012). Körperverachtung oder Phänomenologie der Leiblichkeit? Eine Kritik am Transhumanismus. In M. Eilers, K. Grüber & C. Rehmann-Sutter (Eds.), *Verbesserte Körper—gutes Leben? Bioethik, Enhancement und die Disability Studies* (pp. 287–310). Frankfurt am Main, Germany: Peter Lang.
Musil, R. (1970). *Der Mann ohne Eigenschaften* (special ed.). Reinbek bei Hamburg, Germany: Rowohlt.
Neusüß, A. (Ed.). (1986). *Utopie: Begriff und Phänomen des Utopischen.* Frankfurt/Main, Germany: Campus-Verlag.
Nida-Rümelin, J., & Kufeld, K. (Eds.). (2011). *Die Gegenwart der Utopie. Zeitkritik und Denkwende.* Freiburg, Germany: Alber.
Ritter, J. (1972). Fortschritt. In J. Ritter (Ed.), *Historisches Wörterbuch der Philosophie.* Band 2: D–F (pp. 1032–1059). Darmstadt, Germany: Wissenschaftliche Buchgesellschaft.
Saage, R. (2009). Zur Differenz zwischen dem konvergenztechnologischen Futurismus und der klassischen Utopietradition. In R. Steltemeier, S. Dickel, S. Gaycken, & T. Knobloch (Eds.), *Neue Utopien. Zum Wandel eines Genres* (pp. 116–134). Heidelberg, Germany: Manutius.
Schaper-Rinkel, P. (2013). Politiken des Human Enhancement. Transhumanistische Versprechen und die Analyse von technowissenschatlichen Zukünften. *Technikfolgenabschätzung, Theorie und Praxis*, 22 (1), 16–31.
Schölderle, T. (2012). *Geschichte der Utopie: eine Einführung.* Stuttgart,Germany: Böhlau.
Schummer, J. (2009). *Nanotechnologie. Spiele mit Grenzen.* Frankfurt am Main, Germany: Suhrkamp.
Scully, J. L. (2012). Auf moralisch unsicherem Terrain: Über Embodiment, Enhancement, und Normativität. In M. Eilers, K. Grüber, & C. Rehmann-Sutter (Eds.), *Verbesserte Körper— gutes Leben? Bioethik, Enhancement und die Disability Studies* (pp. 147–163). Frankfurt am Main, Germany: Peter Lang.

Seligman, A. (1988). Christian utopias and Christian salvation: A general introduction. *International Journal of Comparative Sociology*, 39 (1–2), 13–29.
Sloterdijk, P. (2009). *Du mußt dein Leben ändern. Über Anthropotechnik*. Frankfurt am Main, Germany: Suhrkamp.
Taubes, J. (1991). *Abendländische Eschatologie. Mit einem Anhang*. München, Germany: Matthes und Seitz.
Thompson, D. (1997). *Das Ende der Zeiten. Apokalyptik und Jahrtausendwende*. Hildesheim, Germany: Claassen.
Tilly, M. (2012). *Apokalyptik*. Tübingen, Germany: Francke.
Tirosh-Samuelson, H. (2011). Engaging transhumanism. In G. R. Hansell & W. Grassie (Eds.), *H+/: Transhumanism and its critics* (pp. 19–54). Philadelphia, Metanexus.
Tirosh-Samuelson, H. (2012a). Science and the betterment of humanity: Three British prophets of transhumanism. In H. Tirosh-Samuelson and K. L. Mossman (Eds.), *Building Better Humans? Refocusing the Debate on Transhumanism* (pp. 55–82). Frankfurt am Main, Germany: Peter Lang.
Tirosh-Samuelson, H. (2012b), Transhumanism as a secularist faith. *Zygon*, 47 (4), 710–734.
Tirosh-Samuelson, H., & Mossman, K. L. (Eds.). (2012). *Building Better Humans? Refocusing the Debate on Transhumanism. Jenseits des Humanismus: Trans- und Posthumanismus*. Frankfurt am Main, Germany: Peter Lang.
Vinge, V. (1993). *The coming technological singularity*. Available at http://mindstalk.net/vinge/vinge-sing.html.
Watson, J. D. E. (2009). The harm of premature death: Immortality—the transhumanist challenge. *Ethical Perspectives*, 16 (4), 435–458.
Wolbring, G. (2007). New and emerging sciences and technologies, ableism, transhumanism and religion, faith, theology, and churches. *Madang—International Journal of Contextual Theology*, 7, 79–112.

The Cognitive and Practical Relevance of Technological Visions

8

Franc Mali

Introduction

For human beings, the future is never vacuous but an open space filled with various expectations, either in form of hopes or fears. These expectations fulfill an important function because they serve as a reminder that the world does not have to be as it is—that there are other possible (future) worlds. More specifically, utopian dreams have always stimulated human actions with the aim to change the present, the existing world. In modern, future-oriented thinking, progress in science and technology plays a crucial role, even where such progress is seen from a critical perspective. Recently, the magnitude and speed of scientific and technological progress have created a wave of reflections on the role of new and emerging technologies in the building of desirable futures of human societies.

Different imaginaries exist of how humanity will cope with future scientific and technological progress. Each of these imaginaries is based on theoretical assumptions about the future. We can, for example, distinguish between dystopian and utopian visions of the future, between evolutionary and teleological conceptions of technological progress, and between the points of view that technology determines values and that technology is value-neutral. Undoubtedly, the "philosophy" of transhumanism, if compared to most other theoretical reflections on the future, is characterized by great degrees of optimism and activism concerning future technological progress. In transhumanism, activist approaches are combined with frequent references to very distant futures. Although transhumanism is often criticized as overly speculative, its visions and points of view challenge fundamental values and concepts of our current times and societies. Similar to utopian thought, transhumanistic imaginaries about the future may help us better understand our present societies. We can, therefore, ask the following questions: what do far-reaching transhumanistic visions of the future tell us about the present; what is their (implicit) criticism of the latter; and why and how do they want us to change?

In the following, I will first discuss forms of anticipatory governance (Barben et al., 2008) regarding the transhumanistic topology of the future. In recent years, transhumanistic visions have become influential in policy-related activities dealing with the development of so-called human enhancement technologies. Anticipatory technological governance is still in the process of emergence, but experts and policy actors around the world have already developed global scenarios and visions of the future specifically concerning a desirable transhuman society. More needs to be done, however, in particular with regard to future progress in thc area of human-enhancement technologies. I agree with those who argue (see, for example, Coenen et al., 2009; Ferrari; Coenen, & Grunwald, 2012, p. 216) that, at the core of discourse on human-enhancement technologies, is a general, future-oriented goal, namely, human enhancement, rather than a particular set of the new and emerging technologies.

There is a need to take into account more strongly interconnections of technological progress on the one hand and social values and human preferences on the other. To better understand the future prospects of human-enhancement technologies, we need to gain insights into both how social values shape these technologies and how these technologies influence social values. It is, of course, not possible to find a simple and general response to these questions.

In the second part, I discuss the interconnections of technological progress and social values in some detail, specifically with regard to human-enhancement technologies and transhumanistic visions, but also more generally concerning emerging areas of bioscience and technology. For the latter, I will use some cases from the emerging field of synthetic biology. This field is highly relevant in this context for several reasons: it is characterized by tendencies of technoscientific convergence, which appear to shape technological progress (including human-enhancement technologies) strongly in recent decades. Its products are expected to have major or even disruptive impacts on various applied areas of science and technology (such as health, energy, and the environment). As is the case with human-enhancement technologies and other emerging technologies, discourse on synthetic biology is influenced on the one hand by highly speculative expectations concerning the future, which raise "big" questions (for example, regarding notions of life and artificial life) and, on the other hand, by more short-term aspects (for example, concerning the marketization of the results of research and development). In discourse on emerging biotechnologies in general, various dimensions of natural and social life are problematized, for instance, safety and security, questions of human dignity and privacy, and the boundaries between the natural and the technological. In this regard, I emphasize the importance of experts in anticipatory governance of such technologies. As as has been argued elsewhere (Mali et al., 2012), the two established models regulating the relation between experts and policy makers, namely, the *decisionist* and the *technocratic* models, appear now to be outmoded. There is, in fact, a need to create and foster forms of governance of emerging technologies that are socially inclusive and involve all relevant stakeholders (with their various kinds of expertise) as well as the general public. This should, however, not be at the expense of the use of expert knowledge. I argue that experts still have to play a crucial role whenever we aim for rational governance of emerging fields of science and technology.

Technological Visions, Anticipatory Governance, and Social Values

An idea central in transhumanism is the assumption that science and technology will enable us to direct the course of our own evolution as human species. In transhumanism, human beings are understood as entities to be subjected to a radical extension of their capacities beyond their natural limits (see Fuller, 2012a; 2013).

Transhumanist visions sometimes refer to quite distant futures, several decades or even centuries away. They should, however, not be dismissed as mere science fiction. Transhumanist imaginaries about the future are distinct from pure fantasy since they are not merely expressed but also acted on, and "the idea of fantasy carries with it the inescapable connotation of thought divorced from projects and action" (Appadurai, 1996, p. 7). In recent years, transhumanist future visions have increasingly become relevant for policy actions dealing with new and emerging science and technology (see Brown & Michael, 2003; Coenen, 2008; Guston & Sarewitz, 2002). An important example is the so-called NBIC initiative (nano, bio, info, cogno) on converging technologies that started in the early 2000s (Roco & Bainbridge, 2002). The initiative and its policy agenda aim to promote the progress of new and emerging technoscience such as nanotechnology, biotechnology, information and communication technology, and cognitive science that may transform the very constitution of the human species. The initiative has recently been kick-started and has been strongly influenced in the 2000s by transhumanist ideas and networks (Coenen et al., 2009, p. 12). This is but one instance in which transhumanist imaginaries have some cognitive and scientific as well as practical political impact.

If we do not oversimplify, we might even fruitfully compare the far-ranging transhumanist visions with "thought experiments" of the past. Historically, theoretical and practical sciences have often benefitted from thought experiments concerning future technology. As Steve Fuller has stated, "Indeed, they have been productive throughout the history of science, not least when Einstein discovered relativity theory after imagining himself travelling on a light beam" (Fuller, 2012a, p. 43). Although far-reaching transhumanist visions include much speculation and do not exclusively deal with conceptual, heuristic, or "hermeneutic" questions (Grunwald, 2012, p. 203), we have no reason to object to them in principle.

Anthony Giddens (1998) and Niklas Luhmann (1997) have argued that contemporary societies are marked by a much more intense future orientation than those in the past. The premodern relationship to the future was dominated by a theistic orientation toward conserving the present in the future, preserving the natural order of things. The modern world sees a substitution of divine agency by human agency and, with it, the substitution of conservation by dynamics and innovation.

Anticipatory governance of new and emerging technologies evolves against the background of this general topology of the future in contemporary societies. Anticipatory governance can be understood as complex policy action based on a "distributed collection of social and epistemological capacities, such as future imagination and the disposition to learn from trial and error" (Barben et al., 2008, p. 992). It should be driven neither by unconditional prophecies nor by technocratic predictions based on technological determinism. Karl R. Popper distinguished "scientific predictions" from "unconditional historical prophecies" (Popper, 1963, p. 456), arguing that prophecies are unconditional statements about the future ("X will occur"). In Popper's view, representatives of historicism such as Plato, Hegel, and Marx have used such prophecies in social theory, and every historicism draws on the following two postulates: first, it postulates the existence of laws of social development and our ability to detect them; and, second, it presupposes that these laws of social development will allow us to foretell the future. Historicism follows in the footsteps of eschatology because "every version of historicism expresses the feeling of being swept into the future by irresistible forces" (Popper, 1957, p. 160). Historicism looks into the future by means of apodictic prophecies, and such prophecies cannot be refuted. Scientific predictions, on the other hand, are based on hypothetical rather than unconditional knowledge. They have a conditional form ("If the present state of X is such and such and X fits such and such a pattern, then the future state of X will be so-and-so"). Even if the future is different than predicted, Popper argues, we can learn from the prediction something about its conditions. In his view, either outcome, the confirmation as well as the refutation of predictions, is heuristically useful.

If the logic of hypothetic deductive inference is replaced with a logic of dogmatic deductive inference however, the future is seen as entirely "pre-configurated"

(Grunwald, 2012, p. 36), and, at the same time, this imagined future overwhelms the present (Nordmann & Rip, 2009, p. 273). This sort of determinism runs counter to the idea of anticipatory governance. The latter is an inherently forward-looking activity in the sense that expectations concerning the future inform the decision-making processes of today (Guston & Sarewitz, 2002, p. 99), but, in a framework of anticipatory governance, the future is not assumed to be entirely predictable. In determinism, on the other hand, technological progress is deemed "auto-dynamic" (Grunwald, 2012, p. 35) and does not allow for rational human decision making or reflexive ethics (ibid., p. 32).

The main deficiency of technocratic and quasi-eschatological visions of the future is that they neglect the active role of humans with regard to the formation of their future world. Anticipatory governance, to the contrary, makes use of the disposition of individuals and social actors to act with a desirable future in mind. In this context, visions of the future are social constructs, created and "manufactured" by people, groups, and organizations at respectively determined points in time (Grunwald, 2013).

Scenarios in anticipatory governance are based on the assumption that reduction of uncertainty about the future can be reached incrementally by a large number of small but informed decisions. According to Guston and Sarewitz, "This approach favors (but does not ensure) consequences that are manageable and error-correction that is both politically and practically feasible" (2002, p. 99). The main aim of anticipatory governance is, thus, not certainty by means of prediction, but the fostering of a disposition to continuously learn by means of trial and error. This is also the core element of Popper's conception of "social piecemeal engineering" (1962, p. 158). While social piecemeal engineering does, in fact, assume that humans are the masters of their own destiny in the sense that they can change their social life by means of the everyday procedures of trial, error, and social reform, it is not interested in eschatological questions, and it operates without a grand vision concerning the destiny of humankind.

If we look at the transhumanist vision of the future and its goal to use human-enhancement technologies in a massive way, it becomes obvious that, besides forecasting and processes of mutual learning by means of trial and error, anticipatory governance also needs to entail an element of reflexivity concerning the overarching goals of technological progress. Since human enhancement indicates such a goal rather than a particular set of emerging technologies (Ferrari, Coenen, & Grunwald, 2012, p. 216), the intellectual and societal engagement with this topic often involves more reflection on values and worldviews than detailed discussion of specific technological developments.

Topologies of the future can help modern societies establish anticipatory governance of emerging technologies. Around the middle of the twentieth century, various pessimistic theoretical views concerning the social implications of advanced technology were developed (Heidegger, 1977; Jonas, 1979; Marcuse, 1967). These and other highly pessimistic, even dystopian views have contributed to a rise of technophobia, a wariness toward technology that goes beyond reasonable consideration of negative effects of technological progress. It tends to "portray technology as a growing monster of our making that will increasingly determine the course and content of our lives and sap what remains of our humanity" (Hanks, 2010, p. 3). As Drengson wrote concerning neoprimitivist technophobia as an extreme reaction to technological progress, technophobia "can be compared to the disenchantment of early adulthood. One learns that attachments which are centered in romantic and erotic identification can frustrate growth and can generate suffering, pain, grief, and fear of loss" (Drengson, 2010, p. 30). Such technophobia is usually based on technological determinism, understanding technology as a force that transforms the nature of human action. The question here is not what we can achieve with technology or how we can influence technological trajectories, but what technology does with and to us. Such points of view run counter to theories and approaches that focus on social actors who deliberate on future technological trajectories and agree on measures to shape technological developments through society.

Transhumanist visions are much more optimistic with respect to future technological progress. Although some of these visions relate to a very distant time horizon and even sometimes raise extreme, eschatological questions (for example, in the case of Ray Kurzweil's views on the coming singularity or in the vision of a synthetic biology that re-creates life), many transhumanists are heavily engaged with the social shaping of technology, both in motivation and in outlook. They often focus on the progress of human-enhancement technologies. The concept of human enhancement is usually employed as an umbrella term that encompasses a wide range of existing, emerging, and visionary technical interventions in the human body, including attempts to modify cognitive functions (Buchanan, 2011; Coenen et al., 2009). If the theoretical relevance of human enhancement relates to issues of more general philosophical significance (for example, the definition of human nature, personal identity, and so forth), the practical relevance relates to issues such as how to address societal trends and developments aimed at enhancing the capacities of healthy humans, how to solve the numerous ethical conflicts in cases of practical choice, and how to create an adequate consensus that could direct general regulation and control of new technologies (Buchanan, 2011; Coenen et al., 2009).

From such a transhumanist point of view, human-enhancement technologies will help us gain more individual freedom and take responsibility for our lives and technological and societal development. These technologies are seen as a means to increase the plurality of social life and human existence. Steve Fuller has argued in this context in favor of individual freedom, innovation, and experimentation, so that human-enhancement technologies would flourish rather than be limited in their potentials due to an overcautious approach and the exaggeration of possible negative effects (Fuller, 2012a, 2012b, 2013). Future-oriented innovators, or moral entrepreneurs, should be free to use enhancement technology as a means to transform their cognitive and physical capabilities. Their experimentation with enhancement technologies may help us avoid possible societal threats of large-scale transhumanist transformations while, at the same time, allowing us to explore all the possible dimensions of this promising technological area.

Like all technologies, human-enhancement technologies are not value-free. There is an intimate connection between the progress of human-enhancement technologies and social values and preferences. To better understand the possible trajectories of human enhancement technologies, insights need to be acquired about how social values shape these technologies and how they, in turn, influence social values. Europe, for example, defines itself today, at least in the framework of the European Union and its member states' policies, as a community sharing such values and ideas as justice, solidarity, equality, dignity, citizen rights, freedom, and sustainability; and sees these values and ideas as the normative pillars of future technological progress (Göran, 2008; Scott, 2009). At the same time, various ethical dilemmas caused by technological progress have been widely discussed in Europe in recent years. There are national differences, however: in Poland, for example, emphasis has often been put on concerns regarding infertility treatment; in Germany, on prenatal testing and genetic screening; in Spain and Italy, on embryonic stem cell research; and in Ireland, on the administering of new experimental drugs without informed consent. These national differences in bioethical priorities have been pointed out, for example, in empirical studies in which the role and functioning of national ethical committees across Europe have been analyzed (Ahvenharju et al., 2006; Fuchs, 2005; Mali et al., 2012). Apparently, there is no common point of view in Europe—not to mention a common approach—regarding the question of which and how commonly shared values should influence specific technological development applications. And the normative framework is constantly changing. This pluralism of values and bioethical priorities is undoubtedly a source of strength of European sociocultural space but also a source of conflicts. New and emerging technological developments will most probably lead to an increase of conflicts concerning values, their interpretation, and prioritization with regard to technological progress.

Some Challenges on the Way to Anticipatory Technology Governance

Since the 20th century, the social shaping and political regulation of human-enhancement technologies have developed into highly complex challenges. There appears to be a need to go beyond established forms of social reflexivity concerning technological progress. Such established forms had been developed, for example, in the processes of institutionalization of technology assessment, publicly oriented bioethics, and ELSI/ELSA (ethical, legal, and societal implications/aspects) research since the last decades of the 20th century (Fisher, 2005). Recent ethical debates on new and emerging science and technology have problematized much of what had been deemed unquestionable in the past or—to quote Armin Grunwald—"bisherige Selbstverständlichkeiten" (2012, p. 177) in these contexts. To arrive at a good model of governance concerning human-enhancement technologies, potential benefits of these technologies need to be identified, and, at the same time, safeguards against potential risks need to be created; and all this necessarily has to take place under the condition of a very high degree of uncertainty concerning emerging technological developments.

Challenges are already raised by definitional questions. If we see, for example, human-enhancement technologies as merely extensions or as additional applications of existing science and technology, new regulatory regimes are not needed. If, however, we view the use of human-enhancement technologies as a radically new type of human and social activity, there might also be a need for new regulatory regimes. Interestingly, we have recently witnessed the development and extensive use of a strategy of double rhetoric with regard to these definitional aspects. In the case of synthetic biology, for example, the field is very often portrayed as something entirely new, but also very often as old wine in new bottles. At first glance, it is surprising that these two contrary points of view are often expressed by the same experts or stakeholders. The points of view that they express change with the contexts of discussion. If ethical dilemmas or risk issues are debated, synthetic biology is portrayed by those who argue in favor of a rapid development of synthetic biology as old wine in new bottles that should be dealt with in the framework of the regulatory regime for traditional genetic engineering. Scientists, industry stakeholders, policy makers, and others thereby try to persuade the nonexperts that synthetic biology's trajectory is under control; yet, in other contexts, the very same groups of actors emphasize the novelty of the field. To give but one example, in order to avoid eliciting an adverse reaction from the public, proponents of synthetic biology used, for example, the term "self-replicating complex biological entity" instead of "living organism" (Schmidt et al., 2010).

Another challenge to anticipatory governance of new and emerging technologies is to define the balance between caution and progress. Recently, some experts, mainly in the United States, have argued that a third way between "precautionary" and "proactionary" approaches is needed. While the precautionary principle emphasizes circumspection and might even lead us to slow down or stop technological developments, the proactionary principle, which was introduced by transhumanists, is based on a strong belief in the intrinsic value of progress in such areas as human-enhancement technologies. According to the proactionary principle, the assessment of risks should be limited to cases where evidence already exists for negative effects. Thus, the main goal of the precautionary principle is to prevent possible negative outcomes of technological developments, while the main goal of the proactionary principle is to seek the best available opportunities (Fuller, 2012b). As Fuller and Lipinska write, both principles are axiological in the sense that they are concerned with values: "The precautionary says that we are part of a larger whole called 'nature' and the meaningfulness of our lives (not to mention our sheer survival) is based on our appreciating that deep metaphysical point. … In contrast, the proactionary says that we are no mere part of nature; rather our existence gives meaning to an otherwise meaningless nature by serving as means to our ends" (2014, p. 99). According to Fuller and Lipinska, the new approach of "responsible research innovation" (RRI) in the European Union, as for example expressed by René von Schomberg (see, for example, 2012) is based on a precautionary approach, although it is presented as a spur to innovation (Fuller & Lipinska, 2014, p. 100).

Taking into account, however, that the RRI approach aims to shift the focus from preventing risks and other potential negative implications of new and emerging technoscience to the question of their "right impact" (Douglas & Stemerding, 2013, p. 140), we may argue that a dichotomization of a cautious European approach and a more daring one in the United States is exaggerated. Understood in this way, the RRI approach even fits with some of the transhumanist ideas concerning progress.

In a certain sense, the RRI approach has contributed to narrowing the gap between Europe and the United States in terms of the views on risk governance. This gap was, for example, evident in the respective dealings with the progress of modern biotechnology when, in the United States, it was argued for "risk assessment based on sound science" (Jasanoff, 2005, p. 4) while, in Europe, the focus was on a precautionary approach. Two reports from the early 2010s can also be deemed evidence for the narrowing of the gap. In a report of the U.S. Presidential Bioethics Commission on "new directions" in the ethics of synthetic biology and emerging technologies (PCSBI, 2010), it was strongly emphasized that we should develop an approach of "prudent vigilance," to be able to foster innovation and technological progress and, at the same time, take into account safety, security, and environmental aspects; the values of people; and societal needs. The approach of prudent vigilance is presented as an approach to deal with risks, emphasizing the importance of continuous responsible interaction between all stakeholders and societal groups involved in matters of new and emerging technologies. In a report of the European Commission's President's Science and Technology Advisory Council on "science for an informed, sustainable and inclusive knowledge society" (STAC, 2013), it was similarly argued that the new paradigm of "responsible research and innovation" shifts the focus from particular risks to innovation processes at large. Their governance should neither be technology-specific nor solely risk-focused, and it should be based on a transparent, interactive process in which various stakeholders become mutually responsive to each other concerning the (ethical) acceptability, sustainability, and societal desirability of innovation processes and its marketable products. It thus appears that recent approaches on both sides of the Atlantic are converging, at least at the normative level. A pluralistic social epistemology becomes the basis for cooperation of a wide variety of stakeholders, to be able not only to cope with the intrinsic uncertainties of technological progress but also to achieve the "right technological impact" (PCSBI, 2010, p. 65). In a new "art of trans-boundary governance" (Zhang, 2012), the main goal is to facilitate effective interactions between a wide range of social actors in which all of them have the opportunity to express their perspectives and interests at all stages in the pathway of research (ibid., p. 109). Since emerging technoscience will have a global impact, it is a good sign that governance approaches converge internationally. The time, in fact, appears to be ripe for "post-ELSI ethics" (Rabinow & Bennett, 2009, p. 100) and for more encompassing anticipatory governance activities in which the notions of the "right impact" of technologies and of "trans-boundary governance" could play central roles.

Transboundary governance models aim to allow a multitude of actors to imagine, criticize, and thereby shape emerging and future technologies and is, thus, of particular importance with regard to human-enhancement technologies. The latter constitute an area of technoscientific development and application strongly influenced by visions of the future that shape the present (Barben et al., 2008; Ferrari, Coenen, & Grunwald, 2012). To deal with the topic of human enhancement, the regulatory regimes concerning science and technology need to be extended and the range of engagement widened. This may also help avoid acceptance problems that were, to a large extent, created by technocratic approaches (as in the GM food case).

However, the critique of technocratic approaches and the rise of new governance approaches may lead to a flawed understanding of the role of experts in discourse on new and emerging technologies. Scientific experts still play a crucial role in such discourse (Bijker, Bald, & Hendricks, 2009; Weingart, 2001)—in our view, quite rightly so. To realize the goal of an inclusive social shaping of science and technology, society is well advised to continue to tap the intellectual resources of scientific experts, for example, with regard to the development of future scenarios and consensual strategies. As Harry Collins and Robert Evans have argued in their 2007 book *Rethinking Expertise*, expertise builds on skill acquisition that depends on tacit knowledge. The difference between the knowledge of experts and that of laypersons should still be taken into account in new forms of technology governance.

Seen from a more general and traditional sociological point of view, experts have the role of advocates for the rule of reason. With regard to discourse on human-enhancement technologies, the function of expert knowledge is, above all, important in the intermediary space between cognitive and practical values. Expert knowledge may help make visions of the future of human-enhancement technologies better informed and more rational. As long as they do not indulge themselves in hyping or demonizing technological developments, they can help foster reasonable discussions about current and future human-enhancement technologies and counter tendencies toward irrational speculation (Nordmann, 2007; Nordmann & Rip, 2009). Media studies and other research have demonstrated the effects of public rhetoric in discourse on new and emerging technologies (Christidou, Dimopoulos & Kouladis, 2004; Faber, 2006). According to Dorothy Nelkin, "[F]or most people the reality of science is what they read in the press" (1987, 2), and, in recent years, we have witnessed various instances of problematic media reporting on emerging technologies. In the case of nanotechnology, for example, wild visions of self-replicable nanobots running amok have strongly influenced public discourse (Bennett & Sarewitz, 2006; Bowman & Hodge, 2007; Loveridge, Dewick & Randles, 2008). Against this backdrop, policy actions should still be (and increasingly) based on trustworthy knowledge, in order to identify, assess, predict, and regulate new and emerging technologies.

In the case of human-enhancement technologies however, experts themselves have, unfortunately, contributed to the popularization of extreme positions and to ideological debates. The most outspoken proponents and critics of human enhancement have often been very vague as to the specifics of technological developments. Moreover, many proponents of human enhancement tend to see the development of ever-more powerful and safe human-enhancement technologies as inevitable (for example, Bostrom & Sandberg, 2009; Harris, 2007). The tendency to use rhetorical persuasion instead of evidence and arguments is also a feature of the works of many critics of human enhancement (for example, Fukuyama, 2002; Kass, 2003; Sandel, 2007). Technological alterations are rejected by means of appeals to vaguely defined concepts and are portrayed as potential causes of a future disaster for humankind. Some of them have tried to discredit enhancement and transhumanism in toto, by polemical attacks on all who are in favor of any kind of human physical or cognitive enhancement (Buchanan, 2011). Such ideological hardening of positions and polemical discussions, which focus on utopian or dystopian aspects, lead to nothing but a waste of intellectual resources (Pustovrh & Mali, 2013). They arguably constitute the most serious challenge on our way to prudent anticipatory governance of new and emerging technologies. If we have only such discussions, it would probably be better simply to wait until the future has become past.

Of course, such a wait-and-see attitude is not what we recommend. To create new forms of social reflexivity concerning human enhancement and other new and emerging technologies, we should design and experiment with new institutional settings and deliberative procedures (Biegelbauer & Hansen, 2011; Dryzek, 2010; Grunwald, 2013). In empirical research concerning discourse and activities on synthetic biology in Slovenia (Mali & Pustovrh, 2014), we gained insights into the points of view of scientists, policy makers, business representatives, and social activists in this country. Their knowledge concerning new governance and public participation approaches in technology policies was still sparse, and the same can most probably be said of many other countries in Europe and elsewhere. We lack new institutional mechanisms that could help mobilize and involve a wide variety of stakeholders, bringing them together and facilitating a sustainable and truly fruitful mutual communication with regard to visions of the future of technological progress. Short-term thinking still often dominates, in particular due to the time horizons of political decision making and the need for scientists and other stakeholders to adapt to them.

Conclusion

Visions of the future of technological progress are serious work done by serious people. It would be a grave mistake to exclude them in efforts to establish new forms of anticipatory governance. The same applies to transhumanist visions of the future, which should not be treated in toto as idle speculation or as mere fantasies, detached from reality. A crucial challenge that science and technology policies are facing nowadays is to create adequate long-range technological scenarios and strategies, and such scenarios and strategies will necessarily have to take into account the broad spectrum of transhumanist thinking. As regards the future progress of human-enhancement technologies, it is of utmost importance to avoid both the creation of hype around human enhancement and the demonization of any attempt to alter human corporeality. Of course, the benefits and risks of human enhancement cannot be fully understood as long as the relevant technologies and applications are still in their infancy. New governance experiments and settings are, thus, needed to foster rational discourse on technological developments in an approach of prudent vigilance or of similar approaches. In this context, we should also aim for a redefinition of the function of expert knowledge and of its role for risk assessment in the intermediary space between cognitive and practical values. If we want to avoid having the complex relations between values and technologies, between different values, and the competing interpretations of values lead to more conflicts about new and emerging technologies, we need to find new ways to foster rational discourse on these technologies and on the relevant values.

References

Ahvenharju, S., Halonen, M., Uusitalo, S., Launis, V., & Hjelt, M. (2006). *Comparative analysis of opinions produced by National Ethics Councils*. Helsinki, Finland: Gaia Group Ltd.

Appadurai, A. (1996). *Modernity at large: Cultural dimensions of globalization*. Minneapolis, MN: University of Minnesota Press.

Barben, D., Fisher, E., Selin, C., & Guston, D. (2008). Anticipatory governance of nanotechnology, foresight, engagement, and integration. In J. Hackett, O. Amsterdamska, M. E. Lynch, & J. Wajcman (Eds.), *The handbook of science and technology studies* (pp. 979–1000). Cambridge, MA: MIT Press.

Bennett, I., & Sarewitz, D. (2006). Too little, too late? Research policies on the societal implications of nanotechnology in the United States. *Science as Culture, 15* (4), 309–325.

Biegelbauer, P., & Hansen, J. (2011). Democratic theory and citizen participation: Democracy models in the evaluation of public participation in science and technology. *Science and Public Policy, 38* (8), 589–597.

Bijker, W., Bal, R., & Hendriks, R. (2009). *The paradox of scientific authority: The role of scientific advice in democracies*. Cambridge, MA: MIT Press.

Bostrom, N., & Sandberg, A. (2009). Cognitive enhancement: Methods, ethics, regulatory challenges. *Science and Engineering Ethics, 15* (3), 311–341.

Bowman, D. M., & Hodge, G. A. (2007). Nanotechnology and public interest dialogue: Some international observations. *Bulletin of Science, Technology & Society, 27* (April), 118–132.

Brown, N., & Michael, M. (2003). A sociology of expectations: Retrospecting prospects and prospecting retrospects. *Technology Analysis and Strategic Managment, 15* (1), 3–18.

Buchanan, A. (2011). *Beyond humanity? The ethics of biomedical enhancement*. Oxford, England: Oxford University Press.

Burk, D. L., & Lemley, M. A. (2002). Is patent law technology-specific? *Berkeley Technology Law Journal, 17* (4), 1155–1206.

Christidou, V., Dimopoulos, K., & Kouladis, V. (2004). Constructing social representations of science and technology: The role of metaphors in the press and popular scientific disciplines. *Public Understanding of Science, 13* (4), 347–362.

Coenen, C. (2008). *Konvergierende Technologien und Wissenschaften. Der Stand der Debatte und politischen Aktivitäten zu "Converging Technologies."* Büro für Technikfolgen- Abschätzung beim Deutschen Bundestag. Hintergrundpapier Nr.16, March 2008. Available at: http://www.itas.kit.edu/pub/v/2008/coen08a.pdf.

Coenen, C., Schuijff, M., Smits, M., Klaassen, P., Hennen, L., Rader, M., & Wolbring, G. (2009). *Human enhancement study.* European Parliament, DG Internal Policies STOA, Brussels. Available at: http://www.europarl.europa.eu/RegData/etudes/etudes/stoa/2009/417483/DG-IPOL-STOA_ET(2009)417483_EN.pdf.

Collins, H., & Evans, R. (2007). *Rethinking Expertise.* Chicago, IL: The University of Chicago Press.

Douglas, C., & Stemerding, D. (2013). Governing synthetic biology for global health through responsible research and innovation. *Systems and Synthetic Biology, 7* (3), 139–150.

Drengson, A. (2010). Four philosophies of technology. In C. Hanks (Ed.), *Technology and values* (pp. 26–38). Malden, MA: Wiley-Blackwell.

Dryzek, J. (2010). *Foundations and frontiers of deliberative governance.* Oxford, England: Oxford University Press.

Faber, B. (2006). Popularizing nanoscience: The public rhetoric of nanotechnology, 1986–1999. *Technical Communication Quarterly, 15* (2), 141–169.

Ferrari, A., Coenen, C., & Grunwald, A. (2012). Visions and ethics in current discourse on human enhancement. *Nanoethics, 6* (3), 215–229.

Fisher, E. (2005). Lessons learned from ELSI program: Planning societal implications research for the national nanotechnology program. *Technology in Society, 27* (3), 321–328.

Fuchs, M. (2005), *Nationale Ethikräte. Hintergründe, Funktionen und Arbeitsweisen im Vergleich.* Berlin, Germany: Nationaler Ethikrat.

Fukuyama, F. (2002). *Our posthuman future: Consequences of the biotechnology revolution.* New York, NY: Picador.

Fuller, S. (2012a). *Humanity 2.0: What it means to be human past, present and future.* New York, NY: Palgrave Macmillan.

Fuller, S. (2012b). Precautionary and proactionary as the new right and the new left of the twenty-first century ideological spectrum. *International Journal of Politics, Culture, and Society, 25* (4), 157–174.

Fuller, S. (2013). *Preparing for life in humanity 2.0.* New York, NY: Palgrave Macmillan.

Fuller, S., & Lipinska, V. (2014). The proactionary imperative. A foundation for transhumanism. New York, NY: Palgrave Macmillan.

Giddens, A. (1998). Risk society: The context of British politics. In J. Franklin (Ed.), *The politics of risk society* (pp. 23–35). Cambridge, England: Polity Press.

Göran, H. (2008): European values—and others, Europe's shared values: Towards an ever-closer union? *European Review, 16* (3), 373–385.

Grunwald, A. (2012). *Technikzukünfte als Medium von Zukunftsdebatten und Technikgestaltung.* Karlsruher Studien Technik und Kultur 6. Karlsruhe, Germany: KIT Scientific Publishing.

Grunwald, A. (2013). Modes of orientation provided by future studies: Making sense of diversity and divergence. *European Journal of Futures Research, 2* (1), 1–9.

Guston, D., & Sarewitz, D. (2002). Real-time technology assessment. *Technology in Society, 24* (1–2), 93–109.

Hanks, C. (2010). General introduction. In C. Hanks (Ed.), *Technology and values* (pp. 1–6). Malden, MA : Wiley-Blackwell.

Harris, J. (2007). *Enhancing evolution: The ethical case for making better people*. Princeton, NJ: Princeton University Press.
Heidegger, M. (1977). *The question concerning technology and other essays*. (L. Lovitt, Trans.). New York, NY: Harper & Row.
Jasanoff, S. (2005). *Designs on nature: Science and democracy in Europe and the United States*. Princeton, NJ: Princeton University Press.
Jonas, H. (1979). *Das Prinzip Verantwortung. Versuch einer Ethik für die technologische Zivilisation*. Frankfurt, Germany: Suhrkamp Verlag.
Kass, L. (2003). Ageless bodies, happy souls: Biotechnology and the pursuit of happiness. *The New Atlantis: A Journal of Technology & Society, 1* (1), 9–28.
Loveridge, D., Dewick P., & Randles, S. (2008). Converging technologies at the nanoscale: The making of a new world? *Technology Analysis & Strategic Management, 20* (1), 29–43.
Luhmann, N. (1997). *Die Gesellschaft der Gesellschaft*. Frankfurt, Germany: Surkamp Verlag.
Mali, F. (2009). Bringing converging technologies closer to civil society: The role of the precautionary principle. *Innovation: The European Journal of Social Science Research, 22* (1), 53–75.
Mali, F., & Pustovrh, T. (2014). Anticipatory governance of synthetic biology: Some challenges for Slovenia as a country with weak stakeholder networks. Paper at STS Conference Graz, Critical Issues in Science and Technology Studies, May 5–6, 2014, Graz, Austria. Available at http://www.ifz.aau.at/ias/IAS-STS/.
Mali, F., Pustovrh, T., Groboljsek, B., & Coenen, C. (2012). National ethics advisory bodies in the emerging landscape of responsible research and innovation. *Nanoethics, 6* (2), 167–184.
Marcuse, H. (1967). *Der eindimensionale Mensch. Studien zur Ideologie der fortgeschrittenen Industriegesellschaft*. Neuwied, Germany: Hermann Luchterhand Verlag.
Nelkin, D. (1987). *Selling science: How the press covers science and technology*. New York, NY: W. H. Freeman.
Nordmann, A. (2007) If and then: A critique of speculative nanoethics. *Nanoethics, 1* (1): 31–46.
Nordmann, A., & Rip, A. (2009). Mind the gap revisited. *Nature Nanotechnology, 4* (5), 273–274.
PCSBI (U.S. Presidential Commission for the Study of Bioethical Issues). (2010). *New directions: The ethics of synthetic biology and emerging technologies*. Washington, D.C.: PCSBI. Available at www.bioethics.gov.
Popper, R. K. (1957). *The poverty of historicism*. London, England: Routledge.
Popper, R. K. (1962). *The open society and its enemies*. (Vol. 1). London, England: Routledge.
Popper, R. K. (1963). *Conjectures and refutations: The growth of scientific knowledge*. London, England: Routledge.
Pustovrh, T., & Mali, F. (2013). Exploring some challenges of the pharmaceutical cognitive enhancement discourse: Users and policy recommendations. *Neuroethics, 7* (2): 137–158.
Rabinow, P., and Bennett, G. (2009). Synthetic biology: Ethical ramifications. *Systems and Synthetic Biology, 3* (1–4), 99–108.
Roco, M. C., & Bainbridge, W. S. (2002). *Converging technologies for improving human performance: Nanotechnology, biotechnology, information technology, and cognitive science*. Dordrecht, the Netherlands: Kluwer Academic Publishers.
Sandel, M. (2007). *The case against perfection: Ethics in the age of genetic engineering*. Cambridge, MA: Belknap Press of Harvard University Press.
Schmidt, M., Kelle, A., Ganguli-Mitra, A., & Vriend, H. (Eds.). (2010). *Synthetic biology: The technoscience and its societal consequences*. Dordrecht, the Netherlands: Springer.

Scott, N. (2009). Research ethics: European and Asian perspective, global challenges. In M. Ladikas (Ed.), *Embedding society in science and technology policy: European and Chinese perspectives* (pp. 21–39). Luxembourg: Office for Official Publications of the European Communities.

STAC (EC President's Science and Technology Advisory Council). (2013): Science for an informed, sustainable and inclusive knowledge society. Policy paper by President Barroso's Science and Technology Advisory Council, Brussels, August 29, 2013. Available at http://ec.europa.eu/commission_2010-2014/president/advisory-council/documents

Von Schomberg, R. (2012). Prospects for technology assessment in a framework of responsible research and innovation. In M. Dusseldorp & R. Beecroft (Eds.), *Technikfolgen abschätzen lehren: Bildungspotenziale transdisziplinärer Methoden* (pp. 39–61). Wiesbaden, Germany: Vs Verlag.

Weingart, P. (2001). *Die Stunde der Wahrheit? Zum Verhältnis der Wissenschaft zu Politik, Wirtschaft und Medien in der Wissensgesellschaft*. Göttingen, Germany: Velbrück Verlag.

Zhang, J. Y. (2012). The art of trans-boundary governance: The case of synthetic biology. *Systems and Synthetic Biology, 7* (3), 107–114.

Enhancing Machine Nature

From the Myth of the Machine to Sociotechnical Imaginaries of Nonmechanical Machines

9

Alfred Nordmann

Our conversations about human enhancement by technological means are premised on an image of technology: what it is, what it can do, by which of its virtues humans can transcend their present condition. This image of technology, however, is not technical but social. Günther Anders pointed out already that how we valorize technology expresses our sense of deficiency or vulnerability and, thus, *ex negativo*, a conception of a better life, of better humans in a better society (Anders, 1956). Half a century later, Sheila Jasanoff foregrounds sociotechnical imaginaries: all the stories of technological determinism or enablement, including the visionary expectations of emergent technologies, are framed by sociotechnical imaginaries (Jasanoff, 2015). These are imaginaries of how things can work together, and, when it comes to working together, this might be of individuals in a social body or of component parts in technological system.[1] Not surprisingly, therefore, technical ideas of regularity and efficacy become projected into conceptions of the economy or the state, and, inversely, machines can be emblems of cooperation, of a governance of parts.

This general consideration becomes salient in current discourse on emerging technologies, including its trans- or posthumanist concerns. This discourse tends to overlook that the enhancement, empowerment, or transcendence of human nature by technological means is predicated on ideas of the enhancement, empowerment, or transcendence of machine nature by way of sociotechnical imaginaries. Indeed, there is reason to believe that the envisioned technology of the future is entirely social and not at all technical. In other words, sociotechnical imaginaries do not merely valorize known technologies in certain ways but can form an idealized

1 Jasanoff (2015) quotes Charles Taylor on social imaginaries: "I am thinking … of the ways people imagine their social existence, how they fit together with others, how things go on between them and their fellows" (Taylor, 2004, p. 23).

image of a technology that is different and better than any known technology. This is the main thesis of the present essay. We learn to view technology in the image of the social by way of conjuring up imaginaries of a technology that somehow ceases to be technology-as-we-know-it and becomes something else, unheard of, better.

A horn of plenty or cornucopia, lamp of Aladdin, *Tischlein-deck-dich*, or molecular assembler is a wish-fulfillment machine that is the work not of engineers but of fairy tales. Only somewhat less obvious than these, the soft machines of nanotechnology, their ancestors and distant cousins represent not the rationalization of the world but its secular re-enchantment: this strikingly nontechnical machinery is wonderful, indeed. The soft machines to be discussed do not constitute a technological paradigm that then infiltrates biology or social relations. Instead, they are first of all the collective dream of a technology that is unprecedentedly different—almost magically and wondrously so—from anything that was thought to be realizable so far. Soft machines are social machines insofar as they provide an escape from the limited technology currently in existence, an escape into the imaginary world of a technology that offers consolation and hope and is designed to establish a kind of peace throughout society. So soft and gentle are these machines that every hard, rigid, or constraining technology yields to them completely—including our as yet far too rough-hewn regulatory policy tools. So soft and gentle are these irreal devices that they are able to cushion the impacts of all conflicts of interest and social antagonisms and accommodate all "stakeholders" quite comfortably.

What will be pursued in the following, then, is not so much how a notion of the technical comes to be inscribed in different social domains but rather how a socially powerful notion of the technical emerges in the first place. How on earth is it possible—one may end up asking incredulously—for an essentially technophobic concept of technology to provide social legitimacy not only to research funding but to dreams of escaping our current predicaments? The much diagnosed techno-optimism of our age is not simply an expression of confidence in the capacity of future technologies to solve all our problems. As Astrid Schwarz has shown in her discussion of "green nanotechnology," the notion of green technology works as a hollow phrase precisely because it lacks technical meaning, giving it a highly effective symbolic function in public discourse by way of its ability to accommodate various societal visions:

> The openness of discourse on the environment is used, then, to develop green nanotechnology as a space of possibilities compatible with societal wellbeing and sustainable in relation to nature. This is where the image of a kind of nanotechnology arises that is soft—or at least cautious—and that reduces stark oppositions. It draws together what is thought to be irreconcilable, for example, natural history and physical reductionism, application-oriented basic research and scientific progress, the

preservation of natural and cultural resources, limitation and transgression, scarcity and abundance. (Schwarz, 2009, p. 117)

Lewis Mumford's Megamachines

So much for a programmatic introduction that posits a broadened historical context for our current more and less green nanotechnologies along with their soft machines. The next step is to articulate the basic argument and provide examples to support it. The argument is that nontechnical social machines serve as imaginary ideals of both societal and technological development, that they thus become sociotechnical imaginaries.[2] This may sound somewhat paradoxical, at first, and rather abstract, but it provides a distant echo of Heidegger's notion of "enframing" (*Gestell*) and his observation that the essence of technology is not anything technological (Heidegger, 2007, p. 5). For Heidegger, this means, first of all, that the essence of technology has nothing to do with means-ends relations or with technology as a tool or an instrument. References in the following to "nontechnical" or even "technophobic" images of technology invoke magical notions of technology. According to these magical notions, technology is not primarily an ingenious way of extracting as much as possible from the limited resources of a limited world. Instead, nontechnical dreams of technology envision that the world could turn out to be limitless, after all, and that technology can alter even our conceptions of what is technically possible. While Ernst Cassirer draws a strict dividing line between this magical image of technology and the realities of engineering in the context of nature and society (1930, pp. 59–60), present-day discourse tends to gloss over this difference, especially when the phantasm is evoked of a technology capable of transforming our familiar world.

Fantasies of technical control assume that technology can advance "without encountering either technical delays or sobering human inhibitions," thereby turning "our dominant technology itself into the equivalent of scientific fiction." About technology as social and scientific fiction, Lewis Mumford goes on to say, "The one part of the human personality that so far eludes rational control is that

2 This conception adds one word to Jasanoff's definition of sociotechnical imaginaries as "collectively held, institutionally stabilized, and publicly performed visions of desirable futures, animated by shared understandings of forms of social life and social order attainable through, and supportive of, *imagined* advances in science and technology" (Jasanoff, 2015). To be sure, as Jasanoff points out, existing technological infrastructures and established technological achievements can also support sociotechnical imaginaries.

which produces these fantasies" (Mumford, 1970, pp. 223, 290). Writing in the 1960s, Mumford gave pointed expression to one paradoxical aspect of his broader technocritical analysis. Humans, he maintained, technologically transformed the conditions of their lives by calling upon the machine, and they now wish to experience wild, uncontrolled, irrational freedom, they have to imagine new humans. To understand what Mumford means by this, it is worth taking a closer look at the argument he presents in his book *The Myth of the Machine* (1967, 1970).[3]

According to Mumford, "[W]e cannot understand the role that technics has played in human development without a deeper insight into the historic nature of humans" (1967, p. 4). Humans express through technology their latent potentialities "to fulfill more adequately [their] superorganic demands and aspirations" (ibid., p. 8). The needs and desires that are superorganic here are those that—conceived in quite traditional philosophical terms—are associated with human freedom. Accordingly, the "dominant human trait" is the capacity "for conscious, purposeful self-identification, self-transformation, and ultimately for self-understanding" (ibid., p. 10). The self-understanding of humans as historically changeable beings occurs by the very means of technology itself, namely, in the encounter with a world that has been deliberately and consciously transformed by human hands. And what we become aware of when we see ourselves in our technology is the form of societal organization that precedes the reification and the concrete manifestation of technology: thanks to a division of labor that is organized in a specific way, that is, with the help of technology, humans create for one another a world of things in which they recognize themselves—and in which they may be able to transform themselves.

In this deliberately created form of organization, the "megamachine" makes its appearance. To define the megamachine, Mumford quotes from the standard engineering text of the late 19th century, *Kinematics of Machinery* by Franz Reuleaux (1875), which proved to be extraordinarily influential also among philosophers. According to Reuleaux and Mumford, a machine is a combination of resistant parts, each of which has a specific function. It operates under human control to utilize energy and to perform work (Mumford, 1967, p. 191; cf. Reuleaux, 1875, p. 38). Mumford claims that the history of machines that meet this definition goes back at least five thousand years. At that time, the machine was not yet visible as a device but existed solely as a form of societal organization that laid the foundations for the visible machines of the 19th and 20th centuries: "[T]he mechanical agents had first to be 'socialized' before the machine itself could be fully mechanized" (Mumford ,1967, p. 194).

3 The remainder of this section is based on the discussion in Nordmann, 2008, 53–56.

It was this invisible—in a material sense as yet nontechnical—working machine that enabled the pyramids to be built five thousand years ago. Though "made of human bone, nerve, and muscle," the components of the machine were, by definition, "reduced to their bare mechanical elements and rigidly standardized for the performance of their limited tasks. … The secret of mechanical control was to have a single mind with a well-defined aim at the head of the organization, and a method of passing messages through a series of intermediate functionaries until they reached the smallest unit" (Mumford, 1967, pp. 191–192). For Mumford, this social machine is not the societal manifestation of a technical ideal; rather, it exists prior to the construction of the mechanized machine as an engineered device. He envisages the free intellectual invention of a form of organization that already knows how to utilize mechanical forces before it becomes materialized in a fully mechanical labor-saving machine. In the case of the pyramids and a strict, almost totalitarian organization of labor, mechanical forces were still socialized rather than mechanized; as such, they were part of a societal mechanism rather than a technical one. The model for the interplay between cogwheels, screws, winches, and levers comes from the interaction between humans at work.[4] For Mumford, then, when people recognize themselves in the machine, they are not seeing in it their biological or individual nature but the way their society or the megamachine is organized, and, in this organization, they themselves are but one system component. And when they adopt the perspective of liberty as they view themselves as part of a machine culture, they are able to turn against this—only apparently all-encompassing—megamachine by means of conscious self-identification and self-transformation.

Otto Mayr's Centrifugal Governor

Within the philosophy of technology, Mumford's is perhaps the most significant and comprehensive version of the theory that ideas about technology are constituted by society long before the corresponding mechanical device is developed. Moving on from this, we now look at a more narrowly defined social machine, introduced here to act as a foil to "soft machines."

4 This accords well, then, with Joseph Pitt's definition of technology as "humanity at work" (Pitt, 2011, p. 70). However, as opposed to Mumford with his reference to Reuleaux, Pitt does not seek to specify this further.

As curator of the National Museum of History and Technology in Washington, D.C., historian of technology Otto Mayr wanted to write a history of the centrifugal governor, a mechanism for automatic regulation that plays a crucial role in the safety of technical systems. Steam engines produce pressure in a boiler that then drives a transmission system. High pressure in a boiler always brings with it the possibility of too much pressure building up, which would lead to an explosion. This is where the centrifugal governor comes into play. It consists of a few metal "flyweight" balls arranged as a kind of carousel that is propelled by steam pressure. The higher the pressure in the boiler, the faster the carousel rotates, and the more quickly it rotates, the stronger the centrifugal forces acting on the spinning flyweights: these rise higher and higher as the rotational speed increases until they eventually reach a horizontal position. Once the steam pressure in the boiler has reached a critical point, and with it the speed of the rotating flyweights so that the latter have nearly reached a horizontal position, the governor opens up a valve and steam is able to escape from the boiler. This causes the steam pressure to decrease, the carousel to rotate more slowly, and the fllyweights to sink back down. The valve now closes again so that only the right amount of steam escapes. At this point, the whole process can start over again, with the centrifugal governor acting as an automatic system of regulation without which the steam engine and the Industrial Revolution would not have been possible—a system of regulation, moreover, that works with a so-called feedback loop, as its behavior depends on the system of which it is a part, that is, on the system that it is designed to observe and on which it acts in return.

When he embarked on writing the history of this system, Mayr made an interesting observation: centrifugal governors had been described in a famous list of machines dating from the year 60 CE that had been reprinted in the 16th century and had become the blueprint for various mechanical showpieces, not least among them a host of clockwork-driven devices. However, well into the 18th century, there are no feedback mechanisms to be found in continental Europe but only in Britain where the technology had been "cultivated and appreciated." Why was this the case? Mayr's search for an answer led him to conclude that it was not possible to explain the rejection of these systems in continental Europe and their development in Britain by reference to any developmental trajectory inherent to the technology itself:

> I had noticed that, in eighteenth century Britain, the principle of the feedback loop had come into use not only in practical technology but also in abstract arguments, notably in Adam Smith's economic theory. Assuming that this was not coincidental, I tried to establish a connection. I tried to show that the use of the concept in abstract argument had been inspired by practical technology. The attempt failed; in the end, I became convinced that the connection was not direct but that each phenomenon

> independently was the result of some unknown earlier cause. This suggested that I take a closer look at the first question: Why was feedback rejected on the Continent? The question could not be answered in terms of developments in practical theory alone. ... Perhaps the answer was to be sought on another level. Apparently, certain kinds of mechanical inventions were immensely popular on the Continent, while others, including feedback devices, were not: What made the difference? (Mayr, 1986, p. xvi)

Essentially, Mayr's book claims that the technical development of clockwork mechanisms and of feedback systems was due to interactions "between the political, social, economic, or religious ideas dominant in a given society and contemporary preferences and designs of technological hardware" (Mayr, 1986, p. xv). Viewed from this perspective, Adam Smith's economic theory precedes the dissemination of the centrifugal governor in mechanical devices not just in chronological terms: the invisible hand that balances supply and demand finds its counterpart in the automatic action of the mechanism that maintains balance in the steam engine. Thus, Mayr refers to feedback mechanisms as "liberal systems," whereas the clockworks of baroque continental Europe reflect an immutable authoritarian system that is hierarchically organized. To imagine a liberal state is to promote self-regulation; accordingly, liberal economic theories appeared prior to technical refinements of the governor—which, in turn, enabled the governor to become a kind of technical metaphor for the modern state.

> The steam engine governor probably did more than any other agent to publicize the concept of self-regulation among engineers and the general population. The Watt steam engine was greeted as a machine of revolutionary importance and as the herald of a new age. No one would miss an opportunity to see this wonder in operation, and few who saw it would have failed to inquire about the purpose of those rapidly rotating centrifugal weights that were mounted conspicuously over the machine. To explain the concept of self-regulation, from that time on, one only had to point out the steam engine governor. When Norbert Wiener in 1947 christened his new science of cybernetics, he was expressly paying tribute to what he considered the earliest cybernetic device; the word *governor* is derived via the Latin *gubernator*, from the Greek [for] steersman. (Mayr, 1986, pp. 194–195)

A modern-day observer looking at the steam engine designed to operate the water fountain in the royal gardens of Potsdam, for example, will see—despite its rather weak performance by today's standards—a technological miracle housed in an exotic temple, high up on which the bronze governor sits enthroned, a polished and gleaming wise steersman or navigator. And by no means accidentally, atop the regulator as an emblem of enlightened governance sits the crowned Prussian eagle (see Figure 9.1). Likewise, to the present day, the art of government in a liberal state relies on a regulatory system. The material quality of products, the amount and composition of exhaust fumes, and the ingredients in food items are all subject to regulation, in many cases enshrined in law: there are established routines, if not automatic procedures, for observation, and corrective measures are taken once a defined threshold is transgressed. This system is motivated by the notion of balancing the innovative strength of the market against consumers' safety needs: if anything bad were to happen, the measure kicks in and balance is quickly restored.[5]

Fig. 1 Permission granted by photographer Astrid Schwarz.

5 The governor or cybernetic controller of the steam engine is not only a cipher of the "first industrial revolution" and not only the epitome of the technocratic "control and regulation" paradigm that was criticized by Heidegger and Marcuse. It continues to be influential in conceptions of "governance beyond the state" where it refers to self-regulatory mechanisms for the reconciliation of interests or the balancing of goods. Lösch, Gammel and Nordmann, 2009, ask whether perhaps a social machine other than the centrifugal governor might be required as a political model—one more appropriate to current modes of technoscientific and industrial production. The question is all the more urgent in light of the fact that the development of *soft machines* in nanotechnology, synthetic biology, information and communications technologies, and climate engineering eludes classical regulatory procedures. A critique of the *governance* concept on this basis can be found in the final section below.

Jessica Riskin's *Wetware*

Mumford's working machine and Mayr's centrifugal governor have served as examples to show that societies imagine technologies even before the latter actually exist in the form of specific devices or technically realized machines. These machines are at first completely social, materializing only at a later time; they are not inspired by technology. And it can be said that the regulatory control system with its automatic feedback mechanism and its balancing function was a model for the development of society and technology alike.[6]

As complex as the interplay may be between ideas of society and technical devices, it has given rise to a highly tangible technological world. Not only are there now very real factory floors on which work machines are organized into production lines, but also boiler safety technologies with their theoretical underpinnings and engineering practices. However, a more far-reaching claim was put forward at the beginning of this chapter, namely, that societies also find orientation in a quite nontechnical, irreal idea of technology. This will be shown in reference to *soft machines*, but this discussion, too, can be prepared by way of historical precedent. Jessica Riskin describes a short-lived fanciful technology that briefly appeared during the second half of the 18th century and that foreshadows a similarly precarious development in the current age of soft machines.

Only recently in Brussels, window shoppers might have encountered a rather large cat stretched out luxuriantly on its back amid piles of Belgian chocolates. Pretty soon they would have realized that the cat is a mechanical doll performing certain cat-like movements. It strokes its whiskers with its paw as a contented swell rises and falls across its thick underbelly. Jessica Riskin wrote a history of such devices, and, in her words, this doll functions analogically. Inside it is a kind of clockwork that functions like a mechanical model, and that is set up to represent certain visible signs or movements of a cat. The very fact that we recognize it as a toy and understand the representational character of the mechanical model makes it clear that this is an instance of a piece of hardware that technically mediates the gap between a real biological cat and the mechanics of its appearance. This hardware has little in common with the internal living essence of an actual cat, except that it is able to reproduce certain visible signs, albeit using quite different means than those of a real cat. According to Riskin, there were legions of such automata

6 At this point, it would be instructive to consider, for purposes of comparison, a perpetual motion machine (or rather the impossibility of such a machine) as a fictitious technology that served as an organizational form for technology and society and equally for science and industrial production procedures; see Rabinbach, 1992.

in existence in the 17th and especially the 19th centuries. However, between 1730 and 1790, the focus was on other constructions, which Riskin, in contrast to analogical *hardware*, describes as *wetware*—thereby using a very contemporary term from artificial intelligence research and biocybernetics.

Riskin illustrates the difference by describing various writing automata, that is, life-like dolls that move a feather quill across a sheet of paper. In the 17th and 19th centuries, such automata made it seem as if they were writing, moving the quill in characteristic manner through the air above the paper. In the 18th century, however, the aim was to make the apparatus actually write for real. And indeed a writing figure originating from the workshop of Jaquet-Droz was able to write up to 40 programmable words. The inside of this figure looked different as well:

> The Jaquet-Droz automata do not just carry out the processes of writing, drawing, and playing music, they are also anatomical and physiological simulations. Their skeletal structures were likely designed with the help of the village surgeon. Both the Lady-musician and the Draughtsman also breathe. The Draughtsman periodically blows the charcoal dust from his paper and surveys his work, and the Lady-musician sighs in time to the music. Her breathing was what spectators most often commented upon. It made her seem not only alive, but emotional. She appeared moved by the music she played. (Riskin, 2003, p. 102)

So these are not simply mechanisms that model and represent certain behaviors analogically; they are simulations of living behavior—extending as far as the birth machines of Madame du Coudray, which, constructed from pelvic bone, fabric, leather, and moist sponges, served quite literally as wetware so that midwives could rehearse something akin to the process of birth. That these simulations were crude and clumsy does not defeat their purpose—these 18th-century machines are very much like current attempts to learn from simulations rather than from analogical models.[7] Today's biocyberneticians, intelligence and behavioral researchers describe the organic connections in the human brain as "wetware" in contrast to the hardware of a computer. To be sure, one can use the latter to imitate intelligent behaviors or thought processes in a technological medium, yet the fact that an analogy must first be established between organic and electrical processes signifies a gap between model and world, construction and reality. In the simulation of living processes,

---

7 Riskin's distinction cannot be understood from the vantage point of the familiar contrast between digital and analog technologies: her "simulations" are analogical devices as well. Her contrast between simulation and analogy is more like the distinction between icon and index, where the iconic is based on the immediate participation of the sign in the signified (simulation) and an index corresponds to the two-place relation between the representation and that which is represented (analogy).

no such gap should even appear. Thus, a specific demand for knowledge calls for an intimate proximity to things. And this epistemological requirement entails an expectation of the machinery of simulation that this technology cannot actually fulfill.

The automata of the 18th century exhibit clearly how far they fall short of this unrealistic expectation. They are unable to simulate living processes in such a way as to generate theoretical insights or practical skills regarding the physiology of humans and animals in an unmediated way, directly from beholding the machine. The question arises, though, whether today's simulations have really come much farther than that or whether they still gesticulate clumsily toward an unachievable ideal of technology—an ideal device that has been assigned a definite role in the research process and that already functions as a public object of fascination.[8] Riskin comments on this:

> "Wetware" … is the expression of a particular moment, the turn of the twentieth to the twenty-first century. The neologism voices one of the organizing ambivalences of the current moment: we believe that the processes of life and consciousness are essentially mechanistic and can therefore be simulated, and yet we are equally firmly persuaded that the essences of life and consciousness will ultimately be beyond the reach of mechanical reproduction. (2003, p. 97)

As described by Riskin, this ambivalence is an expression of a profound confusion and perplexity that would leave unjustified any expectation of a kind of technology that somehow transcends its mechanistic origins. The simulations that had made an appearance in the 18th century disappeared again in the 19th because the idea of such a technology could not be developed coherently; and they are now returning because, in the transition from the 20th to the 21st century, we find ourselves once again confused and perplexed. This, of course, raises the further question of why these two historical moments, in particular, should be characterized by such confusion and perplexity.

Riskin picks up on this question—"Why were the seventeenth and nineteenth centuries periods of analogy, and the late eighteenth and late twentieth centuries periods of simulation?" (2003, p. 118)—and suggests an answer. Analogies are employed when people consider what machines are and find that animals are a lot like them and can be described in the language of machinery. In contrast, simu-

8 The currently most prominent example of this is the *Human Brain Project* and its American counterpart. The project's aim is to construct a map of the brain and an integrated model of its functioning at the scale very nearly of 1:1. It is quite clear what its scientific contribution will be: the funding is in place, and public expectations are set. It is unclear, however, whether it could ever become technically feasible.

lations respond to perplexity when people are not entirely sure what animals are, or what machines can be, and seek to find out about both by trying to build an animalmachine. With their animalmachine, then, the researchers of the 18th and 21st centuries are imagining a technology without knowing whether it can even exist, that is, whether its very idea may turn out to be incoherent. As such, these machines are an especially vivid example of the fact that technological ideals can be, as it were, nontechnical, indeed almost antitechnical:

> The beginnings of the Industrial Revolution and the beginnings of the Information Revolution were both periods of extreme fluidity in people's understandings of what machines were—and indeed in the nature of machines. Once the industrial period was fully under way, manmade machinery and its relations to living creatures stabilized, replacing the fluidity required by a simulation with two terms that were, for the moment, fixed: "life" and "mechanism." Only when the Information Revolution introduced a new kind of machinery did this fixity give way to a new fluidity, and the possibility of using machinery to simulate life again seemed intriguing. In other words, the modern makers of automata that see, hear, and feel in fact have a great deal in common with the eighteenth-century makers of automata that breathed, spoke, and defecated. They too use machines to simulate life precisely because their conception of machines is no better established than their understanding of life. (Riskin, 2003, pp. 118–119)

With regard to the claim being developed here, then, one might say that societies hope for orientation from merely imagined technology because they understand neither themselves nor technology and because they are troubled, fascinated, and perplexed all at once by the limits and possibilities of technology and by the future promise of an unprecedented new and different technology. This perplexity characterizes the life-simulating wetware of the pre- and postindustrial age and also the "soft" social machines imagined in our own times.

Richard Jones's Soft Machines

The succession of historical examples has now arrived in the present and affords a view of nanotechnology as a prime example of so-called emerging technologies. These technologies feed the hope—against our better judgment—for solutions to nearly all pressing problems. They promise cures for cancer and resource saving, energy-efficient products, interventions in the climate, and the synthesis of novel biological organisms. Nanotechnology is an especially interesting case because at its beginnings stood the idea of an unprecedented machinery, and, since its beginnings, the possible reality or likely irreality of these machines were at issue. Anyone who knows anything about nanotechnology has most likely heard of things called "molecular assemblers" or "nanobots," which may either exist sooner or later or else are completely impossible. These are the devices on which the allure of nanotechnology depends entirely.[9]

The highly differentiated nanomachine discourse cropped up in many different areas of society, including cultural and social science as well as philosophical analysis, but cannot be discussed here (see Drexler 2003a, 2003b; Lösch, 2007; Myers 2008; Nerlich, 2005; Smalley, 2003a, 2003b). While Bernadette Bensaude-Vincent and Xavier Guichet (2007) draw an analytic distinction between Cartesian, complex and concrete machines, the following takes as its starting point the much-discussed book *Soft Machines* by physicist Richard Jones (2004) with its distinction between hard and soft machines. Interestingly, Jones's soft machines are not based on any prior technological model. He introduces them not because he sees an already advancing technical development that demands our attention, but because only the conception of "soft machines" lends meaning to the claim by nanotech visionaries that there might one day be molecular machines. Since molecular machines are not compatible with familiar technical paradigms, a new, quite different kind of technology—wet and soft and modeled on biological processes—needs to be imagined on which these machines could, in principle, be based. Jones presents all this with great intelligence and care—he knows exactly what he is doing. As a physicist, he does not assume any obligation to prove the technical feasibilty of the soft machines he conceptualizes. Indeed, even where it explores the idea of what a molecular machine would have to be from the point of view of physics and biology, his book is ultimately one on the relations between nanotechnology and society.

The question of nanomachinery is salient, according to Jones and numerous other researchers, because it is on this question that societes might refuse to accept nanotechnology's beneficial nature. An understanding of "soft machines," Jones argues, provides a necessary corrective to dangerously simple-minded ideas about the ways in which nanotechnology will disrupt social relations. Those who speak of "molecular machines," "nanobots," " assemblers," and the like are aligning themselves with the visionary rhetoric that was initially introduced by Eric Drexler. According

9 The fact that the allure of nanotechnolgy depends entirely on this dream is confirmed by our sobered awakening. As soon as the promise of these machines vanished from the generally accepted list of nanotechnological promises, nanotechnology lost its magic spell of fascination. It is now an R&D trajectory like any other and is for the most part normalized—no longer a matter of molecular machines and atomic precision but merely of chemically produced nanomaterials. Collective dreams of novel machinery are nowadays projected upon synthetic biology.

to this rhetoric, nanotechnology enables such precise control over individual atoms and molecules that it becomes possible to construct tiny, scaled-down, otherwise ordinary machines that will perform work, for example, by fabricating artefacts, repairing human cells or clearing arteries. For Jones, this notion is dangerously misleading for various reasons. Given that it is not scientifically credible (Jones, 2004; Smalley, 2003a, 2003b), it gives nanotechnology a bad reputation and undermines all its promises. And even though it is not scientifically credible, it might, nonetheless, trigger irrational fears of uncontrollable nanobots.

That this notion lacks credibility, however, opens up the possibility of a thought experiment—one that leads away from the idea of scaled-down conventional machinery at the molecular level and that imagines instead an unconventional kind of machine that, although having fine scientific credentials, is beyond technical reach. Jones asks what kinds of machines could perform work at the nanoscale. These would be molecules among molecules in the bloodstream, for instance. A machine of this sort would have to know how to utilize the particular conditions of a world in which gravitational forces are practically absent and where Brownian motion dominates. A machine that was designed according to principles familiar from the macroscopic world would stand no chance at all against Brownian motion. Instead, what would be needed is a biologically soft machine, for example, a viral or sperm-like machine that is adapted to its environment, that allows itself to be driven by its environment, and whose proper functioning is guaranteed by its environmental conditions. In this way—and to counter the scientifically non-credible nanobot—Jones brings into play a scientifically respectable, albeit merely hypothetical, machine that appears to take its cue from political philosopher Ernst Bloch. Decades ago Bloch dreamt of technology that was allied with nature to produce an "unprecedented insertion" and the "building of human work into the work of nature" (Bloch, 1973[1959], p. 817; cf. also Simondon's "concrete machines" [1958]). In what sense "soft machines" can even be considered "machines" in the first place is not at all clear; neither is it clear whether they can ever be realized in technical terms. Nonetheless, Jones's soft machines promise something new that eludes the opposition of natural and social relations. Instead, these machines are integrated into the physical and social dynamics of human affairs and natural processes—they make for a sensitive, considerate, conducive, friendly technology.

It may come as no surprise, then, that physicist and nano researcher Richard Jones not only wrote this book but subsequently became one of the most prominent and effective nanotechnology communicators in Britain. And that is saying something: influenced by critical scholarship in the field of science and technology studies, Britain has rejected the simple model according to which the communication of scientific ideas serves primarily to promote their acceptance. Here, too, what is intended is "unprecedented insertion"—governance as building or designing human work into the work of science and technology (Kearnes, Macnaghten, & Wilsdon, 2006). The state should no longer be conceived as a hard machine to whose regulatory agencies citizens address their concerns merely by submitting petitions and inquiries, or as a machine with a rigid funding mechanism that is tied to immutable criteria of scientific quality and economic profitability. Rather, processes of technological development should prove to be gentle and soft, open to collective design, accommodating and trustworthy. In this sense, Richard Jones filled his role of designated ombudsman for nanotechnology in an exemplary manner—exemplary not only on account of his analyses, which are readily accessible, carefully differentiated, clearly understandable, and often funny, and not only because he was crucially involved in public consultations aimed at establishing research priorities. Jones also filled his role in exemplary fashion in that he engaged in dialogue with scholars in the social sciences and the humanities (for example, Jones, 2011). As a representative or model citizen of socially constituted research on emerging technologies ("mode 2 knowledge production"; cf. Nowotny, Scott, & Gibbons, 2004), Jones constructed two kinds of soft machinery and two technologies so gentle as to be unreal. His biologically inspired nanoscale machine and the participatory governance machine mutually inform one another, since both share in the social imaginary of a never-to-be-realized but altogether gentle and accommodating technology.

Jones's soft or biological machines will not be considered as machines by some for the simple reason that the blurring of the organic and the mechanical constitutes a category mistake—especially if to the organic is attributed something akin to emergence or spontaneity and, thus, something that eludes predictive control. Jean-Pierre Dupuy, in particular, has emphasized that nanotechnological research programs appear to be dedicated to the paradoxical task of controlling the uncontrollable. In the appopriately muddled words of an anonymous nano researcher:

> The problem is the illusion of control—what we want to do is reverse engineer. We harness self-assembly in a non-linear way to get what we want. To do this at the nanoscale will be a big breakthrough because we can then start to control things, put them in compartments and let them evolve. We don't need the illusion of control. We let the system select what it needs according to its local environment. We can't be an engineer at that level if we want to use bottom up. Nature takes this approach and it works very well. (Quoted in Macnaghten & Kearnes, 2007, p. 17)

The paradoxical brief of controlling without control appears here under the heading of "bottom-up engineering" (implying the use of self-organizing processes such that technological systems can grow and need not be constructed), but it comes also under the heading of "soft machines." Self-organization, contends Dupuy and herein echoes the anonymous researcher, is fundamentally contrary to the art of engineering, and he notes that a dynamic that is sometimes labeled as nonlinear dynamics and sometimes as chaos theory contradicts the idea of reliably functioning technological systems. Engineers, he says, surely have no intention of allowing their constructions to surprise them and surely will not build technical systems so complex that they feature catastrophic tipping points (Dupuy, 2007).

Soft Governance in a Plastic World

The opposite of realism, so it is sometimes said, is not utopian but wishful thinking (for example, Geuss, 2010). Thus, the genie of the lamp or the perpetual motion machine is no utopia either but rather a wish-fulfilling machine—not much different from the steam engine's regulator whose art of governance transforms individuals' interests into the common good, and not much different either from the intelligence of an organized division of labor that allocates each individual a place in the order of things. Accordingly, one could go on from here to produce a critique of the discourse on machine enhancement and attendant concepts of a nontechnical technology that reflect wishful thinking but have no technological precedents. This would be a critique of taking these concepts too seriously, either in funding decisions or in public deliberations. It would seek to expose a confused state of debate.[10]

If, however, ours is an age in which we neither know what humans are and what they can do nor what machines are and what they are capable of, and if we are unsure about the limits of technology, there is no easy way of dismissing wishful thinking and the conception of unprecedented machines. By understanding them as sociotechnical imaginaries of how things might work together, they can neither be valorized for their normative expectation of a better world nor be dismissed as illusory and misleading—the conception of sociotechnical imaginaries cuts across the distinction between utopian and wishful thinking.[11] In sociotechnical

---

10 I have gone this route in an earlier version of the present paper (Nordmann, 2014).

11 According to Jasanoff, "Sociotechnical imaginaries occupy the theoretically uninscribed space between the idealistic collective imaginations identified by social and political theorists and the politically neutered, hybrid networks or assemblages with which STS scholars tend to describe reality." These are visions, then, not of a future that departs from the present but of an idea of the future that reinforces the present. As such, " [u] nlike mere ideas and fashions, sociotechnical imaginaries are collective, durable, capa-

imaginaries, historical experience appears in the horizon of hope and destiny, of what might be possible and what should be necessary. As such, these imaginaries are saturated with experience and "false consciousness," but, in irreal conceptions of unprecedented machines, they also create images of peaceful accommodation, the end of politics, or the infinite plasticity of the natural and social world. These images need to be recognized rather than debunked, rendered amenable for public deliberation. Irony and caricature can help, if only to loosen the hold on the imagination of human enhancement, for example. When the idea of perfecting human nature is discussed alongside that of enhancing machine nature and of enhancing material nature (Nordmann, 2010), these various ideas will gain mutual support by referring to a shared historical reality but, by the same token, will also appear more precarious—symptoms perhaps of normative uncertainty and cultural instability.

It would be a major surprise if it actually proved possible to build transparent wish-fulfillment machines that offered no resistance, that exerted control without exerting control, and that seamlessly expressed the order of nature and a social world.[12] At the same time, it is no surprise that the notion of controlling the uncontrollable inhabits people's imaginations and informs the machinery of social life. The very idea of technical control over historically contingent, dynamic processes offers the promise that politics need no longer be an art but can reliably manufacture social harmony and consensus. Indeed, this is what an imagined technology can facilitate and enable—the depoliticization of politics. In the nonlocal public sphere of soft law, soft regulation, and soft governance, there is no hard machinery of thresholds and regulatory interventions. Its social machinery seeks to exploit the effortless compulsion from self-organizing, almost organic processes of growth.[13]

If one wanted to cite specific reasons for the emergence of such a governance conception of politics and technology, one could do worse than consider Jessica

---

ble of being performed" (Jasanoff, 2015). In the remainder of this conclusion, it will be shown in which sense even Jones's soft machines are collective and durable imagined technologies and that they are capable of being performed.

12 Eric Drexler's "universal assemblers" were quite explicitly to be wish-fulfilling machines and, as such, were able to influence the early years of nanotechnology research in a decisive way. The dream of being able to translate mere thinking into a functioning machine makes so-called *mind-machine interfaces* so fascinating— even if they can only ever be *brain-machine interfaces* and for their functioning require the power of concentration rather than thinking alone.

13 One example of this is the social scientific research conducted to accompany scientific research projects. It recruits and enrolls a critical public in an uncontrolled participatory experiment, aimimg to achieving the self-organization of consensual cooperation in a communal future-oriented project beneath the mantle of "developing nanotechnology in a responsible way" (Jasanoff, 2002; Ferrari & Nordmann, 2009).

Riskin's diagnosis of our time. She has postulated a connection between the fascination with technical simulations of organic processes and people's sense of perplexity regarding the possibilities and limitations of technology. This perplexity can be interpreted as a response to the general notion that there are no limits to the world's malleability—a notion for which nanotechnology and synthetic biology are only the most recent examples). Indeed, this malleable world appears to be dematerialized to such an extent that it has run out of ways to resist technological intervention (Bensaude-Vincent, 2004). The nonphysical, soft social machine, thus, embodies a broadened conception of design that draws together the work of developers and users; technicians, engineers; and scientists; consumers and citizens—if everyone is invited to contribute to the shaping of an infinitely plastic world, conflict will evaporate and environmental problems, for example, "will take care of themselves" (Nordmann & Schwarz, 2010). Mumford, in particular, elaborated the historical significance of a technology that initially existed only in the imagination. This is what we are seeing also today: the soft governance model of a collective social experiment with new technologies suggests the emergence of a social order in which producers and developers voluntarily agree to be accountable, in which consumers willingly act as guinea pigs, in which analytic expertise is spread among all participating citizens, in which monitoring by state agencies is replaced by permanent vigilance distributed over an indefinite number of actors. Subsequently, conflicts of interest become backgrounded in the discourse of responsibility, while "ethics" serves as an international *lingua franca*. All this, at any rate, is part of the dream of soft machines as emblems of environmentally benign, socially integrative technology.[14]

References

Anders, G. (1956). *Die Antiquiertheit des Menschen*. München: Beck.

Bensaude-Vincent, B. (2004). *Se libérer de la matière? Fantasmes autour des nouvelles technologies*. Paris, France: Institut national de la recherche agronomique.

Bensaude-Vincent, B., & Guichet, X. (2007): Nanomachine: One word for three different paradigms. *Techné, 11* (1), 71–89.

Bloch, E. (1973[1959]). *Das Prinzip Hoffnung*. Frankfurt, Germany: Suhrkamp. In English: *Principle of hope*. (3 vols.) (N. Plaice, S Plaice, & P. Knight, Trans.). Cambridge, MA: MIT Press, 1986.

---

14 I would like to thank Katherine Cross for her translation into English which provided the basis for this revision of Nordmann (2014).

Cassirer, E. (1930). Form und Technik. In E. Cassirer, *Symbol, Technik, Sprache* (pp. 39–92). Hamburg, Germany: Meiner, 1985.
Drexler, E. (2003a). Drexler counters. *Chemical & Engineering News, 81*, 40<–41.
Drexler, E. (2003b). Open letter to Richard Smalley. *Chemical & Engineering News, 81*, 38–39.
Dupuy, J.-P. (2007). Some pitfalls in the philosophical foundations of nanoethics. *Journal of Medicine and Philosophy, 32* (3), 237–261.
Ferrari, A., & Nordmann, A. (Eds.). (2009). *Reconfiguring responsibility: Lessons for nanoethics—Part 2 of the report on deepening debate on nanotechnology.* Durham, England: Durham University.
Geuss, R. (2010). Realismus, Wunschdenken, Utopie. *Deutsche Zeitschrift für Philosophie, 58* (3), 419–429.
Heidegger, M. (2007). *Die Technik und die Kehre*. Stuttgart, Germany: Klett Cotta.
Jasanoff, S. (2002): Citizens at risk: Cultures of modernity in the US and the EU. *Science as Culture, 11*, 363–380.
Jasanoff, S. (2015). Future imperfect: Science, technology, and the imaginations of modernity. In S. Jasanoff & S.-H. Kim (Eds.). *Dreamscapes of modernity: Sociotechnical imaginaries and the fabrication of power* (pp. 1–47). Chicago, IL: University of Chicago Press.
Jones, R. (2004). *Soft machines: Nanotechnology and life.* Oxford, England: Oxford University Press.
Jones, R. (2011). What has nanotechnology taught us about contemporary technoscience? In T. Zülsdorf, C. Coenen, A. Ferrari, et al. (Eds.). *Quantum engagements: Social reflections of nanoscience and emerging technologies* (pp. 13–26). Heidelberg, Germany: AKA.
Kearnes, M., Macnaghten, P., & Wilsdon, J. (2006). *Governing at the nanoscale: People, policies and emerging technologies.* London, England: Demos.
Lösch, A. (2007). Nanoroboter und Mini-U-Boote—Mediale Vermittlung nanomedizinischer Visionen. In D. Korczak & A. Lerf (Eds.), *Zukunftspotentiale der Nanotechnologien: Erwartungen, Anwendungen, Auswirkungen* (pp. 127–145). Kröning, Germany: Asanger.
Lösch, A., Gammel, S., & Nordmann, A. (2009). *Jenseits von Regulierung: Zum politischen Umgang mit der Nanotechnologie.* Heidelberg, Germany: AKA.
Maasen, S. (2010). Converging technologies—diverging reflexivities? Intellectual work in knowledge-risk-media-audit societies. In M. Kaiser, M. Kurath, S. Maasen & C. Rehmann-Sutter (Eds.). *Governing future technologies: Nanotechnology and the rise of an assessment regime* (pp. 303–319). (Sociology of the Sciences Yearbook 27). Dordrecht, the Netherlands: Springer.
Macnaghten, P., & Kearnes, M. (2007). DEEPEN deliverable 4: Working paper—Scenario planning and draft design of focus groups. Unpublished. Durham, England.
Mayr, O. (1986). *Authority, liberty and automatic machinery in early modern Europe.* Baltimore, MD: Johns Hopkins University Press.
Mumford, L. (1967). *The myth of the machine: Technics and human development (volume 1).* New York: Harcourt Brace Jovanovich.
Mumford, L. (1970). *The myth of the machine: The pentagon of power (volume 2).* London, England: Secker and Warburg.
Myers, N. (2008): Conjuring machinic life. In *Spontaneous Generations, 2* (1), 112–121.
Nerlich, B. (2005, December 22). From nautilus to nanobo(a)ts: The visual construction of nanoscience. *AZoNano—Online Journal of Nanotechnology.* Available at www.azonano.com/article.aspx?ArticleID=1466. doi: 10.2240/azojono0109.
Nordmann, A. (2008). *Technikphilosophie zur Einführung.* Hamburg, Germany: Junius.

Nordmann, A. (2010). Enhancing material nature. In K. L. Kjølberg & F. Wickson (Eds.), *Nano meets macro: Social perspectives on nanoscale sciences and technologies* (pp. 283–306). Singapore: Pan Stanford,

Nordmann, A. (2014). Sanfte Technik: Vom Mythos der Maschine zum Mythos nicht-maschineller Maschinen. In A. Gelhard & A. Kaminski (Eds.). *Zur Philosophie informeller Technisierung*. Darmstadt, Germany: Wissenschaftliche Buchgesellschaft.

Nordmann, A., & Schwarz, A. (2010). Lure of the "yes": The seductive power of technoscience. In M. Kaiser, M. Kurath, S. Maasen, & C. Rehmann-Sutter (Eds.), *Governing future technologies: Nanotechnology and the rise of an assessment regime* (pp. 255–277) (Sociology of the Sciences Yearbook 27). Dordrecht, the Netherlands: Springer.

Nowotny, H., Scott, P., & Gibbons, M. (2004). *Wissenschaft neu denken: Wissenschaft und Öffentlichkeit in einem Zeitalter der Ungewißheit*. Weilerswist, Germany: Velbrück.

Pitt, J. (2011). *Doing philosophy of technology: Essays in a pragmatist spirit*. Dordrecht, the Netherlands: Springer.

Rabinbach, A. (1992). *The human motor: Energy, fatigue and the origins of modernity*. Los Angeles, CA: University of California Press.

Reuleaux, F. (1875). *Theoretische Kinematik. Grundzüge einer Theorie des Maschinenwesens*. Braunschweig, German: Vieweg. Translated as *Kinematics of machinery: Outlines of a theory of machines* (A. W. B. Kennedy, Trans. & Ed.). London: Macmillan and Co., 1876.

Riskins, J. (2003). Eighteenth-century wetware. *Representations, 83*, 97–125.

Schwarz, A. (2009). Green dreams of reason: Green nanotechnology between visions of excess and control. *Nanoethics, 3*, 109–118.

Simondon, G. (1958). *On the mode of existence of technical objects*. London, England: University of Western Ontario.

Smalley, R. (2003a). Smalley responds. *Chemical & Engineering News, 81*, 39–40.

Smalley, R. (2003b). Smalley concludes. *Chemical & Engineering News, 81*, 41–42.

Taylor, C. (2004). *Modern social imaginaries*. Durham, NC: Duke University Press.

Part III
The Transhumanist Imagination in Context

Perfecting European Democracy

Science as a Problem of Technological and Political Progress

10

Brice Laurent

Introduction

The European Union is the outcome of a process that is still very much in development. This process was conceived from its inception as a matter of perfecting a collective of (at least) three identities: a geographical space torn apart by devastating wars and longing for peace, a market that could benefit from the extension of the circulation of economic goods, and a political entity where liberal democracy would finally triumph. The idea of progress was at the core of the European project and closely articulated with the rapid economic growth experienced by Western European countries in the postwar decades. Progress, there, was as much economic as it was political and moral, as it associated the extension of markets with the development of liberal democracy and a collective concern for lasting peace.

The imaginary of progress was part of a landscape made of the decades of economic growth in Western Europe combined with a Cold War situation within which what mattered was the extension of liberal democracy. Science and technology were to be contributors to the progress and topics for the European coordination to work with. A telling example is that of Euratom. Created as a vehicle for the development of "atoms for peace" within a cooperation framework associating European countries, Euratom quickly faced the issue of the crucial role of nuclear energy for national policy choices (Barry & Walters, 2003). In this case, phrasing the problem of cooperation for nonmilitary use of nuclear energy meant problematizing the sovereignty of participating states eager to maintain control of what they considered sensitive areas, indeed integral components of their *grandeur* (Hecht, 1998).

This chapter argues that European construction is a process whereby science and democracy are jointly problematized as matters of progress and perfectibility. In turn, framing science and democracy as matters of progress participates in the making of Europe as a political, economic, and moral entity worthy of public support. For science and technology within European institutions are issues of public support and require the enrollment of a citizenry that is not pregiven. That technological development is also a project of political imagination, defining authority, power relationships, and channels of democratic legitimacy has been demonstrated by STS scholarship (for example, Barry, 2001; Ezrahi, 1990; Jasanoff, 2004, 2005; Jasanoff & Kim, 2009). Euratom, for that matter, is a relevant example, whereby the public to be enrolled is made of a heterogeneous set comprising member states negotiating their sovereignty and publics imagined as consumers in need for cheap and accessible energy (yet potentially reluctant to nuclear energy).

The articulation of technological development and political imagination reaches an even deeper dimension as recent technological developments act on human beings themselves. Biotechnology and, more recently, nanotechnology promise to make technological intervention hinge upon living organisms themselves. By transforming "life itself" into an area of human intervention, they articulate scientific and political progress in that the world they construct is thought anew—made of new technoeconomic objects, institutional arrangements, markets, and eventually human beings (see Jasanoff, 2005; Rose, 2007). These domains of technological development have been at the core of the European science policy. As this paper will argue, the European programs in these domains are specific, in that they are as much (if not more) devoted to the perfectibility of the European collective than to that of the individual.

This movement is inscribed in the gradual Europeanization of science policy. Research became a component of the European Commission's activities with the 1986 Single Act. This important institutional evolution occurred as issues about the European public of science were emerging. As the first framework programs for the conduct of European research policy were introduced, European-wide opinion polls were also applied to science to measure the European public's opinion. Since then, the concern for the development of European research has been tightly connected with a concern for the public support from European citizens (Felt, 2010). Refinements of science communication initiatives at the European level, including the more recent participatory exercises, show that ensuring technological progress in Europe is also a matter of turning passive European publics into active scientific citizens. Eventually, and as this chapter will explore, the narrative of joint progress, both technological and democratic, participates in the making of Europe itself. Analyzing this requires taking this narrative not as a given but as an outcome of processes that need to be accounted for. This means that narratives of progress emanating from the institutions of the European Union cannot be taken at face value. But it also means that critical perspectives on the effective realization of this progress cannot be taken for granted either.

For instance, scholars have discussed the incomplete transition from the "deficit" to the "dialogue" model (Hagendijk, 2004) or shown that public dialogue, by *not* questioning the overall framing of public issues, did not manage to use public engagement in a transformative way (Wynne, 2006). Others regret that the "opportunity" for extended public participation that new technological developments would offer is not truly realized (Macnaghten, Kearnes & Wynne, 2005; Kurath & Giesler, 2009). These critical interventions, while of clear interest for any student of democratic practices in Europe, participate in the construction of narratives of democratic progress and share an objective of democratization through dialogue, only under a negative mode ("the democratic objective has not been reached *yet*"). Both the positive and the critical mode of the discourse of joint technological and democratic progress share the same way of phrasing the problem in need of addressing. They both consider that there is a known direction toward which technological and political development should go. In contrast, what is of interest for us here is, in Michel Foucault's words, the *problematization* of European science, European polity, and eventually Europe itself, in the terms of progress and perfectibility. Problematization, for Foucault, points to the set of processes whereby acceptable public problems and a range of expected solutions are defined (Foucault, 1984). These processes pertain to the manufacturing of individuals, collectives, and concerns solidified in various instruments and discourses. This chapter proposes to analyze the problematization of European science as an issue of joint technological and democratic progress. To do so, I identify some of the mechanisms by which the narrative of joint progress takes shape. I discuss three of these mechanisms: the experiment, the narrative of failure, and the integration of values into research policy.

In the following, I start with empirical sites where perfectibility is problematized, and explicitly so because it conflicts with extra-European definitions of the problems to solve. These sites are related to nanotechnology, which was framed as a "European experiment." I then turn to two other groups of sites where joint technological and democratic progress is made a European objective. Biotechnology—genetically modified foods (GMOs), in particular—has been described as a failure of public acceptability, market harmonization, and public engagement. Making such a narrative of failure is an operation that deserves analytical attention, as it actively defines the expected direction of scientific and democratic progress. Finally, I turn to the recent call for "responsible research and innovation" (RRI) within European research policy and show that it illustrates attempts at integrating the "values and needs" of the European public in ways that are consistent with a general objective of "excellence."

Experimenting with nanotechnology, writing a narrative of failure about GMOs, and calling for integration through RRI are three operations that make science a problem of both technological and democratic progress. These mechanisms are by no means neutral. They shape certain imaginations of the European future at the expenses of others. As the European polity is imagined in the terms of continuous democratic progress, it is also meant to be purified from unorderly publics or alternative technological trajectories. As it makes excellence a key term and the accountability toward a general European public a central concern, it also disqualifies other forms of politics. It is, thus, extremely important to be able to characterize the type of European ordering that three operations I describe below result in. Doing so will offer analytical material to identify what the European narrative of joint perfectibility produces.

Experimenting with Nanotechnology

Defining Nanotechnology as an Experiment

In the mid-2000s, two publications of the European Commission—"Towards a European strategy for nanotechnology" (EC, 2004) and the "Nanosciences and nanotechnologies" action plan (EC 2005)—made nanotechnology a priority of the European science policy, which would be conducted in connection with "publics" and "stakeholders." While the American nanotechnology programs insisted on the "improvement of human performance" (Roco & Bainbridge, 2003), the European program would be pursued for the sake of an "inclusive" European society. This difference was made explicit in a report commissioned by the services of the Directorate-General for Research to a group of scholars and for which philosopher of science Alfred Nordmann acted as a rapporteur. This report argued that nanotechnology was an opportunity to "shape the future of European societies" thanks to close connections between technological development and the expectations of the European public (Nordmann, 2004). In doing so, the differences between the American imagination of technological development and the opposition to its transhumanist undertone were made explicit. Rather than perfecting the individual, European converging technologies would perfect the collective. They would participate in the integration of European societies.

This ambitious objective meant that innovative instruments were to be designed, so that joint technological and political progress could be realized. This required that nanotechnology be constituted as an experimental moment, a "European experiment"—this expression was Nordmann's (2009)—through which both technological development and the democratic involvement of European publics in the making of desirable futures could be rethought. Nanotechnology was then described as an "extraordinary opportunity for the social sciences" (Macnaghten, Kearnes & Wynne, 2005) that could participate in the orientation of technological development.

The European experiment was developed into new regulatory instruments targeting scientific research. A good example is the "Code of Conduct" produced by the Directorate-General for Research (DG Research), which encouraged European researchers to limit the exposure to nanomaterials, restrain from engaging in "human enhancement" and engage in dialogue with lay publics (EC, 2008). The code did not define such crucial terms as "nanomaterials" and "human enhancement." It was purely voluntary, though was also considered a necessary condition for research projects to receive European funding.[1] It attempted to introduce ways of anticipating potential safety risks without constraining public regulation, while being conceived as a platform for a continuous dialogue involving European institutions, member states, researchers, and European publics (von Schomberg, 2009).

Through this initiative, scientific development was conceived as a matter of technological progress as well as an issue of democratic perfectibility. As the code inscribed the reluctance of the European institutions to display objectives related to "human enhancement" (as in the United States), it also proposed to make nanotechnology an experiment in the transformation of the democratic legitimacy of scientific research. What was to be enhanced, then, was not the individual who would be transformed by technology but the whole European polity, expected to reflect collectively on a technology that was, in turn, expected to be acceptable for all. Using a voluntary instrument such as the code is characteristic of an intervention that aims to transform individual researchers, research institutions, and member states in an indirect manner, by suggesting rather than imposing and using detours such as introducing conditions for benefiting from research funding. Success, then, was explicitly considered in an experimental manner, depending on whether this initial test would have to be refined and adapted, possibly to broader domains than nanotechnology. Such an intervention is, in part, an outcome of an interpretation of the subsidiarity principle, according to which ethics is not a primarily area of regulatory action of the European Commission. It also defines both the direction of democratic progress and a practical path to follow it.

This articulation of the support for technological development with the democratic progress of Europe was further conducted in science communication initiatives. One of the first of them was a European project called *Nanodialogue*, initiated in 2006

1 Interview, DG Research, February 2010.

and which gathered a few European science museums.[2] Among its objectives was the test of new formats of representation in the science museum in order to meet a "democratic ambition" (this expression was used during the project's meetings). This "democratic ambition" related to the expected transformation of the role of the science museum in the involvement of European publics in science policy. As sociologists and historians of science have demonstrated, the construction of the "informed citizen" through scientific communication is at the core of the publicization of science in the museum (Bensaude-Vincent, 2000; 2003; Hilgartner, 1990). Yet for the participants in the *Nanodialogue* project, the democratic nature of science museums representing nanotechnology for the sake of European publics was not primarily related to the scientific information transmitted to the visitor. For them, the "democratic ambition" meant that science museums had to represent the "ethical, legal, and societal aspects" (ELSA) of nanotechnology, while making the exhibit an opportunity for dialogue between the European institutions and European publics.

Thus, the circle-shaped *Nanodialogue* exhibit presenting the technical components of nanotechnology was meant to expose the ELSA associated with each of these components, including the safety risks of nanotechnology objects or privacy issues related to the use of nanoelectronics in chips. To implement "dialogue" itself, *Nanodialogue* involved sociologists specializing in public participation in science, including Simon Joss, a notable contributor to the field (cf. Joss & Durant, 1995). These sociologists were in charge of organizing focus groups in the various museums, based on which the opinion of the "European public" about nanotechnology was presented at the European Commission at the end of the project.

Museum professionals and sociologists did not consider *Nanodialogue* entirely satisfying. There were long discussions among the partners about the relevant format through which ELSA were to be represented and occasional disagreements about whether they should hold a significant space within the exhibit. The sociologists thought that the focus groups provided a "snapshot" of European public opinion about nanotechnology that could not pretend to scientific robustness.[3] In the end, the "opinion of the European public" was not very surprising: it was mainly framed as a hope to see nanotechnology realize "benefits," particularly in

2 The empirical material about *Nanodialogue* is based on interviews with several participants in the project (including the coordinator) and archives from the *Grenoble Centre de Culture Scientifique, Technique et Industrielle*, which was one of the project partners.

3 Interview with S. Joss, London, April 2009. Because of limited funding, the practical organization of focus groups was delegated to the museums. The sociologists in charge only distributed guidance booklets.

the medical domain, while controlling the potential risks. It would not be difficult to demonstrate how the instruments used to measure this "public opinion" actively shaped it.[4] But the description of the construction of public opinion through tools like focus groups is less interesting here than the experimental format of *Nanodialogue*. The project tested ways of communicating science in museums and, perhaps even more important, a way of conceiving of the nature of European democracy wherein it would be politically legitimate because of nonelectoral mechanisms of representation targeting a "general public" and "European" because of the various locations of the sites where samples would be collected.

From the Nanotechnology Experiment to a European "Technical Democracy"

The test was relatively successful. After it was completed, *Nanodialogue* was used as a topical example in training programs in public participation conducted by sociologists (including some of those who participated in *Nanodialogue*) and funded by the European Commission.[5] After the conclusion of the project, the participants in *Nanodialogue* were actively involved in the work conducted within the European Commission about the future orientations of the public communication of nanotechnology (EC, 2007). These discussions would eventually lead to a report written by Matteo Bonazzi, a civil servant at the Nanotechnology and Converging Technologies unit of the DG Research (Bonazzi, 2010).

Bonazzi's report defined European priorities regarding the communication of nanotechnology and framed the future calls for projects in this area. As *Nanodialogue*, it made the communication of nanotechnology a matter of dual representation—of nanotechnology ELSA on the one hand and of European publics on the other. Thus, Bonazzi's report was less interested in the content of what was to be communicated than in the forms of exchange. It said nothing about nanotechnology but only proposed a short list of simple (almost simplistic) messages:

---

4 The guidance booklets distributed to the focus group moderators, thus, invited them to question participants about the "benefits" and "risks" of nanotechnology—a framing that inherently constrains the opinion eventually produced by the focus group.

5 In the United States, science communication professionals saw a European specificity in the role devoted to science museums. During an interview, the official in charge of a network of American science museums engaged in nanotechnology communication explained that the production of a public opinion expected to inform policy makers was for her not a concern of the American science policy institutions. Mentioning *Nanodialogue*, she viewed this as a crucial difference between European and American science museums.

- nano is *not* magic;
- nano is *a new phase of technology* exploiting nanoscale effects;
- it deals with new *beneficial applications and markets*, impacting *health, safety, privacy, ethics, and the socioeconomic divide*;
- it *must and can* be controlled and driven conscientiously. (Bonazzi, 2010, 106; *emphasis in original*)

For all their simplicity, these messages pointed to a particular direction: "impacts" were to be communicated, particularly those related to the "benefits" but also the health, ethical, and socioeconomic effects of technological development. In the perspective suggested here, technological development is not thought of as an autonomous force but as a topic of conscious collective intervention. The *Communicating Nanotechnology* report indeed proposed to connect the representation of ELSA and the measure of public opinion, making technological progress a matter of gathering publics for ensuring substantial collective support by providing information about expectations and concerns. Science communication, then, was less about the representation of science for the sake of a passive public than about gathering public opinions.

This shift was summed up by Matteo Bonazzi as an evolution from "public understanding of science" to the "scientific understanding of the public" (2010, p. 20). This shift was a test for European democracy, now required to transform science communication and its connections with research policy. Eventually, the report proposed to make Europe a "technical democracy." Used without reference to the work of scholars of science, technology, and society (STS) who introduced the expression to point to situations where the constructions of issues and concerned publics go hand in hand (Callon, Lascoume & Barthe, 2009), Bonazzi's technical democracy is close to market studies: "The same way we conduct market surveys to understand trends in public opinion" (Bonazzi, 2010, p. 66), the future European technical democracy will be able, so it is argued, to identify trends in the evolution of public opinion, describe the varieties of European publics, and adapt both its communication strategy and its priorities in technological development. Technological and political progress could then be realized thanks to sophisticated instruments through which European public opinion would be "monitored on a continuous basis" (Bonazzi, 2010, p. 152).

This last proposition is an attempt at fleshing out the democratic experiment that nanotechnology was expected to lead to. This experiment eventually occurred at several levels. Nanotechnology was construed as an opportunity to test a large-scale science policy that would anticipate potential negative externalities of technological development and the controversies that these externalities could induce. Based on an instrumentation (such as the code of conduct or focus groups) expected to demonstrate the European specificity of technological development, this experiment is uncertain, as it may or may not transform European research policy or correctly anticipate the controversies related to technological development. It is, in any case, an operation that makes science a matter of joint technological and political progress, under modalities expected to be tested before potential extension to the whole of European research policy.

Narrating GMOs as a Failure

Solving a "Confidence Crisis"

Talking about "experiment" as a way of problematizing technological development in Europe echoes a series of works analyzing European policies as examples of "experimental governance" (Szyszczak, 2006) or "democratic experimentalism" (Dorf & Sabel, 1998; Eberlein & Kerver, 2004). These works point to the use of non legally binding coordination devices (such as the open method of coordination) through which general regulatory choices can be adapted to local particularities, at the level of individuals, companies, or member states. They make experimentalism a form of policy making, ensuring a higher level of democratic legitimacy and efficiency. In this body of work, experimentalism is conceived as an answer to a situation construed as a failure—that of the rigidity of constraining law or the poor democratic legitimacy of existing regulatory interventions.

The nanotechnology experiment was also expected to respond to a situation of failure. The *Communicating Nanotechnology* (Bonazzi, 2010) roadmap postulated the existence of a "confidence crisis" requiring that the relations between science and European publics be rethought. This diagnosis was not isolated. The 2011 *Innovation Union Competitiveness Report*, for which Commissioner for Research Máire Geoghehan-Quin wrote a preface, explained that as "the percentage of European citizens that trust science and technology to improve their quality of life decreased over the last five years," there was "a genuine expectation for science to reorient its efforts to contribute to addressing the societal challenges of our time" (EC, 2011, p. 12). While the diagnosis of crisis fueled the need for experimenting (as in the case of nanotechnology), it also contributed to the problematization of science as a matter of joint technological and democratic progress—"democratic" in that it would provide answers to practical concerns voiced by European citizens.

The diagnosis of a "trust" or "confidence crisis" has roots in the evaluation of public controversies related to technology. It stems directly from a set of works devoted to the public opinion of science in Europe, mostly based on the *Eurobarometer*. The *Eurobarometer* is an opinion study conducted by polling institutes in all member states, commissioned by the European Commission, and is analyzed by public opinion specialists participating in the design of the instrument. *Eurobarometers* devoted to science and technology, including particular areas such as biotechnology, are regularly conducted, with the aim of measuring both the level of public trust and that of public knowledge.

The *Eurobarometer* quantifies European public opinion through the use of macro categories. With it, there is no hope of grasping the subtle variations in individual reasoning about technology. Qualitative studies indeed demonstrate the complexity of public perceptions of controversial technologies such as GMOs (Marris, 2001), which the *Eurobarometer* fails to account for since it prevents lay publics from voicing sophisticated political positions about technological change. This limitation is not surprising considering the effects of polls on the construction of public opinion (Blondiaux, 1998). For our concern here, the *Eurobarometer* is of interest as a performative instrument, constituting both the public and the problems of science communication in Europe (Jasanoff, 2005; Law, 2009). In doing so, it solidifies a general problem of "trust" at the price of eliminating the subtle construction of local perceptions. By the same token, it also necessitates experimentation with the public uptake of science to reconstitute trust on the one hand and to characterize what went wrong in particular domains on the other.

Making GMOs a Case of Failure

Through the *Eurobarometer*, the diagnostic of a trust crisis is connected to a narrative of failure about technological progress in particular domains. Biotechnology, and GMOs in particular, has been one of these domains, perhaps the most explicit. Starting from the late 1990s, the *Eurobarometer* displayed a falling rate of positive perception of technological development in general and of biotechnology in particular. But the "failure" was even deeper since traditional interpretations also failed to explain the trust crisis. Thus, and as the commentators of the *Eurobarometer* regretted that the European public had a "surprisingly limited" level of understanding of science (EC, 2000b), the trust crisis could have been explained in the terms of a deficit of understanding—what sociologists have labeled the "deficit model." But the coordinators of the first *Eurobarometer* devoted to biotechnology (1991) were among the first critics of the deficit model,[6] and the successive *Eurobarometers* eventually undermined the explanatory power of the model. Thus, the results of the 1996 *Eurobarometer* were used by sociologist George Gaskell, a specialist in the public perception of science and coordinator of many of the biotechnology *Eurobarometer* studies, to demonstrate the absence of a correlation between the level of technical knowledge and the degree of acceptance of technology (Gaskell et al., 1996).

The narrative of failure was further developed as dialogue mechanisms targeting biotechnology were organized at the European level (Bengtsson & Klintmann, 2010) or within member states (Boy, Kamel & Roqueplo, 2000; Heller, 2003). While the *Eurobarometer* identified better perceptions of science after 2010, it still made GMOs the "Achilles' heel of biotechnology" as "the wider picture is of declining support across many of the EU Member States—on average opponents outnumber supporters by three to one, and in no country is there a majority of supporters" (Gaskell et al., 2010, p.7). Finally, "GM food is seen as unnatural and makes many Europeans 'uneasy'" (ibid.). This unease was seen as a sign of the failure to make GMOs acceptable to Europeans. But the failure of acceptance was only one among many. Another was the failure to ensure to harmonize the European public space. The *Eurobarometer* displayed strong differences in the perception of GMOs across European member states. This can be seen as parallel to a regulatory fragmentation, which, since the late 1990s, had made member states adopt different positions regarding the authorization of production and trade of GMOs—which eventually forced the European Commission to authorize member states to use safeguard clauses for them to be able to refuse GMOs, potentially against the opinion of the European Food Safety Agency (EFSA) (Wickson & Wynne, 2012).

Thus, the narrative of failure drew together various elements. It made biotechnology and GMOs an issue of failed public acceptance, failed explanatory power of the deficit model, failed attempts at pacification through "dialogue" or "public engagement," and, as a result, failed harmonization of the European public space. Within this general narrative of failure, the development of GMOs in Europe failed to engage European publics and failed to produce a consistent polity in Europe. It made it impossible for the European democracy to make technological development a matter of collective discussion about the future. Thus, in the context of this general narrative of failure, both European officials and social scientists see nanotechnology as an opportunity to transform the relations between technological development and

6 For instance, John Durant, a specialist in the public understanding of science and co-author of the first report based on a *biotechnology Eurobarometer*, was one of the first critics of the deficit model (Evans & Durant, 1995).

European publics. The value of the European experiment that is nanotechnology is grounded in the hope that failure may be avoided: experimenting is conceived as an answer to a situation diagnosed as failed technological and social development.

The narrative of failure is shared by both the institutions of the European Union and their critics. From the viewpoint of the European Commission, science communication initiatives, including under the dialogue format that attempted to replace more traditional pedagogical initiatives, failed to ensure the general acceptance of GMOs and the harmonization of the European political and economic space at both regulatory and public perception levels. For scholars interested in GMO-related controversies, the attempts at introducing dialogue failed to question the overall framing of GMO issues outside of risk considerations that do not envision any significant transformation of the agroindustrial sector (Levidow & Carr, 2009; Wynne, 2006), whereas safeguard clauses at member-state levels do not question the uniqueness of centralized scientific expertise produced at EFSA (Wickson & Wynne, 2012).

However, narrating the public treatment of GMOs in Europe as a failure is an operation that is not to be taken for granted. One could think of alternative narratives, in which the fragmented European situation would be the outcome of successful social mobilization making GMOs a topic of collective concern, and which, by making coexistence (between GM and non-GM, between various national approaches) a central objective, would propose an original democratic construct (cf. Jasanoff, 2013). Narrating the GMO case as a failure has a direct effect in determining the direction of expected progress for technology and European democracy. It implies that perfecting the European polity is about harmonizing markets and following certain technological paths rather than others. By qualifying particular technical and political trajectories as failures, it also defines the expected and necessary directions of progress.

Integrating values through Responsible Research and Innovation

Integrating Values in European Research

There are sites where the narratives of the GMO failure and that of the nanotechnology experiment meet. Consider, for instance, a report written in 2013 by a group of experts for the DG Research of the European Commission, titled *Options for Strengthening Responsible Research and Innovation* (EC, 2013). These experts were academic scholars and civil servants who had been involved in reflections about the

ELSA of nanotechnology. In this report, they mentioned the experiments undertaken by the European Commission with nanotechnology's publics (EC, 2013, p. 11). For them, these experiments offered a way forward, which they contrasted with the GMO case, construed once again as a European failure. They discussed the notion of "responsible research and innovation" (RRI) as a label for a European science policy that would ground technological choices on the expectations and concerns of European publics and would anticipate controversies related to technological development. Throughout this 2013 report, the code of conduct for nanotechnology communication projects appeared as tests through which the failure of GMO could be fixed and European research policy transformed.

Within RRI, technological development is problematized as a matter of joint technological and political progress through an operation that is neither the isolated experiment of nanotechnology nor the narrative of failure of the European GMOs, but extends to the whole of European research policy. RRI has, indeed, been heralded as a general objective by DG Research (EC, 2012a). Still a vague concept, it is explicitly invoked and discussed in practical terms by researchers working at the boundary between academic research and intervention in science policy. The request to do so stems from the European Commission itself. Thus, the official in charge of the European Research Area (ERA) at DG Research spoke to experts in these terms during a workshop devoted to RRI:

> We need your help to define responsible research and innovation. After several years of research on the relation between science and society, we evidenced that we need to involve civil society very upstream to avoid misunderstanding and difficulties afterwards. . . .We cannot guarantee the social acceptability for anything but the more we have dialogue the easier it is to understand the potential obstacles and to work on them. . . . Your advice is important to help us build a policy for the years to come, notably for the Common Strategic Framework that will begin its life in 2014 and for the European Research Area.[7]

As authors of the EU 2013 report, the experts to whom this request was addressed had been involved in transforming nanotechnology into a democratic experiment. Some of them were academics who had seen in nanotechnology an opportunity for social science to transform European research policy (Macnaghten, Kearnes & Wynne, 2005) and engage in an approach that would make "responsibility" a component of European science (Owen, Macnaghten & Stilgoe, 2012). Others were European civil servants and had multiplied public intervention-supporting instruments such as the voluntary code of conduct (von Schomberg, 2009; 2012; 2013). In these explorations, what emerged of RRI, beyond the plurality of what was gathered

---

7 Quoted in Owen, Macnaghten & Stilgoe, 2012, p. 752.

within this broad programmatic discourse, was the objective of integration, that of the values and needs of the public within the trajectory of innovation (Owen, Macnaghten & Stilgoe, 2012).

Whether or not this objective was (or promises to be) satisfactory for the actors involved is not what matters here. I am more interested in what it says about how science is problematized as an issue of joint technological and democratic progress in Europe. And progress within RRI connects the concern for the integration of public values and needs in science and technology with the organizational integration of technological and public communication issues. Consider the institutional evolutions within the DG Research of the European Commission. After 2010, the budget of the "Science and Society" domains was significantly reduced,[8] and the Science in Society unit of DG Research disappeared shortly afterward.[9] Members of the late Science in Society unit presented this evolution as a component of the general reorganization of the European science policy under RRI:

> "Science in Society" disappeared as such but… "Responsible Research and Innovation" emerges as a governance concept and cross-cutting requirement.[10]

That RRI was presented, in the previous excerpt, as a "governance concept" becoming a "cross-cutting requirement" displays, in the DG Research technocratic language, the extension of the problem of the relations between science and the European public to the whole of European science policy. For members of the DG Research, this made it necessary that program officers intervene in purely scientific as well as science communication projects. This institutional evolution is an outcome of the experiment conducted with and by nanotechnology[11] and rephrases the problem of

8 From 51.5 millions euros in 2010, the budget was reduced to 45.69 million euros in 2012.

9 Within the Directorates General of the European Commission, a "unit" is a team of about 20 people. That Matteo Bonazzi did *not* belong to this unit is significant: he considered that as a sign of the "integration" of science communication within the technicalities of research policy (Interview, May 2010).

10 Presentation "Responsible Innovation," Science in Society unit, June 2012.

11 In this description by the former members of the Science in Society unit, nanotechnology was mentioned as a crucial step, ensuring the transition from previous attempts at bridging the gap between science and European publics, to the definition of RRI as a central component of the whole European science policy. Thus, a chronology leading to RRI was presented as such by the civil servants of the soon-to-disappear Science and Society unit: "2000 Science, Society and the Citizen in Europe; 2001 Science and Society Action Plan; 2002 Science and Society (FP6); 2007 Sciences in Society (FP7); 2008 Code of Conduct Nanotech, 2011 Proposal for Horizon 2020; 2014 Horizon 2020 starts." This chronology started in 2000 with the follow-up of the Lisbon strategy targeting the

joint technological and political development within operations of integration—an integration that unifies the whole of European science policy under the banner of "responsibility."

Integration for the Sake of Excellence

Integration is, of course, a common trope within European institutions and a general objective of European construction. Within RRI, integration connects technological development with the progress of European societies by considering that the problem of the public government of research and that of the identification of public values and needs are one and the same. As such, it echoes previous attempts at using science and technology as engines of integration in Europe, while also introducing important differences.

One of the most significant previous cases of the use of science for integration objectives was the 2000 Lisbon strategy and the subsequent programs aiming at developing the "European Research Area" (EC, 2000a). The Lisbon strategy hoped to make Europe a "knowledge-based economy" able to compete in the international economy thanks to the high level of its scientific research. It made science and technology engines of political and economic integration in ways that treated them as a collective undertaking on the part of all member states. All of them were required to reach a target level of R&D spending. This indicator became both the sign of the uniform European transformation that it aimed for and a visible means of controlling its success. Success, however, proved elusive: the targets were not reached in many of the member states. After the mid-2000s, the outcomes of the Lisbon strategy were regarding as disappointing by the European Commission itself (EC, 2005).

RRI, and the version of integration it proposes, is inscribed within the post-Lisbon European research policy, formalized in a program called *Europe 2020* (EC, 2010). *Europe 2020*, and its component called *Horizon 2020* defining the European plans for funding research projects, targets priority domains of public investments and thereby transforms the objective of European integration that was present in the Lisbon strategy (EC, 2010; Lundvall & Lorenz, 2011). Rather than attempting to harmonize the level of public investment in research across the entire Union, *Europe 2020* proposes to guide the development of technologies in selected areas. Thereby, this new European strategy expects to meet the expectations and concerns of European publics while also reserving limited public funding in the post

relations between science and European publics. It went on with the successive funding programs for scientific research in Europe, marked a step with the code of conduct, and eventually reached "Horizon 2020" that replaced the previous framework programs.

financial-crisis era to a small number of domains considered crucial (EC, 2011). Thus, European science policy is defined in *Europe 2020* as contributions to major "societal challenges" that are expected to represent the concerns of the European public. Among these challenges are "health and well-being," "safe, clean and efficient energy," and "smart, green and integrated transportation." *Horizon 2020* uses these "challenges" to define the expected topics of the research projects that apply for European funding.

The integration that RRI proposes is inscribed within the general rephrasing of European research policy, whereby targeted domains of intervention are chosen to ensure both technological results and democratic legitimacy. Used as a vague objective based on the integration of public values and needs in European research, RRI can encompass a variety of meanings, which might even verge on the contradictory, as some expect the European public to have a direct say on technological development and others conceive "responsibility" mainly as a way of ensuring public support for innovation (De Saille, 2013). But this very characteristic makes RRI a useful concept for European science policy, as it captures both the expected democratic progress of the European polity and the targeted use of funding for the sake of economic growth. RRI is an instrument through which the anticipation of promising scientific domains is joined to political anticipation of the expectations and concerns of European publics. This is precisely the role of the "challenges" around which the European science policy is being redefined. This version of "integration" is not to be taken for granted—and the difference with the Lisbon strategy is a telling indicator of that. Consistent with the notion of "excellence," it proposes to target limited amounts of resources to selected (geographical and thematic) areas of public intervention. It grounds democratic legitimacy not on the game of electoral politics but on the monitoring of a European public opinion in need of technical solutions for collective problems.

Conclusion

In January 2013, the European Commission announced that the Human Brain Project (HBP) had been selected as a "flagship project" and would be supported with 1.9 billion euros. Expected to publicly display the "excellence" of European research, the HBP proposed no less than the development of a computer- simulated human brain. This project attracted the praise of transhumanist movements and was criticized in the media for its transhumanist undertones.[12] After the skeptical position of the nanotechnology-related European reports had construed transhumanism as an extra-European philosophy, this later initiative seems to be a clear contrast. Yet it shared with the examples discussed in this chapter a will to make science a problem of both technological and political progress. For the problematization of science within the HBP is not a matter of engendering innovation for the sake of perfecting a particular technological future. Rather, the HBP itself reflects the project of perfecting European research, and so too the progress of the EU: it is a technoscientific instrument for ensuring the integration of European society through the development of solutions for common problems.

As the HBP example shows, there may yet be many other sites where science is problematized in Europe as an issue of joint technological and political progress. This paper has not attempted to be exhaustive but has instead demonstrated that emerging technologies proposing to intervene in "life itself" are entangled with crucial political questions pertaining to the identity of Europe. While these technologies offer the ways and means to perfect human life, European institutions situate them in a problematization of progress that extends far beyond the individual. As the examples discussed in this chapter have shown, European initiatives make them simultaneously matters of identifying desirable technological trajectories and perfecting European democracy. Through operations such as the experimental interventions in the European government of nanotechnology, the production of narratives of failure about past technologies, and the integration of "public values and needs" into European research policy, the problem of perfecting the technological future appears as that of producing of a politically perfected European collective, a collective within which a European polity would at last be supportive of science, and technology would, in turn, be responsive to public expectations. Problematizing science and technology as such is far from neutral. The three operations I have described propose to legitimate certain technological trajectories and certain forms of democratic organization at the expense of others. Within the later *Europe 2020* strategy, they privilege a few technoscientific challenges that are imagined to respond to the needs of a general European public, a public whose political representation thereby ceases to be a problem of (perfecting) electoral politics.

---

12 For instance, the "All about transhumanism" website (http://abouttranshu.tumblr.com/), a "congregation of news, events, websites, and popular culture involving transhumanism" prominently displayed the promotional video of the HGP. A paper in a French newspaper devoted to the Human Brain Project was titled "Transhumanistes sans gêne" (Transhumanists without self-restraint).

The three operations identified in this chapter probably do not cover the entire range of mechanisms crafting the European concern for joint technological and democratic progress. But they do offer insights for an analysis of technological development that do not take the expected direction of change for granted. Problematizing science and democracy in the terms of progress has effects on individuals and collectives, and the interest of the European case is precisely that it illustrates processes whereby European construction goes hand in hand with the crafting of narratives of technological and democratic progress. Studying progress in the making, then, appears as a way of accounting for the gradual shaping of large-scale political entities.

References

Barry, A. (2001). Political machines: Governing a technological society. London, England: Athlone Press.

Barry, A., & Walters, W. (2003). From EURATOM to "complex systems": Technology and European government." *Alternatives: Global, Local, Political, 28* (3), 305–329.

Bengtsson, B. & Klintmann, M. (2010). Stakeholder participation in the EU governance of GMO in the food chain. In K. Bäckstrand, J. Kahn, A, Kronsell, & E. Lövbrand (Eds.), *Environmental Politics and Deliberative Democracy* (pp. 105–122). Cheltenham, England: Edward Elgar Publishing. Available at Elgaronline.

Bensaude-Vincent, B. (2000). *L'opinion publique et la science: à chacun son ignorance.* Bruxelles, Belgium: Synthelabo.

Bensaude-Vincent, B. (2001). A genealogy of the increasing gap between science and the public. *Public Understanding of Science, 10* (1), 99–113.

Blondiaux, L. (1998). *La fabrique de l'opinion: Une histoire sociale des sondages.* Paris, France: Seuil.

Bonazzi, M. (2010). *Communicating nanotechnology: Why, to whom, saying what and how.* Brussels, Belgium: European Commission DG Research.

Boy, D., Kamel, D. D., & Roqueplo, P. (2000). Un exemple de démocratie participative: La "conférence de citoyens" sur les organismes génétiquement modifies. *Revue française de science politique, 50* (4–5), 779–810.

Callon, M., Lascoume, P., & Barthe, Y. (2009). *Acting in an uncertain world: An essay on technical democracy.* Cambridge, MA: MIT Press.

De Saille, S. (2013). Innovating innovation: Operationalising RRI in the ERA. Paper presented at the Devices of Responsibility Workshop, Paris, Mines ParisTech, September 12–13.

Dorf, M. C., & Sabel, C. (1998). A constitution of democratic experimentalism. *Columbia Law Review, 98,* 267–473.

Eberlein, B. & Kerwer, D. (2004). New governance in the European Union: A theoretical perspective. *Journal of Common Market Studies, 42* (1), 121–142.

EC (European Commission). (2000a). Toward a European research area: Communication from the Commission to the Council, the European Parliament, the Economic and Social Committee and the Committee of the Regions, Brussels, 18.1.2000 COM (2000) 6, final.
EC (European Commission). (2000b). *The Europeans and biotechnology—Eurobarometer 52.1.* Luxembourg: Office for Official Publications of the European Communities.
EC (European Commission). (2004). Towards a European strategy for nanotechnology. COM2004(338). Available at: ftp://ftp.cordis.europa.eu/pub/nanotechnology/docs/nano_com_en.pdf.
EC (European Commission). (2005a). Nanosciences and nanotechnologies: An action plan for Europe 2005–2009. Luxembourg: Office for Official Publications of the European Communities. Available from ftp://ftp.cordis.europa.eu/pub/nanotechnology/docs/nano_com_en.pdf.
EC (European Commission). (2005b). Working together for growth and jobs: A new start for the Lisbon strategy, COM (2005) 24. Brussels, Belgium: Official Journal of the European Communities. Available at http://cordis.europa.eu/documents/documentlibrary/73674591EN6.pdf.
EC (European Commission). (2007*). Strategy for communication outreach in nanotechnology.* Brussels, Belgium.
EC (European Commission). (2008). Commission recommendation of 07/02/2008 on a code of conduct for responsible nanosciences and nanotechnologies research. Bruxelles, C (2008) 424, final.
EC (European Commission). (2010). *Europe 2020: A strategy for smart, sustainable and inclusive growth.* Communication from the Commission, Brussels, 3.3.2010 COM (2010).
EC (European Commission). (2011). *Innovation union competitiveness report.* Available at ec.europa.eu.
EC (European Commission). (2012). Responsible research and innovation: Europe's ability to respond to societal challenges. KI-31-12-921-EN-C. Available at ec.europa.eu.
EC (European Commission). (2013). Options for strengthening responsible research and innovation: Report of the expert group on the state of art in Europe on responsible research and innovation. KI-NA-25-766-EN-C. Available at ec.europa.eu.
Evans, G., & Durant, J. (1995). The relationship between knowledge and attitudes in the public understanding of science in Britain. *Public Understanding of Science, 4* (1), 57–74.
Ezrahi, Y. (1990). *The descent of Icarus: Science and the transformation of contemporary society.* Cambridge, MA: Harvard University Press.
Felt, U. (2010). Vers la construction d'un public européen? Continuités et ruptures dans le discours politique sur les cultures scientifiques et techniques. *Questions de communication, 17,* 33–58.
Foucault, M.(1984). *L'usage des plaisirs.* Paris, France: Gallimard.
Gaskell, G., Stares, S., Allansdottir, A., Allum, N., Corchero, C., Fischler, C., Hampel, J., Jackson, J., Kronberger, N., Mejlgaard, N., Revuelta, G., & Schreiner, C. (1996). *Public perception of biotechnology in Europe: Eurobarometer 52.1.* Brussels, Belgium: European Commission. Available at http://ec.europa.eu/public_opinion/archives/ebs/ebs_244b_en.pdf.
Gaskell, G., Stares, S., Allansdottir, A., Allum, N., Castro, P., Esmer, Y., Fischler, C., Jackson, J., Kronberger, N., Hampel, J., Mejlgaard, N., Quintanilha, A., Rammer, A., Revuelta, G., Stoneman, P., Torgersen, H., & Wagner, W. (2010). *Europeans and biotechnology in 2010: Winds of change.* A report to the European Commission's Directorate-General for

Research. Brussels, Belgium: European Commission. Available at http://ec.europa.eu/public_opinion/archives/ebs/ebs_341_winds_en.pdf..
Hagendijk, R. (2004). The public understanding of science and public participation in regulated worlds. *Minerva, 42* (1), 41–59.
Hecht, G. (1998). *The radiance of France.* Cambridge, MA: MIT Press.
Heller, R. (2003). *GM nation? The findings of the public debate.* London, England: UK Department of Trade and Industry.
Hilgartner, S. (1990). The dominant view of popularization: Conceptual problems, political uses. *Social Studies of Science, 20* (3), 519–539.
Jasanoff, S. (Ed.). (2004). *States of knowledges.* London, England: Routledge.
Jasanoff, S. (2005). *Designs on nature: Science and democracy in Europe and the United States.* Princeton, NJ: Princeton University Press.
Jasanoff, S. (2013). Epistemic subsidiarity: Coexistence, cosmopolitanism, constitutionalism. *European Journal of Risk Regulation, 4* (2), 133–141.
Jasanoff, S., & Kim, S. H. (2009). Containing the atom: Sociotechnical imaginaries and nuclear power in the United States and South Korea. *Minerva, 47* (2), 119–146.
Joss, S., & Durant, J. (Eds.) (1995). *Public participation in science: The role of consensus conferences in Europe.* London, England: Science Museum with the support of the European Commission Directorate General XII.
Kurath, M., & Gisler, P. (2009). Informing, involving or engaging? Science communication, in the ages of atom-, bio- and nanotechnology. *Public Understanding of Science,18* (5), 559–573.
Law, J. (2009). Seeing like a survey. *Cultural Sociology, 3* (2), 239–256.
Levidow, L., & Carr, S. (2009). *GM food on trial: Testing European democracy.* London, England: Routledge.
Lundvall, B.-A., & Lorenz, E. (2011). From the Lisbon strategy to Europe 2020. In N. Morel, B. Palier, & J. Palme (Eds.), *Toward a social investment welfare state? Ideas, policies and challenges* (pp. 333–351). Bristol, England: Policy Press.
Macnaghten, P., Kearnes, M., & Wynne, B. (2005). Nanotechnology, governance, and public deliberation: What role for the social sciences? *Science Communication, 27* (2), 268–291.
Marris, C. (2001). La perception des OGM par le public: Remise en cause de quelques idées recues. *Economie Rurale, 266*, 58–79.
Nordmann, A. (2004). *Converging technologies: Shaping the future of European societies.* European Commission Research Report. Available at: http://ec.europa.eu/research/social-sciences/pdf/ntw-report-alfred-nordmann_en.pdf.
Nordmann, A. (2009). European experiments. *Osiris, 24* (1), 278–302.
Owen, R., Macnaghten, P., & Stilgoe, J. (2012). Responsible research and innovation: From science in society to science for society, with society. *Science and Public Policy, 39* (6), 751–760.
Roco, M. C., & Bainbridge, W. S. (Eds.). (2003). *Converging technologies for improving human performance: Nanotechnology, biotechnology, information technology and cognitive science.* Dordrecht, the Netherlands: Kluwer Academic Publishers.
Rose, N. (2007). *The politics of life itself: Biomedicine, power, and subjectivity in the twenty-first century.* Princeton, NJ: Princeton University Press.
Szyszczak, E. (2006). Experimental governance: The open method of coordination. *European Law Journal, 12* (4), 486–502.

von Schomberg, R. (2009). Organizing collective responsibility: On precaution, codes of conduct and understanding public debate. Keynote lecture at the meeting of the Society for the Study of Nanoscience and Emerging Technologies, Seattle, WA, September 11. Available at nanotechproject.org.
von Schomberg, R. (2012). Prospects for technology assessment in a framework of responsible research and innovation. In M. Dusseldorp & R. Beecroft (Eds.), *Technikfolgen abschätzen lehren: Bildungspotenziale transdisziplinärer Methoden* (pp. 39–62). Wiesbaden, Germany: Vs Verlag.
von Schomberg, R. (2013). A vision of responsible research and innovation. In R. Owen, J. Bessant, & M. Heintz (Eds.), *Responsible innovation: Managing the responsible emergence of science and innovation in society* (pp. 51–74). London, England: John Wiley.
Wickson, F., & Wynne, B. (2012). The anglerfish deception: The light of proposed reform in the regulation of GM crops hides underlying problems in EU science and governance. *EMBO reports, 13* (2), 100–105.
Wynne, B. (2006). Public engagement as a means of restoring public trust in science: Hitting the notes, but missing the music? *Public Health Genomics, 9* (3), 211–220.

Technologies of Transcendence at Singularity University

11

Margarita Boenig-Liptsin and J. Benjamin Hurlbut

The concept of "technological singularity," a point in time in which artificial intelligence will supersede human intelligence that was made famous by computer scientist and leading transhumanist Ray Kurzweil, is frequently uttered in the same breath with "transhumanism."[1] However, the Singularity University (SU), cofounded by Kurzweil and entrepreneur Peter Diamandis is, according to people both inside and out, not about the Singularity and not a university.[2] It is certainly not specifically for transhumanists, and it does not teach transhumanism as such.

The project of Singularity University is, however, like the idea of the Singularity itself, grounded in a faith in the inevitability of radical, technology-driven social transformation. If the project of transhumanism could be summed up in a phrase, it would be "to break with the boundedness and limits of humanity,"[3] to transcend constraint through technological means. The same description applies to SU, though the boundaries that it aspires to transgress are articulated not in terms of human ontological self-transformation but in the language of technological solutions to global grand challenges to engender large-scale social change.

The rhetoric of "technological solutions" is ubiquitous in contemporary knowledge societies. It reflects a faith in the socially transformative potential of technology. From engineering agriculture to "feed ten billion" to modifying the atmosphere to

---

1 There are disagreements about how the two concepts relate (for example, *Max More: Transhuman and the Singularity*, 2012), but the fact that both work toward similar goals of envisioning and enabling a different kind of human life with technology is undisputed.

2 This idea is frequently uttered at the Singularity University, and it is also written about in popular accounts of the University; see, for example, Rowan (2013, May 6).

3 This idea is central both to foundational texts in the history of transhumanism, such as Pierre Teilhard de Chardin's *The Phenomenon of Man* (1965) and Julian Huxley's "Transhumanism" (1957), as well as to contemporary transhumanist thinkers.

mitigate climate change, technological innovation is widely imagined as a source of solutions to problems that society and government appear impotent to address. The promise of technological innovation is a promise of transcendence: of transcending the problems of the present by realizing a better world in the future. Yet this is transcendence, with a small "t," is articulated in a secular idiom and imagined as in the service of society's secular institutions. Innovation is undertaken in response to society's needs and desires and engendered through policy and market—those liberal institutions capable of translating present preferences into future material orders. There is a consonance between societies' secular faith in innovation to bear us beyond scarcity and constraint and the more explicitly eschatological imagination of technological transcendence that animates transhumanism. Importantly, they are consonant in part because they are both recognizably secular: both narrate transcendence in an idiom of human agency, built on advances in technical capacity and control. As such they invite questions about commonplace constructions of secular knowledge, religious belief, and the boundaries between them (Hurlbut, forthcoming).

This essay explores this consonance by attending to the narratives of transcendence that are incipient in both the transhumanist imagination of technological progress and in the wider culture of innovation in Silicon Valley. SU offers a provocative site for interrogating these continuities and disjunctions. It is located at a kind of midpoint between a widely shared imagination of technological innovation as driving social progress and transhumanist visions of technologically driven transcendence. SU's vision is very much a reflection of Silicon Valley's innovation culture, in particular of the notion that entrepreneurial start-ups and hackers tinkering in a garage have the power to transcend the limits of what is possible, solving social problems with technology and thereby transforming the world.

One aim of this chapter is to draw attention to the ways that a sociotechnical imaginary of technological transcendence also invites reframings of responsibility and agency. The theoretical framework of sociotechnical imaginaries foregrounds how shared visions of technological futures function as loci of collective moral imagination in the present, thereby shaping future trajectories of sociotechnical change.[4] By attending to imaginaries that come to be culturally dominant, we can

4 Our analysis in this essay is informed by the theoretical concept of "sociotechnical imaginaries." Sociotechnical imaginaries are "collectively held, institutionally stabilized, and publicly performed visions of desirable futures, animated by shared understandings of forms of social life and social order attainable through, and supportive of, advances in science and technology" (Jasanoff, 2015, p. 6). This paper is an outgrowth of a larger research program developed in the Harvard STS program that interrogates imagined technological futures as profoundly normative articulations of shared understandings

see how practices of collective imagination produce asymmetries in power and shape social, political, and technological futures.

The notion that technology reshapes society privileges the innovator as an essential custodian of human well-being. In imagining those who create new technologies to be in the position to envision and generate a better future, it likewise authorizes them to declare imperatives of action in the present, for instance, what legal and economic configurations are required to allow the benefits of innovation to flow unimpeded to society. Thus, imagining beneficent futures—a prerogative of democratic governance—becomes a matter of making technical judgments about what social needs technology can meet and what allocations of authority and responsibility are necessary to bring about requisite forms of innovation.

The imaginary of technological transcendence that we explore below cuts through some of the contradictions that are incipient in common notions about the relationship between technological innovation and social life. Common accounts of innovation as a source of solutions tend to construe technological change as governable and in the service of society because it is undertaken in the name of an already articulated public good. Governance becomes a matter of ensuring adequate incentives and minimizing political sources of friction to technological progress;[5] take, for instance, the specter of "overregulation" that is constantly invoked in contemporary innovation communities. At the same time, technological change (and the innovators who produce it) is widely characterized as reshaping society such that law and policy inevitably "lag behind" and are constantly in need of reform to respond to technology's impacts and consequences.

This idea that technological change inevitably outstrips existing social and legal regimes is a commonplace in Silicon Valley culture. Importantly, it is a belief that is by no means limited to celebratory accounts of the revolutionizing effects of technology. Skeptics of the promised benefits of innovation also tend to attribute to the technologist a uniquely powerful role in shaping social life, a role whose excesses are captured in the notion of "playing God."[6] Scientific and policy elites

---

of what configurations of social life are desirable, right, and good. Cf. Jasanoff, chapter 3, and Nordmann, chapter 9, this volume.

5 This is the predominant attitude toward governance at SU, as we will discuss in more detail in relation to the concept of disruption. It is also shared by many entrepreneurs in the Silicon Valley; see, for example, Schumpeter (2014, January 24).

6 The phrase "playing God" is generally used as a critique of a technology innovator who is perceived to have overstepped or transgressed a socially acceptable limit, such as in the case of creation artificial life or, alternatively, as a dismissive gloss of putatively irrational critiques of technology. For but one example among many, see American Scientist . . . , 2010, May 21.

tend to dismiss the notion of "playing God" as a vague and irrational reaction to technological change, one that also fails to abide by norms of secular public reason. Yet the assumed asymmetry between expert and public knowledge of the technological future is itself the basis for dismissing it. One of the standard responses to such critiques is to circumscribe legitimate public scrutiny of the technological future to technologies that are deemed by experts to be realistic and on the immediate horizon, thereby excluding imagined futures that are significantly different from the way things are now (for example, the President's Commission, 1982; Presidential Commission, 2010; cf. Evans, 2002). This move has the effect of disciplining public moral imaginations to engage with only a circumscribed range of scenarios while preserving an imagination of science as a cornucopia of beneficial, if unpredictable, technological innovations. Put simply, the notions that the technological future can only be foreseen by those who create it and that culture necessarily lags behind are made to justify limiting public authority over technological trajectories to control (primarily in the role of consumer) the uses of products of innovation. This is a form of political deference: it allows for problems of responsibility that would otherwise be confronted as matters of political judgment to be sidestepped. Deferring to an imagination of technological innovation as an inevitable source of public benefit suppresses warrant for collective scrutiny of the technological future by limiting discussions of risk to immediate and familiar forms of technological capacity. It narrows questions of the good to questions of what is technically possible, deferring to the accounts of the technologists who are in a position to anticipate the future (Hurlbut, 2015). After all, to paraphrase a popular Silicon Valley adage, those best able to predict the future are those who invent it. Thus, imaginations of technological futures also encode normative commitments, including to the political and ethical identities of scientists and technologists.

SU affirms and celebrates the notion that innovators are responsible for human futures, as well as the politics of deference that this notion entails. SU ratchets up and renders explicit commitments that are mostly tacit (but very much present) in contemporary imaginations of innovation as a means of advancing social progress. In the routines of contemporary governance, innovation is characterized as in the service of society and governed by society's institutions even as it is seen as "revolutionizing" society. SU's rejects this contradiction, instead attributing far greater autonomy to technological innovation and to the innovator as agent. It constructs innovation not as an instrument of government but as the primary and most potent mode of governing the future, exposing social institutions of governance as weak and obsolete. Technological capacity, made available to the planet's billions, is seen as revolutionary, emancipatory, and an agent of transcendence: technological change perturbs regimes of power and authority and engenders forms of agency that allow the individual to transcend the circumstantial constraints of time and place. SU imagines a global community of individuals with myriad technologies at their fingertips, bound together no longer by political identity or geographical constraint but through networks of information and opportunity that ostensibly allow individuals to make of themselves what they will. Thus, whereas mainstream governance tends to reject vehemently the notion that the technologist is "playing God" and is thus ungovernable, SU affirms it.[7] As such, SU offers a rich case study for interrogating how imagined technological futures imply reconfigurations of governance that are, in turn, rooted in an imagination of technological transcendence.

This paper is based on fieldwork conducted by the authors at Singularity University intermittently from 2009 to 2013.[8] As participant-observers, we followed the development of SU and its visions, discourses and rhythms of work. Our analysis draws on these experiences supplemented by extensive research into publicly disseminated information about the activities of the university.

We begin with an analysis of the transhumanist imagination, focusing in particular on the ideas of transcendence—founded on notions of technological inevitability and exponential evolution—that underwrite Ray Kurzweil's account of the Singularity. We then introduce SU, exploring how these ideas translate into practice within this institution. We observe that SU grounds an ethical vision of responsibility and justice in a narrative of exponential technology, constructing "disruptive technologies" as agents of social progress and innovation as a moral imperative. We argue that "disrupting" the status quo, especially of social and normative orders, is seen as a responsibility of the innovator. As such, it is an ethical posture that is underwritten by the notion that the innovator is an agent of transcendence. Finally, we explore implications of SU's notions of disruption, progress, and social order for imaginations of governance and for the political institutions authorized with the task.

7 This is consistent with the view of Salim Ismail, former executive director of Singularity University, who affirmed Stewart Brand's declaration "We are as gods and might as well get good at it" (Portola Institute, 1969; Socrates, 2011, September 11).

8 Boenig-Liptsin was a student in the GSP-09 program and a teaching fellow for the policy, law, and ethics track in the GSP-11, while Hurlbut lectured in the policy, law, and ethics track in GSP-11, and both returned to the SU campus for various events during GSP-12 and GSP-13.

A Technocratic Transcendence

"Transhumanism" variously refers to a body of philosophical ideas, a social movement, and a stage of human evolution. Its multiple versions are unified by a particular, shared imagination that reflects two central commitments: first, the notion that technology is the primary force of historical change and the defining feature of human life, and, second, an optimistic anticipation of the progressive transcendence of human constraint by transforming the human body and the world through technological means.

Many expressions of transhumanism reflect a kind of self-styled radicalism, an iconoclastic vision of the human future that smashes conservative sensibilities and rejects human essentialism in the name of the dynamic remaking of self and world through ever-advancing technological capacity. The notion that technological change precipitates radical disjunctions between present and future is a central feature of the transhumanist imagination. This volume draws attention to the teleological-cum-eschatological dimensions of this imagination. One aim of this essay is to examine how such radical visions are "drawn down" to the level of ethical posture and practice, in particular to interrogate how they become consonant with more quotidian, secular ideas of progress and transcendence achieved through technological innovation. As we demonstrate below, the Singularity University is exemplary of this dynamic. However, before we turn to the case study, we here briefly discuss how the recognizably eschatological dimensions of transhumanism draw upon a discourse and repertoire of imagination that is more widespread in contemporary society, one that privileges innovation—and thus technological futures—as a locus of moral imagination and an agent of transcendence.

For purposes of our analysis, we draw attention to three salient features of the transhumanist imagination. First, envisioning radical technological futures is the bread and butter of transhumanism. From extensive modification of the human body to a world permeated by artificially intelligent machines and from hedonist technological utopia to dystopian apocalypse, transhumanist imaginations narrate the human future in terms of technological transformation. Second, in privileging technology in accounts of the human future, transhumanism simultaneously privileges it in accounts of what human life is. Transhumanist ideas notoriously reject the modern, humanist project of articulating a human quintessence as the ground of reason and normativity (Tirosh-Samuelson, 2012). They advance a dynamical rather than essentialist imagination of the human, characterizing technological progress as the defining feature of human history and identity. Technological innovation is what defines human beings as a species, and human life is always already embedded in a technological world. In this view, technology is what makes us human, and, insofar as there is a human nature, it is to endlessly tinker with, transform, and thereby transcend the technological present. Third, technology is also privileged within an ethical posture that is evident in certain threads of transhumanism. This ethical posture treats technological innovation as human essence, but also as human *telos*, as the defining purpose of human life. Innovation in this imagination becomes an obligation and a responsibility to acknowledge and claim ownership over technological power and to forge ahead into the future in the name of transcending the limitations of the present.

There is an important continuity between these three key elements of the transhumanist imagination and a commonplace imagination of technology in contemporary knowledge societies. Although transhumanists do tend to imagine radical departures from the state of human life in the present, beneath such radical visions is a cultural commonplace whereby technology is seen as a primary agent of historical transformation and social progress. As transhumanists are often quick to point out, their ideas of technological progress are coherent with mainstream sensibilities about the urgency of technological innovation in the face of human frailty and insecurity. They simply extend the logic of these sensibilities to more distant technological horizons (Solomon & Bostrom, 2005) or to macroscale, synthetic visions of the social "impacts" of technological systems (Kurzweil, 2005; Stock, 1993). Thus, the transhumanist imagination reflects a culturally common imaginary of human progress and perfectibility: the human condition is progressively ameliorated through advances in technological capacity. Furthermore, the pursuit of technology-driven amelioration is perceived to be an obligation of any responsible society.

The logic that links wider cultural sensibilities about technological change and the transhumanist imagination is particularly evident in Ray Kurzweil's ideas. For Kurzweil, the transhumanist ethic follows from dispassionate scientific observations about the dynamics of technological change. The sweep of human history—and, thus, human life—he argues, is best understood in terms of the technological trajectories that define it. These trajectories reveal law-like regularities that both predict future trajectories and reveal the irrelevance of particular social contingencies to the inexorable progress of technology. Kurzweil has quantified rates of technological change to demonstrate that technology is subject to the "law of accelerating returns" (Kurzweil, 2001, March 7). The law holds that evolutionary processes entail positive feedback mechanisms that increase in efficiency with each iteration of the process. Therefore, evolutionary processes are exponential, not linear. Evolutionary systems are also progressive. They bootstrap themselves by drawing upon a repertoire of resources for innovation that become increasingly powerful with each generation. According to Kurzweil, the law applies equally to biological and technological evolution, but it is only with technology that the time between generations becomes sufficiently short that the exponential rate of change can be experienced within a human lifetime. With the advent of human beings who "reach beyond [their] limitations," biology gives way to technology as the driving force of evolutionary change.

According to Kurzweil, "the law of accelerating returns" is a highly regular process that can be modeled and predicted. Yet to see the past (and future) in these terms requires an enlightened mind. Kurzweil declares that most people are ignorant of these dynamics, despite contributing to them. This is, according to Kurzweil, because most people have an "intuitive linear" view of history; time is measured by linear passage of years, not by the exponential process of the technological transfiguration of the world. This false intuition, according to Kurzweil, prevents us from seeing the future with clarity and appreciating the rapidity and inevitability of progress.

The "intuitive linear view" stands in contrast to the more enlightened "historical exponential" view. The historical exponential view measures history (and future) according to the dynamics that drive it, as opposed to the mere passage of time. These dynamics are, of course, figured most fundamentally by rapidly changing technological capacity. Measured in terms of historical trends at the macroscale, technology drives history by driving itself, and people are, in effect, along for the ride. Or, more accurately, the aggregate activities of human beings generate innovation according to law-like patterns such that specific patterns of technological change are not a choice so much as a natural and inevitable evolutionary expression of human nature. They are both nature and destiny. However, we must note in passing that this picture applies at the macroscale, but, at the same time, this account of technological progress does depend upon the agency and creativity of innovators who harness existing capacity to achieve transformative effects on the world, effects that are felt through the diffusion and uptake of technological innovation. This mapping of technological change posits a profound asymmetry between the producers and consumers of technological innovation, constructing the former as shaping the future and the latter as passively along for the ride. This is because the former can both see and create the future, whereas the latter can do neither. We return to this point in the next section.

For Kurzweil and his followers, the distinction between linear intuitive and historical exponential understanding of technology is the key distinction between an outmoded humanism and an enlightened transhumanism. Kurzweil's assertion that "the essence of being human...is to reach beyond our limitations" (Kurzweil, 2005, p. 311) is first a historical observation and only secondarily an ontological and ethical appraisal. The "law of accelerating returns" reveals (on Kurzweil's account) a natural fact about the human past and future that must be acknowledged and deferred to in constructing moral philosophical accounts of what we are and what we ought to be (Kurzweil, 2001, March 7). Thus, in his *Age of Spiritual Machines* (1999), in which he imagines a (utopian) future in which human beings are transcended by new forms of disembodied intelligence, Kurzweil characterizes this narrative as merely a playing forward of a version of Moore's law on a larger scale than the microchip. Transcendence, in this imagination, becomes a function of inevitable technological trends, and transhumanity is a natural and inevitable waypoint in a process of technological-cum-spiritual transformation. The Singularity is at once an eschatological fulfillment of historical destiny and an inevitable outcome of mundane processes of technological change. For Kurzweil, therefore, transhumanism is not a choice or ideology or faith but a rational acknowledgment of particular historical dynamics and of their implications for how we should understand ourselves and aspire to a better future. A dispassionate appraisal of the dynamics of technological change reveals a picture of human nature and human destiny.

The centerpiece of this picture is the notion of technology as enabling human abilities. Without their technological prostheses, human beings are "dis-abled": constrained by the limitations of their bodies and their social environments. We use the concept of dis-ability to highlight how technological capacity is made the "measure of man." As technology transcends its prior iterations, so too human life transcends its prior constrains, rendering what was previously the limit of ability into dis-ability vis-à-vis newfound capacities. Technology enhances agency and thereby endlessly establishes new baselines for human agency, freedom, and fulfillment. Once again, this construction of human fulfillment finds easy consonances with wider cultural sensibilities. It is worth noting that Kurzweil became a public figure in part for his contributions in developing a technology to assist the disabled in overcoming the constraints of their bodies, namely, text-to-speech technologies for the visually impaired and people with learning disabilities (American Foundation for the Blind, 2004). For Kurzweil, there is a complete conceptual continuity between his technologies for the disabled and his more general vision of technological transcendence—a vision that, as we shall see, profoundly inflects the project of SU.

In Kurzweil's thought and in the transhumanist imagination more generally, we find the trope that technological change is the frame through which to mark history and to reflect upon the human condition. This is a technocratic vision in the most basic sense of the word: *techne* is the primary force that governs human life. Yet it is also an imagination of who is in the know: of who can see the future clearly and is, therefore, in a position to prepare for it, produce it, and govern it. The future is known to—and, thus, belongs to—those enlightened experts who understand the dynamics of innovation. And ethical postures that invite transformations of human life are recast as simply accepting—and preparing for— the lives to come. This narrative of inevitability constructs technology as the locus of agency and the driver of history. It allows students of the dynamics of innovation to claim a privileged position from which to see the future and thereby to diagnose what ideas and institutions are obsolete, or, worse, contradicting and inhibiting the unfolding of technological progress.

This imagination of inevitability and predictability also offers the basis for a kind of moral clarity and ethical remit. It is a common trope that we are thrown into a condition of power by the putatively inevitable unfolding of technological progress, that questions of "what should be done" with our newfound powers are somehow consequences of, rather than contributions to, regimes of technological innovation. The power of technology to reshape and redeem human life, thus, becomes an ethical idiom: what must be done becomes a question of what technological futures are possible and who can and will bring them about.

In summary, we have drawn attention to the consonance between transhumanism's (ostensibly radical) construction of the future and the common conviction that technological innovation drives progress and enhances human well-being. In both views, technological change drives history and transcends the constraints—the dis-abilities—of the present. In mainstream discourses of innovation, this view is less explicit and bares less of the historicism and eschatological trappings of the transhumanist imagination; but, for both, technology is treated as a privileged site for imagining social transformation, progress, and justice. In the next section, we examine these ideas in practice.

Singularity University

The Singularity University was founded in 2009 by Ray Kurzweil and the technology entrepreneur Peter Diamandis. Kurzweil is an accomplished inventor and a public figure who is best known for popularizing the idea of the Singularity (see the introduction to this volume). Heralded as a prophet by a sizable contingent of techie groupies, he is a well-known figure. He was awarded the National Medal of Technology by President Bill Clinton, was called one of 16 "revolutionaries who made America" by a PBS documentary, and has been repeatedly and extensively honored by a number of elite universities and technical institutes, including MIT and Cal Tech. Peter Diamandis founded the XPRIZE foundation, which has pioneered a model of supporting "moonshot" innovation projects through competitions in which the team that achieves some specified technological feat wins a multimillion dollar prize. XPRIZE competitions have the dual function of engendering innovation in a specified domain and creating a public spectacle. The competitions, in effect, simultaneously declare what the technological future will look like and take steps to bring it about, albeit for a specific and circumscribed technological domain. It is a model that has been widely celebrated as an effective means to solve defined problems through innovation and has been taken up by certain agencies within US Federal government, including DARPA and the White House (Ackerman, 2012, October 3; DARPA, 2013; Dorgelo, 2015, September 5). The XPRIZE model reflects Diamandis's conviction that creative capacity for innovation exists in the world and that, to engender transformative technological change, innovators must simply be drawn out, given resources (or incentives), and set free to create the future.

The Singularity University bears the fingerprints of its creators. Its mission is to "assemble, educate and inspire a new generation of leaders" capable of driving the development and use of "exponential" and "disruptive" technologies to "address humanity's grand challenges." SU's core offering is an annual two-month-long graduate training program, known as the Graduate Studies Program (GSP), in which promising young innovator-entrepreneurs are instructed in the entrepreneurial strategies, creative techniques, and social (and ethical) visions that SU sees as critical for technological innovation.

SU is located on the NASA-AMES research campus in the heart of the Silicon Valley in Mountain View, California. Also known as Moffett Field, NASA-AMES was an important site of research in aeronautics, space, and military technology during the Cold War. Its most visible landmark is Hangar One, a massive airplane hangar that looms over all the other structures in the area and encloses nearly eight acres of land. It was built in the 1930s to house the navy's 785-foot-long dirigible, the *U.S.S. Macon* (Veronico, 2006). For the small group of locals who turned out to see the Macon moor in Hangar One, the scale of these technologies reflected the emergence of the United States as a high-modern state and a global military power (Grossnick, 2013).

The decaying vestiges of this period are scattered around the NASA-AMES campus, with Hangar One the most prominent among them. In recent years, the hangar's asphalt skin has been stripped away, leaving the steel bones of the structure exposed. Like an ancient whale on display in a natural history museum, the skeletal remains of the structure stand as a monumental memorial to the early days of the military-industrial complex. For the Singularity University and the Silicon Valley innovation culture in which it is embedded, it also marks an epoch in the history of technology that has been displaced by a new era with new configurations of power. The Singularity University teaches that the lumbering leviathan of the high-modern technological state is on the decline, transcended by the small and nimble start-up, operating in an increasingly globalized arena. Across from Hangar One, colorful banners with the SU logo attached to streetlights around the parade grounds flap in the wind. During SU's inaugural session, the banners posed questions that harkened to the newfound power of the techno-elite to change the world: "What are the laws and ethics of progress?", "How will machine intelligence touch our world?", "How can small devices solve big problems?"

The SU faculty includes some of Silicon Valley's most successful engineers, venture capitalists, and high-tech entrepreneurs. Its corporate founders include Google, Genetech, Autodesk, and Cisco (Singularity University, 2015b). Many of the students in the GSP, of which there are roughly 80 per year,[9] hold a Ph.D. or equivalent, most from elite institutions. The majority have started at least one company. Between the faculty and the student body, SU represents a powerful social network and pool of talent with links to reservoirs of capital that are now sufficiently large to underwrite risky innovation of a sort that has historically depended on government (especially military) R&D investment.

The transhumanist imagination informs SU's educational curriculum, but it is not explicitly presented as such. In fact, the word "transhumanism" is rarely heard in the SU classroom. If it is mentioned, it is with a certain caution and skepticism because the faculty and students at SU, though they believe in the tremendous potential of technology to transform the human body and the world, consider the abstract philosophical ideas associated with transhumanism speculative and potentially detrimental to the SU image. Transhumanism's self-styled radicalism is seen as a liability for the credibility of SU.[10] Rather, SU represents itself as an educational institution that teaches forms of entrepreneurial, strategic, and ethical thinking that are, on its account, necessary drivers of innovation. Its mission is to transform the world by giving a vanishingly small number of would-be innovators the know-how and resources to innovate exponential technologies.

The SU project is to build capacity for engendering technological innovation to address globally recognized threats to human well-being and to improve human

---

9 The first GSP class had 40 students but doubled in the second year to 80 students and has remained at this number since.

10 Careful policing of the SU brand name is an important task of university administrators. SU's acquisition in 2012 of the Singularity Summit, an annual conference on AI founded by the Singularity Institute, which was itself founded by Ray Kurzweil and venture capitalist Peter Thiel, was an effort to control the activities that go on under name of "Singularity." The article that announced the purchase contains a statement by Luke Muehlhauser, executive director of the Singularity Institute, that the word "Singularity" has "broadened" since the institute's founding in 2000 (SU, 2012, December 6). Arguably, the activities of the SU focused on technological innovation in the here-and-now have been the main contributor to the expanding definition of the concept.

lives. Its stated mission is not to produce speculative visions of the future world but to transform the world through innovation (though, in practice, there is much of the former and little of the latter.) If, as is frequently said, the Singularity University is not about the Singularity and is not a university, what then is it, and what is it about? What is the relevance of what takes place there to projects of "perfecting the human"?

Ray Kurzweil announced the creation of SU in a Ted Talk in 2009: "A university for the coming Singularity" (Kurzweil, 2009, February). Most of his presentation was dedicated to making the case for the exponential nature of technology and the inevitability of the Singularity. He described the regularity and predictability of exponential increases in computing power, observing that the same rules apply to "any technology where we can measure the underlying information properties" (ibid., mins. 01:25–01:30). Three key themes that are central to SU's pedagogical mission stood out in his comments. First was the predictability and inevitability of exponential technological change and its immunity to social contingency and perturbation. With a graph of increases in computing power displayed behind him, he declared, "What's even more amazing, really, than this fantastic scale of progress, is that—look at how predictable this is...through thick and thin, through war and peace, through boom times and recessions. The Great Depression made not a dent in this exponential progression" (ibid., min. 02:17). Second was the notion that technological capacity transcends forms of social and political power. It engenders possibilities for individual agency that tend to (literally) revolutionize the structure of societies, and these revolutions track exponential technologies—and, thus, are foreseeable to those with an enlightened "historical exponential" view. "[Exponential communication technologies] are deeply democratizing. I wrote, over 20 years ago...when the Soviet Union was going strong, that it would be swept away by this growth of decentralized communication" (ibid., mins. 04:30–04:44). Third was the prediction that exponential technologies will inevitably and irredeemably transform human life itself:

> We ultimately will merge with this technology. It's already very close to us. When I was a student it was across campus; now it's in our pockets. What used to take up a building now fits in our pockets. What now fits in our pockets would fit in a blood cell in 25 years. And we will begin to actually deeply influence our health and our intelligence, as we get closer and closer to this technology (ibid., mins. 06:01–06:25).

Thus, the idea of the Singularly is present not only in the name of SU but also in the animating vision that lies behind it. Yet the Singularity is rarely discussed at SU, except in a kind of quasi-disavowal. However, while the grand speculative vision of the Singularity is put aside, the idea of exponential technology and the (inevitable) trajectories of transformation that they engender figures centrally in the SU curriculum. Indeed, SU was, in Kurzweil's words, created as a place to study exponential technologies "with an emphasis on their interactions, and to help guide the process for the benefit of humanity and its environment" (Kurzweil, 2009). SU's tagline, displayed above a logo that is more than a little reminiscent of Superman's "S", is "preparing humanity for accelerating technological change."

Soon after Kurzweil's presentation, the call for applications for the first GSP went out. SU received over 1,200 applications for 40 spots in the nine-week-long pedagogical experiment (Johnson, 2009, September 2). Students were selected for their expertise and leadership in one of the ten academic tracks: (1) future studies and forecasting, (2) policy, law, and ethics, (3) finance and entrepreneurship, (4) networks and computing systems, (5) biotechnology and bioinformatics, (6) nanotechnology, (7) medicine, neuroscience, and human enhancement, (8) AI and robotics, (9) energy and ecological systems, and (10) space and physical sciences (Singularity University, 2009, July 30).[11] The founders also considered their potential to contribute to what Diamandis called the "benign conspiracy": "benign" because the project was supposed to be good for humanity and "conspiracy" because the group was planning to do "big things" that would change the world (Rein, 2009, June 29).

These selection criteria reflected the foundational aim of SU: to engender a culture of thinking "big" and doing "good" with technology—a culture infused with faith in the power of innovation to transcend constraint and transform human life. In this spirit of technologically scalable beneficence, the SU curriculum is organized around the 10^9+ (read: "ten to the nine plus") Team Project. The challenge of the 10^9+ project is to develop an innovation "to positively affect a billion people in the next 10 years using accelerating technologies" (Singularity University, 2015c).

The leadership has taken steps to make the "global" a defining feature of SU. The SU curriculum is organized around the "global grand challenges" (GGC): food, water, poverty, environment, health, energy, education, space, and security (Singularity University, 2015c). The GGCs identify the social problems that SU

---

11 For the upcoming GSP-15 program, the number of tracks has been reduced from ten to six. These are biotechnology and bioinformatics; energy and environmental systems; networks and computing systems; AI and robotics; medicine and neuroscience; nanotechnology and digital fabrication (SU, 2015a). Notably are missing the "social" tracks (future studies; policy, law, and ethics; and finance and entrepreneurship). This change reflects a shift in the pedagogy of SU over the years where subjects concerning the governance of science and technology are not discussed in their own right but incorporated into the "technical" tracks. This reflects not so much the secondary nature of these issues for the university, but rather the view that social questions arise only in relation to specific technological developments that invariably lead the way.

believes its approach to innovation can address. Students are taught that GGCs are exponential problems, addressable only through exponential technologies. The student body reflects this concern for the "global." During the opening ceremony for GSP- 12, students paraded into the auditorium waving the flags of their countries of origin. Thirty-six nations were represented (Saenz, 2012, June 16).[12] The diversity of the students acts to ensure not only that SU draws upon a breadth of cultures and experiences but also that it represents the global population—the "billions" on behalf of whose interest each student acts.[13]

At SU, this vision of technological change is also a vision of social justice. Indeed, one of SU's foundational commitments is to creating a more just world through exponential technologies. (This was also a condition of receiving the support and largess of key Silicon Valley funders at its founding.) During the keynote address at SU's fifth anniversary celebration, Larry Brilliant, a physician famous for leading Google.org, the philanthropic arm of Google, called on the recent graduates of the summer program and all alumni of the five years of the Singularity University's existence to "jump on, grab ..., twist ..., and bend" what Martin Luther King referred to as the "arc of the moral universe" so as to force it towards "justice."[14] By transforming Martin Luther King's famous quotation about the overall tendency of morality into an imperative to seize and manipulate, Brilliant affirmed the SU view that technology—and the technologist—is an agent of social transformation.

Thus, this vision of technological innovation is also a vision of responsibility. The power to shape global justice— indeed, to declare what is (un)just—is claimed by those who are (putatively) in a position to create the future: to align the exponential curve of technological change and the arc of the moral universe. Problems of scarcity, infirmity, and illiberty are, in the SU imagination, all technologically remediable; it is the responsibility of technology to "improve the lives of a billion people." Importantly, this is a power relationship. It asserts a profound asymmetry between the innovator-benefactor and the benefiting billion. Little attention is given at SU to how one might ascertain what would constitute improved lives for

12 By contrast, in the GSO-09 class, 13 countries were represented (SU, 2009, July 30). The number of countries represented in the GSP classes has increased with each year.

13 This idea that a student at SU is a representative of his or her country or of some seemingly universal global citizen is echoed in the online fundraising campaigns that students who are accepted at SU engage in. To raise money for the GSP tuition, some students launch crowdsourcing campaigns on sites such as IndieGoGo with slogans such as "Help me help 1 billion people," reflecting the idea that the student acts on behalf of her constitutents-investors (Coolidge, 2013).

14 Brilliant delivered the same message to graduating students of the Harvard School of Public Health in his 2013 commencement address (Harvard, 2013, May 30).

the billion, let alone understanding the culturally situated moral imaginations that inform notions of human flourishing. Instead, innovation itself is seen as indisputably necessary and good. It, like the technological future it brings into being, is unimaginable to the billion until it is already before them. Hence, the thinking goes, there is little to be gained (and much to be lost) from making innovation contingent on first ascertaining collective desires or seeking society's permission.

Responsibility, therefore, is seen a function of the *ability* to act, rather than the duty to do so. The enlightened innovator is in the position of agency because only he can imagine and enact a beneficent future. But he has no obligation to do so. Rather, responsibility is understood as "response-ability"—as the innovator capitalizing on his unique ability to reshape the world, to offer a response to human needs. Innovation promises to rectify injustice and redeem human life, but it is the innovator's choice whether to act, and action is supererogatory by definition.[15]

During the opening ceremonies of GSP-13, one of the senior SU faculty gave a brief exposition on the idea of responsibility that informs SU. He told the story of scuba-diving with a companion. His companion failed to calculate how much oxygen he would need. With this lack of foresight, he found himself in a position of scarcity: short on air and in danger of drowning. The SU faculty member had had the wherewithal to see into the future and, thus, had an abundance of oxygen to spare. He rejected the notion that he had an obligation to supply his spare oxygen to his companion. It was "not my responsibility." But it did present an "*opportunity*" to leverage his position to solidify his friendship. This, he declared, is the relationship between SU and the world. The innovator elite has no *obligation* to save humanity from its own self-imposed destruction, but he has the *opportunity*—the ability to respond in a way that the rest of humanity is incapable of. At SU, the speaker declared, "we seize opportunity," and we get rich in the process.

15 With "response-ability," we do not intend to reference Donna Haraway's (2008) use of the term, as we came to it independently. Our term emerged from our observation of the language and concerns of the SU community's characterizing responsibility as "opportunity" and privileging technologically enabled "responses" to global problems. Haraway uses the term to reference an ethical comportment that acknowledges being engaged in relationships of response and "becoming with" beings that exceed (and problematize) the category of the rational human subject. Whereas, for Haraway, attunement to "becoming with" opens space for new modes of ethically deepened relationality, in our case, the basis for imagining ethical interrelations is the instrumental techniques through which those capable of acting (that is, the innovators) aspire to reconfigure the subjectivities and intersubjectivities of those upon whom they act. The response-abilities that, for Haraway are the wellspring of enriched ethical attunements are, in our account, already configured as an expression of a sociotechnical imaginary.

The problem of poverty poses an "opportunity" rather than a "responsibility." The GGCs are regularly marked as business opportunities at SU. As a consequence, "improving the lives of a billion people" is a secondary benefit of *interested* entrepreneurial action. This notion that the world's greatest challenges are "opportunities" comes in part from the faith at SU in the possibility of individual human beings to use technology to right wrongs. This optimistic vision of the possibility of "doing well by doing good," seizing "opportunities," and bringing about a world of "abundance" is intimately connected to a view of people as "dis-abled," as constrained by the limitations of their bodies and their social environments, and of technology as enabling.

Talk of the "global" is ubiquitous at SU, in this respect tracking a common discourse in contemporary neoliberal imaginations. Importantly, however, SU's "globe" is imagined as knit together by technology. Technology transcends cultures and borders, and technological futures affect and embrace humanity as a whole. For SU, innovation is not (and ought not be) politically or geographically circumscribed. Exponential technologies escape the grasp of social and political institutions and inevitably percolate into the wider world. Indeed, they are seen as agents of political transformation rather than as means of political control. (The example of social media in the Arab Spring revolutions is often invoked.) They offer forms of agency to individuals that were unimaginable in the recent past. As Peter Diamandis recently remarked to former President Bill Clinton, "A Maasai warrior in Kenya on a smartphone has more access to knowledge and information than you had when you were the president twenty years ago"(*President Bill Clinton . . .* , 2014). Implicit in this statement is the idea that exponential technologies empower without regard for national identity or boundaries.[16] They are, in this imagination, agents of globalization that extend to, connect, and unify into new solidarities people who have historically been segregated by national identity. SU celebrates microfinance and communications technologies as examples of the emancipatory and agency-engendering power of exponential technologies, though these are seen as representing only the tiny tip of the iceberg. For SU, the global is not an imagined

16 There is, of course, much to object to in these characterizations, for instance, in the politically configured variations in the technology of the Google search itself, with constraints on search in China, with the "right to be forgotten" in Europe, and so forth. However, our aim is to explicate rather than "fact check" the imaginary that underwrites the SU project, in part because, while specific claims do demand qualification, the general narrative that SU advances has acquired significant credibility in contemporary American culture, including within those institutions whose practices contradict the imaginations of technological empowerment that SU advances. Indeed, Clinton's response to Diamandis's comment was, "That's a frightening thought!"

community in Benedict Anderson's sense (Anderson, 1991; cf. Taylor, 2003), but an *imagined* imagined community: a unified global community is, in the SU vision, an anticipated but inevitable consequence of a future that will be ushered in through technological change and corollary disruptions of social, economic, and political orders. Indeed, the idea of global solidarity is an aspirational normative dimension of the technological visions that SU celebrates, visions in which ideas of development and political emancipation figure centrally as achievements of technological innovation.

In this sense, SU's "global" is a flattening concept. The globalizing force of exponential technologies is seen as obviating political and cultural differentiation, not by challenging differences so much as rendering them irrelevant. In the SU vision, exponential technologies generate a densely interconnected world whose character is determined more by those technologies than by culture or politics. As technologies become "better, faster, and cheaper," they diffuse from center to periphery, unifying cultures and reducing the distinction between the rich and the poor. Indeed, SU faculty are fond of saying that, in the future, the world will no longer be divided into "haves" and "have-nots" but into haves and have-a-lots, where the former benefit from technology-derived abundance and the latter are the innovator-entrepreneurs who reap the rewards of bringing benefit to the rest of society (Diamandis & Kotler, 2014).

Importantly, this is as much a vision of political transformation as it is of technological diffusion. During a public event at SU, one of the present authors had a conversation with several GSP students from Latin America about political sovereignty and the issue of transnationally conflicting visions of whether a technology is socially desirable, for instance, when a society rejects a particular technology through its political institutions, as has happened with genetically modified crops. For these students, postcolonial legacies of interventions made in the name of development were felt concerns. A senior member of the SU faculty happened by, and we engaged him in the conversation. He declared with disgust that "people don't think," implying that those who did would see no values conflict. At the same time, he discounted the ultimate significance of such resistance and the political institutions through which it is expressed, declaring that a high enough "value proposition" overcomes any resistance.

In this formulation, the "value proposition" is intrinsic to the technology itself: technology as a bearer of benefit eventually and inevitably displaces those repositories of values that stand in the way of its diffusion. There is, of course, ample evidence to challenge this simplistic narrative, but we wish to draw attention specifically to the imaginary of technological change that underwrites this particular way of thinking and talking about cultural, political, and moral configurations and relationships. This imaginary flattens the global into a homogeneous space that is defined by human needs and desires for which cultural particularity is seen as largely irrelevant. Humanity comes to be defined by the putatively universal benefits that particular forms of innovation can offer: the global community, as imagined at SU, is unified by the shared grand challenges for which innovation offer solutions. The billion for which the one claims response-ability is without place or identity. It is a generic unit of global humanity, for whom the value of a technology derives from that technology itself and not from any historically or culturally situated values.

The notion of response-ability, configured as a generic relationship between the response-able few and the dependent many, represents the ethical core of SU. SU students are characterized as "leaders" because they are the elect few who understand the worlds' challenges as exponential problems that demand exponential technological solutions. Questions of ethics do come up at SU, and one of its academic focus tracks was (up until 2014) dedicated to ethics, policy, and law. But ethical problems tend to be understood as "bad" applications of "good" technologies—abuses of technological capacity to do ill. This notion is underwritten by the assumption that what is good and bad is self-evident to the innovator, since, after all, the value proposition is as much a creation of the innovator as the technology itself. Furthermore, the solutions to such problems tend to be also understood in technological terms: as a matter of engineering risk containment mechanisms into risky technologies or building systems that are more robust against dangerous applications. Where law or regulation are invoked, they tend to be seen as instruments for enforcing risk containment measures to prevent abuses of technology, with the corollary notion that otherwise unconstrained uses are beneficial by definition; the intrinsic "value proposition" needs no justification and, thus, ought not be constrained by externally imposed rules.

Therefore, the power—the response-ability—to reshape the world for the better is imagined as lying in the hands of the innovator. The asymmetry between the one and the billion is an expression of this power of the one and the impotency of the billion to address humanity's grand challenges. The right response to dis-ability is expanded technological capacity, conceived and constructed by the one, and delivered to the many.[17]

---

17 It is worth noting that this vision is not simply a neoliberal one. It is distinct from the notion that the market can harness individual self-interest to bring about collective benefit. While the market is seen as a necessary and important modality of dissemination, transformation in the SU vision comes not from market forces but from individual innovators whose creations create and transform markets, including by serving needs and possibilities that market demand has not yet imagined. Thus, the imagination of

This account of the innovator as an agent of social justice is grounded in an imagination of technology as shaping and reshaping social and moral order. At SU, visions of the exponential impacts of technology, of their promise of social justice and of corollary notions of an ethic of response-ability, are unified in an imaginary of transcendence wherein progress is a consequence of the disruptive effects of technology on social life.

Exponential Technology and the Dynamics of Disruption

Although technological innovation is the central focus of SU, students are not taught technical skills for designing particular technologies. Instead, they are exposed to trends in technology sectors and to the accounts of players in those sectors about what is on the horizon, all in the name of mapping the world-transformative potential of exponential technologies. Students are taught to think "exponentially" about these trends. "Thinking exponentially" is the central project and pedagogical core of SU. The SU faculty regularly repeats the slogan that "thirty linear steps is thirty steps, but thirty exponential steps is a billion." In thinking exponentially, SU promises its students a course that will render them leaders and world transformers, since, after but a few exponential steps, the rest of the world is necessarily miles behind. Indeed, this notion of thinking exponentially, and thereby bringing about exponential futures, is the basis for several asymmetries in the SU vision: between the present and the future, between the one and the billion, and between the enlightened few who understand and embrace the technological future and the ignorant many who ignore or fear it. These asymmetries underwrite the idea of response-ability and serve as the warrant for undertaking technological projects that "disrupt" the status quo. The SU faculty repeatedly affirms how difficult, unintuitive, and unconventional it is to be an exponential thinker.

Students are challenged daily to "think exponentially," that is, to imagine how technologies, industry, and society would change as a result of exponential transformations (decreases in price, increases in capacity and power) in technology and to train their sights on the business opportunities that they thereby discern. Indeed, the 10^9+ question—"How will you improve the lives of 1 billion people?"—is a learning exercise in exponential thinking. Since technological development is seen at SU to be the result of exponential increases in capacity and decrease in the price

improving lives is grounded not in an invisible hand but in the visible and directive hands of the technologists who construct the future.

of information (all technologies discussed at SU are presented as being foundationally informational technologies and therefore subject to exponential growth of the information technology sector), thinking exponentially is presented as *the* way to anticipate where technology is headed and, thus, what opportunities exist to do good (and, thereby, to do well).

Yet, at the same time as students learn to think exponentially and are asked to incorporate exponential projections into their 10^9+ projects, they are also required to produce tangible business plans, product prototypes, and convincing investor pitches by the end of the summer program. The students must accommodate the radical, necessarily extrapolative, and sensational future to the mundane challenges of constructing a credible business plan. This places students in a difficult position, requiring them to draw down exponential thinking into a pitch for innovation projects that are plausibly achievable, in short, to bridge a radical futurism and a realistic pitch.

At this intersection of radical future and realistic pitch, the idea of disruption takes shape. The discourse of disruption is common in Silicon Valley and ubiquitous at SU. The idea that exponential technologies are socially, economically, and politically "disruptive" is one of SU's central tenets. Talk of disruption abounds and is almost universally celebratory. Importantly, "disruption" has a different meaning at SU than it does in business schools, where the idea of disruptive entrepreneurship was first conceived. For example, in the influential work of Harvard Business School professor Clayton Christensen (2011), disruptive innovation refers to firms that find a way to compete in a particular market space by developing faster and/or lower-cost methods of serving a substantially similar market niche. At SU, disruption implies something more: it refers not only to market destabilizations but to social, political, and moral destabilizations precipitated by technological innovation. A disruptive technology disrupts because it transgresses—indeed, transcends—prior limits and the corollary sensibilities that emerged around them. It breaks the rules by remaking the rules. Thus, whereas Christensen's notion of disruption depends on capturing existing demand, and thus on existing forms of value, for SU, disruptive technologies transcend and transform existing desires and values. They give something unprecedented and thereby shatter established orders, offering something totally novel and transformative and engendering new market-reshaping desires in the process.

For instance, the iPhone is frequently held up at SU as exemplary of a disruptive technology. It displaced phones, cameras, flashlights, computers, and a host of other technologies by unifying them in a single piece of hardware (Diamandis & Kotler, 2014, p. 150). More importantly, however, it created desire (on a global scale) for an object that people did not yet know they needed and offered forms of capacity that were not yet within people's imaginations.[18] And it created a platform of innovation itself, certainly for the myriad apps that have been produced for it but also for the forms of social connectivity, the habits of communication, and the new forms of agency that people enjoy through this platform. With a smartphone, Diamandis's Maasia warrior not only has (ostensibly) more access to information than an American president did at the turn of the millennium, but he is potentially linked into a system of global finance through which he can access capital and become an entrepreneur himself, perhaps contributing to the churn of innovation that has emerged around this single piece of hardware. Disruption in this vision is emancipatory: it allows new access to information and new tools through which a larger number of individuals can become "makers."

The ethos of disruption is well captured in Larry Brilliant's "revision" of the famous saying by Martin Luther King at the SU Night of Impact celebration. Brilliant's revision encourages students to boldly take the future into their own hands—to actively bring about with technology the vision of the future that is consistent with their understanding of what is good and just. This attitude is "disruptive" because it challenges the existing social order and seeks to transform it using violent (jumping on, grabbing) and technological (twisting, bending) means. Two examples from the GSP-13 program illustrate the central and at the same time somewhat ambiguous position that the ethos of disruption occupies at SU.

During the course of the GSP-13, a member of the SU faculty invited Cody Wilson to discuss his projects at SU. Cody Wilson is the founder of Defcad (now Defense Distributed), a search engine for 3D printable models. Wilson became famous after the release of a controversial video in which he voiced his mission for Defcad as a site for "subversive" 3D printing that began with an example of 3D printing guns using open-source files (Wilson, 2013). Wilson was invited because, as one SU faculty member put it, "he is someone who has an idea for a revolution." The subversive tones of Wilson's idea, his desire to go around government efforts to tighten gun control with the help of a technology identified as "exponential" at SU, seems on its face to resonate with the SU ethos of disruption. However, Cody Wilson's presentation at SU was one of the most controversial events of the summer. A number of students and faculty vigorously disagreed with this self-proclaimed "crypto-anarchist's" project of disruption. One prominent faculty member got into a heated exchange with Cody, saying that SU is not about creating social disorder through technological anarchy, because the world cannot be orderly in anarchy. The

18 The ability of the technology innovator to generate desire for a technological object and the lifestyle it represents makes up the popular mystique of the innovator. This ability is commonly attributed today to legendary innovators like Steve Jobs and Henry Ford.

Cody Wilson case highlights the difference between technologically driven disorder and disruption in the SU vision and the anti-anarchical posture of SU toward innovation. For SU, as in the transhumanist imagination, disruptive technology is an agent of social reordering by creating capacity for new forms of autonomy and more effective means of governance, including, especially, self-governance. Disruption does not mean destruction but transcendence of the constraints of dis-ability.

By contrast, our second example is exemplary of SU's visions of technological disruption. Glowing Plant is a project of GSP-11 alum Antony Evans, a techno-entrepreneur originally from England and now living in San Francisco, and Omri Amirav-Drory, a biochemist whom Evans met at SU. The project grew out of the SU network that Evans developed while a student, in particular, the relationship to the DIY "bio-hacker" space called BioCurious that is run by Raymond McCauley, an SU faculty member and chair of the biotechnology and bioinformatics track (Dzieza, 2013, August 18). Evans sought funding for his project on the crowd-funding site Kickstarter.com. His goal was $65,000, but he raised a remarkable $484,000 to make glowing plants that would, in his long-term vision, replace electrical lighting. For pledges over $40, Glowing Plant promised to send out packages of glowing plant seeds in May 2014 (Evans, 2013, April 23).[19] Glowing Plant elicited significant public attention and controversy (Cutler, 2013, April 23; Pollack, 2013, May 7). Two environmental organizations, Friends of the Earth and ETC (Erosion, Technology, and Concentration) wrote to Kickstarter and the USDA with concerns about the potential release of genetically modified seeds into the environment (ETC Group, 2013a and 2013b; Leach, 2013, May 7). Kickstarter told the groups to take up their complaints directly with Evans and with the relevant public authorities, but, in July 2013, it modified its rules such that genetically engineered plants and animals could not be funded, placing them in a category that already included alcohol, drug, and gun-related projects (Geere, 2013, August 2). SU-affiliated BioCurious was sued over its agreement to provide Glowing Plant laboratory space. SU subsequently refrained from publicly supporting or defending Glowing Plant, seeking to sidestep the controversy; but the people, lab spaces, and ideas associated with the project all trace back to SU. Despite the controversy surrounding Glowing Plant, Evans remains an important model of the SU entrepreneurial approach. For example, during GSP-13, Evans was invited to speak about his project as a model for how to run a highly successful crowd-funding campaign.

Glowing Plant captures multiple dimensions of the idea of disruption that animates the SU project. First, it is a hacker project turned start-up that has had a

---

19 At the time of writing, the glowing plant technology is still in development, and the seeds have not yet been sent.

significant impact on society. SU, like the wider Silicon Valley culture of which it is part, celebrates the hacker space—the garage—as a birthplace of world-transforming innovation. (Apple, Google, and Hewlett-Packard all began in garages in Palo Alto, California.) The garage-hacker-turned-entrepreneur is exemplary of the asymmetry between the innovator and the world, the one and the billion. In the "benign conspiracy," a few innovators in a garage are able to change the world, perturbing material, social, and normative orders through technological innovation.

Second, the Glowing Plant project embraces technological mastery, in this case over living systems. SU celebrates the emerging project of synthetic biology to apply to tools and methods of computing to biology, in effect, approaching life as an information technology.

Third, though the Glowing Plant project was nominally conceived as an effort to disrupt the technology of electric lighting (something it is not even remotely close to achieving), the disruptions that have been noted and celebrated by the SU community have more to do with embarrassing existing regulatory institutions and normative sensibilities and attracting attention (and resources) by being willing to develop a controversial technology. Indeed, the project was controversial in part because it seemed to slip through the cracks of the existing regulatory regime by using a technology to make its plants that is not subject to USDA regulation of agricultural biotechnology.[20] The project is exemplary of the putative power of the individual innovator to push the limits of technological possibility, thereby exposing the impotence and obsolescence of regimes of governance in the face of exponential technological change.

Finally, the Glowing Plant project also demonstrates the evangelical element—zealous advocacy for an idea—in projects of "disruption." The project was highly visible, and its success, quite literally, was predicated on the controversy it engendered. Evans tracked the correlation between Kickstarter pledges and press reporting, and he found that negative press—articles and blog posts that criticized the project as irresponsible— each generated somewhere between \$6,000 and \$8,000 in additional pledges. The Kickstarter campaign enrolled nearly 8,000 people as investors in the project and became known to many more via media attention—all without having actually produced the technology. SU faculty member Andrew Hessel affirmed this dimension as the project's most positive impact: "Where the project really shines, though, is in seeding interest and discussion. It has convinced almost 8000 people to chip in a few bucks. It also received millions in media attention. It has been tweeted

20 For a recent discussion of how companies circumvent regulation by performing certain bioengineering techniques, see (Pollack, 2015, May 7).

and 'liked' and commented on....It's fueling conversations on biosafety, patents, liability, open vs. corporate science, legislation, and more" (Hessel, 2013, June 6).

Glowing Plant is counted as a success at SU *not* because it offers a technology that disrupts an existing market space, since, for that, the technology would have to be real and realized. Instead, it has disrupted imaginations of what the future holds and of who will realize—and govern—that future. It positions the individual innovator as empowered to chart the course into the technological future and shows that the rest of us are inevitably caught by surprise and can only react post hoc. It also represents a challenge to existing regimes of governance, revealing the inadequacy of those regimes given their inability to anticipate or respond to emerging technological capacity. It is notable that many authorities within these regimes have bought into this critique, notwithstanding the fact that there is as yet no technology to regulate. Antony Evans has been invited to speak to the National Academies and has been identified by prominent academics as a demonstration case of responsible innovation (Maynard, 2014; National Academies, 2013). A National Science Foundation-funded effort to map the environmental regulatory challenges of synthetic biology identified Glowing Plant as a defining case (Drinkwater et al., 2014). Whether or not an actual glowing plant is in the future, Glowing Plant has achieved a privileged position in imaginations of what the future holds—and what must be done to prepare for it. Thus, disruptive technology in SU's imaginary scandalizes: it exposes culturally dominant ways of thinking about and governing human life as naïve and outmoded. We are not suggesting that this result is achieved, but that this is a central feature of SU's vision of how technology acts—and should be made to act—in society and history.

Conclusion

We conclude with two observations. First, there is a notion of agency at work in the imaginary of technological transcendence that is fundamental to both the SU project and to the transhumanist imagination. SU conceives of human agency in terms of technological capacity. What we are is what we can do—to matter, to information, to our bodies, in our intersubjective social relations, in our built world, and so forth. On one level, this vision is profoundly individualist, since the value of technological capacity is understood in terms of the forms of freedom it engenders: in effect, how technologies emancipate the individual from prior limitation and constraint. Yet agency in this sense is engendered for the user by the technology. The ability to exercise new capabilities and forms of control is promised as a consumer product.

On another level, however, and this is our second point, the imaginary of technological transcendence promises an additional form of agency, one that accrues to the innovator: the power to govern the future. Through the disruptive and exponential potential of the world-transforming 10^9+ Project, the innovator capitalizes on response-ability to assume responsibilities of governance: to declare what future is right, desirable, and good. This imaginary attributes to technology—and, thus, to its creators—powers of mastery and control that, through technological means, transcend the constraints of the present and, in a messianic mode, "bend the arc of the moral universe toward justice."

This essay has sought to tease out the threads that connect the radical technological futures of the transhumanist imagination with a wider cultural imagination of technological innovation as a secular source of transcendence. For the latter, innovation promises to solve social problems, to address humanity's grand challenges, and to guarantee security, liberty, and abundance. Yet the specific futures that tend to be imagined in this idiom do not seem to represent radical departures from human life as we know it. Instead, they promise to increase the yields of our fields, the healthfulness of our lives, and the robustness of our industrial economies. This essay has sought to draw attention to the ways in which those modest—and not unvirtuous—visions can reflect an underlying logic that privileges *technological* imaginations of the right and the good as a primary locus of moral imagination. In imagining technology as an agent of historical change, such visions occlude the tacit ethical postures that underwrite them. The Singularity University serves as an example of similar technological visions organized around an explicit commitment to an otherwise more tacit ethical posture. It embraces a deterministic imagination of technology as agency: as acting upon, disrupting, and transforming society. And within this imaginary, it constructs a vision of responsibility and governance that entails profound asymmetries in power, asymmetries that are justified—indeed, naturalized—by reference to the redemptive futures that exponential technologies promise.

SU's promissory technological futures are as yet only imagined. But they are futures that capture the public imagination, and this, in turn, allows the people who speak for those futures to be more easily heard. While SU purports to train leaders who will lead by transforming the world through technological change, in practice, they lead by speaking on behalf of the future, declaring what is possible, beneficial, and good. In promising technological transcendence of present insecurity, they simultaneously claim the authority to declare which of society's anxieties are misguided and what modes of governance stand in the way of the future. In a society that venerates innovation as a source of solutions, this is a powerful mode of diagnosing the dis-abilities of moral and political imagination that contradict the technological future and, thus, inhibit progress and human flourishing. Though SU is itself only a start-up, struggling to become visible and relevant, it warrants attention because it is repeating a story of emancipation, redemption, and transcendence that is becoming increasingly compelling in our contemporary moment. Equally important is what is silenced in this story: the global grand challenge of imagining ourselves as a vulnerable and mutually dependent human collective. This grand challenge demands that we cultivate technologies of humility even as we aspire to technologies of transcendence (Jasanoff, 2003). And it demands a politics of responsibility wherein the one stands not apart as redeemer of the billion but among them in servitude and solidarity.

References

Ackerman, S. (2012, October 3). Darpa offers a million dollars for crowdsourced amphibious tank.*Wired.co.UK*. Available at http://www.wired.co.uk.

American Foundation for the Blind. (2004). Ray Kurzweil: An oral history interview. Available at http://www.afb.org.

American scientist who created artificial life denies 'playing God'. (2010, May 21). *The Telegraph* (UK), News.

Anderson, B. (1991). *Imagined communities: Reflections on the origin and spread of nationalism*. Rev. ed. London, England: Verso.

Christensen, C. M. (2011). *The innovator's dilemma: The revolutionary book that will change the way you do business*. New York, NY: HarperBusiness.

Coolidge, G. (2013, April 23). Help me help 1 billion people." *IndieGoGo*. Available at: https://www.indiegogo.com/projects/help-me-help-1-billion-people.

Cutler, K.-M. (2013, April 23). The first ever synthetic biology kickstarter is about growing "glowing plants". *TechCrunch*. Available at http://techcrunch.com.

DARPA. (2013). Darpa robotic challenge trials 2013. Available at http://archive.darpa.mil.

Diamandis, P. H., & Kotler, S. (2014). *Abundance: The future is better than you think*. New York, NY: Free Press.

Dorgelo, C. (2015, September 5). Challenge.gov: Two years and 200 prizes later. The White House, Office of Science and Technology Policy. Available at http://www.whitehouse.gov.

Drinkwater, K., Kuiken, T., Lightfoot, S., McNamara, J., & Oye, K. (2014). *Creating a research agenda for the ecological implications of synthetic biology*. Joint Workshops by the MIT Program on Emerging Technologies and the Wilson Center's Synthetic Biology Project. May 7. Available at http://www.researchgate.net.

Dzieza, J. (2013, August 18). Plants that glow in the dark spark heated debate. *The Daily Beast*. Available at http://www.thedailybeast.com.

ETC Group. (2013a, May 30). Kickstopper! Putting a stop to synthetic biology pollution. Available at http://www.etcgroup.org/kickstopper.

ETC Group. (2013b, May 30). Kickstopper: Stopping synbio pollution. *IndieGoGo*. Available at https://www.indiegogo.com. Evans, A. (2013, April 23). Glowing plants: Natural lighting with no electricity. Kickstarter. Available at https://www.kickstarter.com.
Evans, J. H. (2002). *Playing God? Human genetic engineering and the rationalization of public bioethical debate*. Chicago, IL: University of Chicago Press.
Geere, D. (2013, August 2). Kickstarter bans project creators from giving away genetically-modified organisms. The Verge. Available at http://www.theverge.com.
Grossnick, R. A. (Ed.). (1987). Kite balloons to airships: The navy's lighter-than-air experience. Washington, D. C.: Deputy Chief of Naval Operations (Air Warfare) and the Commander, Naval Air Systems Command.
Haraway, D. (2008). *When species meet*. Minneapolis, MN: University of Minnesota Press.
Harvard University, Office of Communications. (2013, May 30). Commencement 2013: Larry Brilliant's address. Available at http://www.hsph.harvard.edu/news.
Hessel, A. (2013, June 6). Grow your own glowing plant. *Kurzweil Accelerating Intelligence*. Available at http://www.kurzweilai.net.
Hurlbut, J. B. (2015). Reimagining responsibility in synthetic biology. *Journal of Responsible Innovation, 2* (1), 113–116.
Hurlbut, J. B. (Forthcoming). Technologies of imagination: Secularism, transhumanism, and the idiom of progress. In D. Yerxa (Ed.), *Religion and Innovation: Antagonists or partners?* London, England: Bloomsbury.
Huxley, J. (1957). Transhumanism. In *New bottles for new wine, essays* (pp. 13–17). London, England: Chatto & Windus.
Jasanoff, S. (2003). Technologies of humility: Citizen participation in governing science. *Minerva, 41* (3), 223–244.
Jasanoff, S. (2015). Future imperfect: Science, technology, and the imaginations of modernity. In S. Jasanoff & S.-H. Kim (Eds.), *Dreamscapes of modernity: Sociotechnical imaginaries and the fabrication of power* (pp. 1–47). Chicago, IL: University of Chicago Press.
Johnson, B. (2009, September 2). A school for changing the world. *The Guardian*, Technology. Available at http://www.theguardian.com/technology/2009/sep/02/technology-school-singularity-university.
Kurzweil, R. (1999). *The age of spiritual machines: When computers exceed human intelligence*. New York, NY: Penguin Books.
Kurzweil, R. (2001, March 7). The law of accelerating returns. *Kurzweil Accelerating Intelligence*. Available at http://www.kurzweilai.net.
Kurzweil, R. (2005). *The singularity is near: When humans transcend biology*. New York, NY: Penguin Books.
Kurzweil, R. (2009, February). *Ray Kurzweil: A University for the Coming Singularity. TED*. 8:47 mins. Available at http://www.ted.com/talks.
Leach, A. (2013, May 7). Glow-in-the-dark plants make activists see red. *The Wall Street Journal*, TechEurope.
Max More: Transhuman and the singularity. (2012). https://www.youtube.com.
Maynard, A. (2014). Responsible innovation. Talk at Arizona State University, April 1.
National Academies. (2013). Forum on synthetic biology. Washington D.C., October 21. Available at http://sites.nationalacademies.org.
Pollack, A. (2013, May 7). A dream of trees aglow at night. *New York Times*, Fehler! Linkverweis ungültig. & Environment, B-1.

Pollack, A. (2015, January 1). By "editing" plant genes, companies avoid regulation. *The New York Times*, Energy & Environment, B-1.
Portola Institute. (1969). *Whole earth catalog.* Menlo Park, CA: Portola Institute, Inc.
President Bill Clinton and Peter Diamandis at CGI 2014. (2014, September 24). *YouTube.* Available at www.youtube.com.
President's Commission for the Study of Ethical Problems in Medicine and Biomedical and Behavioral Research. (1982). *Splicing life: A report on the social and ethical issues of genetic engineering with human beings.* Washington, D.C.
Presidential Commission on Bioethical Issues in Biomedical Research. (2010). "New directions: The ethics of synthetic biology and emerging technologies." Washington, D.C. Available at: http://bioethics.gov/sites/default/files/PCSBI-Synthetic-Biology-Report-12.16.10_0.pdf.
Rein, L. (2009, June 29). Singularity University—day one. *Humanity+.* Available at http://hplusmagazine.com.
Rowan, D. (2013, May 6). On the exponential curve: Inside Singularity University. *Wired UK.* Available at http://www.wired.co.uk.
Saenz, A. (2012, June 16). Singularity University 2012: The graduate studies program is near and we have insider coverage. *SingularityHUB.* Available at http://singularityhub.com.
Schumpeter. (2014, June 24). The not so golden state. *The Economist.* Available at http://www.economist.com.
Singularity University. (2009, July 30). SU names GSP-09 students. *Singularity University Blog.* Available at http://www.kurzweilai.net.
Singularity University. (2012, December 6). Singularity University acquires Singularity Summit. *Singularity University Blog.* Available at http://singularityu.org.
Singularity University. (2015a). Core curriculum. Available at http://singularityu.org.
Singularity University. (2015b). Corporate founders. Available at http://singularityu.org.
Singularity University. (2015c). Graduate studies program. Available at http://singularityu.org.
Socrates [Nikola Danaylov]. (2011, September 11). Salim Ismail on Singularity 1 on 1: We are already gods, we might as well start acting as such. Available at http://www.singularityweblog.com.
Solomon, M., & Bostrom, N. (2005.) Transhumanist dream. *Foreign Policy, 146*, 4.
Stock, G. (1993) *Metaman: The merging of humans and machines into a global superorganism.* New York, NY: Simon & Schuster.
Taylor, C. (2003). *Modern social imaginaries.* Durham, NC: Duke University Press.
Teilhard de Chardin, P. (1965). *The phenomenon of man.* (B. Wall, Trans.). New York, NY: Harper & Row.
Tirosh-Samuelson, H. (2012). Transhumanism as a secularist faith. *Zygon,* 47 (4), 710–34.
Veronico, N. A. (2006). *Moffett Field.* Charleston, S.C: Arcadia Publishing.
Wilson, C. (2013). *Defcad.com.* Available https://www.youtube.com.

Scientific Humanisms and Technological Utopias 12

Situating the Transhumanist Imagination

Nasser Zakariya

In this essay, I will first briefly relate transhumanist imaginaries to genealogies and critiques of utopia-minded technological determinisms and technological sublimes. I will also relate such imaginaries to conceptions of transcendence active in the linked traditions of scientific humanism, naturalism, and cosmic religion. Against this backdrop, I will focus on some of the ways that transhumanism tends to stitch together conventionally secular and theological grammars, suggesting elements of instability in those designations and certain corollaries that may come with that instability. I will then relate transhumanism to some of the recent literature on technoscience and technology assessment, suggesting the breadth of interlinked visions of technologically inflected futures. Emphasizing the commonplace nature of central underpinnings of the transhumanist soteriology suggests how that soteriology can come to appear less controversial and more compelling. Companion visions of technology help identify how much of that which appears characteristic of transhumanism is, in fact, not specific to it, while putting into relief the specificities of its imaginary—specificities that can, nevertheless, be reconciled with more pervasive, mainstream attitudes to technology.[1] This triangulation suggests, in broad strokes, the extent to which the technological imaginary of transhumanism draws from technological deterministic and futurist sensibilities distinct from older forms of transhumanism; that is, these current imaginaries are neither the only child nor only the child of an older "transhumanism." I hope that this genealogical exercise sheds light on what enables tranhumanist discourse to project technological futures in both secular and nonsecular keys and how it is and is not religiously

1 Here, I will not repeat the many thorough readings of transhumanism employed in this and other volumes, though my analysis depends on them along with examinations of popular and academic transhumanist texts.

construed—an oscillation that is not unique to this discourse but is, nevertheless, central to its modes of authorship and authority.

Secular Substitutes? Technological Determinisms, Sublimes, and Utopias

Among the most prevalent of prominent understandings of technology are the broad claims that technologies proceed in some sense according to their own logic and that they have a preeminent position in culture. In the first case, technological development is treated as an unstoppable and predictable line of descent, increasing power and efficiency: computer capabilities will expand at exponential rates, technically guided medical interventions will be ever more pervasive and penetrating (for example, the promise of nanotechnologies and genomic medicine), robots increasingly more human or animate, humans ever more cyborg-like, the material ever more organic, the organic ever more material, and so on. Such an account provokes the sometimes misleading term "technological determinism" but in its least misleading sense.[2] Endorsement of such technological imminence is entirely consistent with accepting, repudiating, or having no position with regard to transhumanism, implicit or otherwise.

In the second case, the culture of a period is understood to be grasped through technologies or technological systems, often by reference to prominent or glamorous examples such as the steam engine; modes of locomotion such as automobiles, aviation, and spacecraft; the factory system as a whole; nuclear bombs and power; social media; or 3D printing. The act of listing is a tired one, partly because of how such technologies or technological systems repeatedly have been held up as figures for specific eras. Technological artifacts reveal the essence of a generation, in which individuals are subordinate to the needs of technology rather than vice versa, submitting to technocratic norms that give primacy to technological efficiency and efficacy. This specific belief in technological preeminence, called "technological determinism's most familiar face" by Bruce Bimber, does not generally evoke a transhumanist vision, linked as it is to wider present-day anxieties and dilemmas. For these visions of technology, assessing the present is a matter of assessing the present and future of technology.

---

2 Bruce Bimber refers to this vision of technological determinism as the "nomological" account. For this, and the two other categories that Bimber defines and critiques, the "normative" and "unintended consequence" account, see Bimber (1994).

Regardless of the breadth of such theorizations of technology and the comparative narrowness of contemporary transhumanism, the former may help locate the latter. To ask what sustains the status of technology in the present might come close to diagnosing modernity—if, at times, more specifically American modernity—given the breadth of relevant literature and events. Within this variegated cultural field, notions of a "technological sublime" and projects of "technological utopianism" inflect contemporary extremes of futurist optimism suffusing some technological discussions.

Scholars such as David Nye have historicized this newer sublime, charting how it is provoked not only by natural visions but also increasingly by technological artifacts and measured by a consistent affect of astonishment. Nye's claims depend on a "substitution thesis":

> Americans have long found the sublime more necessary than Europeans, so much so that they have devised formations of the sublime appropriate to their pluralistic, technological society. Precisely because American society is so pluralistic, no single religion could perform that function. Instead, ever since the early national period the sublime has served as an element of social cohesion. (Nye, 1994, p. xiv)

According to Nye, then, technology will serve where religion will not.

Nye's claim is in some sense consonant with aspects of Charles Taylor's genealogies of secularism and elements of the conversations these genealogies have provoked. Taylor traces a process of secularization wider and older than the history of pluralism in the United States. His diagnosis of that process is not primarily on the level of the construction of God-free public spaces and practices (what he refers to as "secularism 1") or as a matter of the loss of active belief in God (secularism 2), but through the conditions for such belief (secularism 3): "the change I want to define and trace," he notes instead, "is one which takes us from a society in which it was virtually impossible not to believe in God, to one in which faith, even for the staunchest believer, is one human possibility among others." This change, then, is in the broader "context of understanding" (Taylor, 2007, p. 3).

Like Nye, and in relation to the sublime, Taylor ultimately tends toward a kind of compensatory thesis: "I have been placing sublimity, and the moral meaning of wilderness, within the context of the felt inadequacies of modern anthropocentrism, and the need to recover contact with a greater force." As such, the sublime is "opposed to a sense of life which is too shallow and human-centered—shallow because human-centered" (Taylor, 2007, pp. 341–342). Taylor provides an almost causal account of the emergence of the sublime after the decline of the enchanted medieval order:

> The moral meaning of the sublime, and much of the post-eighteenth century cosmic imaginary, can be seen as a reaction to this shallowness and anthropocentrism. But it can't just react by moving backwards. The sense of depth cannot be found in an eternity which is no longer a felt reality. It is now found in the vastness of space and the abyss of time. (Taylor, 2007, p. 343)

However, in contrast to Nye, Taylor tends to emphasize the comparative difficulty in making contact with the "fullness" of experience accompanying the change to the secular and argues against any simple measure of change with secularization. Taylor intimates that the possibility of fullness has been compromised, a loss that cannot be compensated by any sublime, technological or otherwise. These latter sublimes he also diagnoses through the rise of secularization, the dynamics of which are explained by his notion of a "buffered self"—the bounded atomistic self of modern experience that Taylor tends to read as intrinsically and fundamentally self-assured. Such a self can occupy the safe positions required by, for example, Burke and Kant's formulations of the sublime, without ultimately being overwhelmed.[3]

Be that as it may, the consonances and contrasts between the positions evident in the historical work of Nye and the more conceptual work of Taylor already begin to suggest the difficulty in establishing any stable contrast between secular and nonsecular experience, either in the self-same period or between past and present. This possible elision raises the question of whether technologically inflected experience can provide that fullness so often linked to religion.

At the same time, Nye and Taylor's differences of position begin to suggest the invocation of certain stable or at least recurrent markers for the divisions between overt secular and religious experience: on the one side, sobriety and domesticity, on the other, astonishment and wilderness. The confinement of the former and the unboundedness of the latter point to contemporary paradigmatic elements of what the scholar of religion Talal Asad terms grammars of the secular and the religious, respectively. These affinities do not exclude a secular sublimity or a spiritual constriction and might, under the genealogical lens of Asad, seem like evidence of the ever-incomplete process of universalization and demarcation at play in the rise of secular spaces in the West.[4] Instead, these affinities or associations serve to

3 It might be noted here that the intellectual historian Peter Gordon (2008), in a critical review of Taylor's work, has observed that the disenchanted modern is often characterized instead as disempowered, marginal, and vulnerable in comparison to his or her ancestors. Taylor might perhaps retort that the context of understanding that underpins such an imagination would, nevertheless, rely on the metaphysical assertiveness he attributes to the buffered self.

4 See, for example, Asad (1992, for example, p. 6).

explain why excesses of technological optimism can provoke religious diagnoses, and depths of religious bourgeois sensibility can provoke secular diagnoses, respectively— even if, by a certain light, each is also respectively a logical extension of conventionally secular or religious patterns.

But the example of transhumanism becomes more pointed and more interesting if Taylor's argument is followed a little further, less as an indication of its validity as such, but rather more as a marker of a current strand of belief. For Taylor, secularism and "exclusive humanism" arise together: "a secular age is one in which the eclipse of all goals beyond human flourishing becomes conceivable; or better, it falls within the range of an imaginable life for masses of people" (Taylor, 2007, pp. 19–20). (Here we should keep in mind the distinction between "fullness," often linked to the transcendent, and "flourishing," often linked to the immanent.) Religion in the Western tradition ultimately means for Taylor a multidimensional "transcendence" defined by opposition to each of his forms of secularization: the displacement of a transcendent God from the social life, from individual belief, and from a condition of experience that requires a dimension beyond the immanent. If transhumanism is understood as a seeking of transcendence through some notion of exclusive humanism, its very terms seem to circumvent the contrasts on which Taylor establishes his own notions of transcendence and, therefore, confute the secular/religious divide. Regardless, to diagnose transhumanism as religion or even as having a religious dimension is to bind the apparent extremes of exclusive humanism and religion, to deny and to assert the transcendent scope of the transhuman imaginary at once.

I will return to these points momentarily in considering the conceptual genealogies generally understood as rooting transhumanism, but a brief consideration of technological utopianism will help clarify these formulations. Technological determinism has often entailed an optimistic conclusion, one met in American and European efforts to conceive such utopic visions. In sketching the history of technological utopianism, the historian Howard P. Segal has contrasted the American technological inflection with the extended Old World tradition of utopianism, including figures as diverse and dispersed as Francis Bacon, the Marquis de Condorcet, Auguste Comte, Robert Owen, Karl Marx, and Friedrich Engels. Segal claims in this regard that "none of the Europeans made technological advance their panacea, as did all the American technological utopians"— a contrast upheld despite the fact that, as Segal further observes, "American technological utopianism never constituted a self-conscious movement and ideology" (Segal, 2005, pp. 2–3).

The contrast is arguably more strained in the 20th century, which sees in Europe not only such futurist technological reflections as those of J. B. S. (John Burdon Sanderson) Haldane or J. D. (John Desmond) Bernal but also the activity linked to enthusiasms for scientific social planning among the social relations of science movement and the more politicized variations of European positivism. The argument for national differences here might turn on uneasy distinctions between futurism and utopianism, between scientific planning and technological materiality. Regardless, for Segal, optimism as characteristic of American conceptions of technology and statecraft is more the point, rather than any institutionalization of utopian ideas.

More recently, however, Segal finds a decline in this optimism, which he attributes to technological assaults on the environment, to the history of disappointments with technological promise, and to "diminishing faith in experts as objective analysts" (Segal, 2005, p. 165) linked to Vietnam-era politics. In contrast to the former emphasis on the technological as solving wide social and infrastructural problems, Segal claims that "in recent decades technology has increasingly come to be associated with the fulfillment on a small-scale of individual needs and desires" (Segal, 2005, p. 166). In a position indirectly consonant with Taylor's buffered self, Segal notes that such technology "enhances individual sanctity and provides choices in an ever more digitized, programmable world" (Segal 2005, p. 166). He holds that, as a result, "the consumer mind…wonders what a specific item will do for the individual rather than the society" (Segal, 2005, p. 166)—the grounds for what might be called a "consumerist transcendence."

In this vision, then, the actual widespread institutionalization of technologies catering to the individual has undercut more collective utopic pictures; the authority hovering over or underlying technological change appears to turn less on government regulation than it does on the tastes of the market. Expertise fails to assess the impact of change, government fails to track it, and the disempowerment that flows from such diffusive technological agency appears only to reinforce the picture of technological determinism: we (as governments and individuals) cannot prevent ourselves from gratifying ever more buffered tastes, which are ever more powerfully gratified by customized, consumerist technologies. Utopia becomes less about communal bonds and joint welfare and more about individual sanctity and activity—and so, on conventional construals of the utopic, conceives no utopia at all.

By these lights of pervasive technological and deterministic imaginations, the character and even degree of contemporary transhumanist reliance on technology in itself are not singular, even in relation to subject construction. If Segal is right concerning recent disappointments with technological achievement, transhumanists are perhaps more insistently optimistic than many of those who presently look to technology as the dominant cultural force. Seizing upon the conditions in the present, much of the transhumanist imagination embraces the future, both with and against the grain of earlier American utopianism. Transhumanism goes with the grain insofar as new technologies raise the question of utopia as a possibility and goes against it insofar as it embraces apparently new technologies, despite and perhaps partly because of the very trends that Segal sees as linked to a widening pessimism undermining belief in collective utopianism.

Individual gratification and collective social welfare are not, within transhumanist imaginaries, necessarily imagined as mutually exclusive (see Boenig-Liptsin & Hurlbut, chap. 11 this volume). The mark of that improved welfare can instead be the summation across the gratified desires of all the individual consumers—though to neither Taylor nor to Segal, nor perhaps to Nye, would the assembly of untouched, buffered, consumerist selves amount to a utopic condition. If there is a further theological dimension to transhumanism itself, it is far from decisively individualist. It might hold open the promise of more collective and meaning-making self-construction through cosmic reorientation and a euphenic reworking of the flesh—the possibility of that very fuller experience apparently frustrated by a disenchanted and disoriented present. And if so, does such an orientation extend to other paradigmatic elements of theological constructions of the self, through human ritual and moral discourse? This brings us to conceptual genealogies attested in relation to transhumanism.

Immanent Scientific Humanisms

To some transhumanists themselves, as well as to their commentators and critics, philosophizing and theological speculation point to Julian Huxley. However, it is unclear how revealing an invocation this is, particularly if understood as a straightforward genealogy. Not only does Huxley infrequently use the word "transhumanism" in publication, but he also adopts related terms including "scientific humanism" and, later, "evolutionary humanism," terms that are at times appealed to by other public figures. These other scientific humanisms signal alternative cultural markers more distinctly than does transhumanism, evoking not only technologically empowered self-directed evolution but also 19th-century debates opposing the centrality of science to humanism at a time when (as J. Z. Smith suggests in *Imagining Religion* [1982, 102–103]) human sciences might still be contrasted with divine sciences.

Huxley appears to use "transhumansim" publically for the first time in lectures in 1951. There, he claims that "[t]he new central incentive of human life will be fulfilment through the realization of potentiality" (1951, p. 139). Making overt the semantic contrast he has in mind, he notes that "[s]uch a broad philosophy might perhaps best be called, not Humanism, because that has certain unsatisfactory connotations, but Transhumanism" (1951, p. 139). In his lectures, both in 1951 and later, Huxley emphasizes a "oneness" that helps clarify the religious feeling he attributes to his transhumanism:

> [Man] need no longer feel insulted by the fact of death or by his own insignificance. Once he can grasp that he is a part, and an operative part, of an enduring process, he need not continue to compensate for his isolation by mere personal ambition or by frantic striving for superior status or superior achievement. Once he grasps that his job is the maximum realization of the possibilities of his own development, and that this, if well and truly accomplished, will at the same time inevitably also facilitate and contribute to the realization of the possibilities of society, and of mankind, and of the cosmic process as a whole, he will acquire a new confidence and a new sense of oneness with the rest of existence. (Huxley, 1951, p. 139)

The sentiments of Huxley's later more famous essay "Transhumanism" are more consonant with such a picture, but are less insistent that "humanism" informs, albeit insufficiently, "transhumanism." Rather, the essay emphasizes more the possibility of the "post-human"—a future existence "as different from ours as ours is from that of Pekin[g] man" (Huxley, 1957, p. 17)[5] and less the religiously resonant unity of existence. Much of the thick context and genealogy of Huxley's own intervention is discarded, then, when a thin line is drawn from him to the present.

In relation to the views of Huxley's transhumanism—what constitutes its imagined center or its paradigms—the past presents as many contrasts as it does continuities. The terms "scientific humanism" and "evolutionary humanism" of Huxley and others are in some respects nearly synonymous with Huxley's use of the word "transhumanism." In either case, humanity has a duty to take control of its own evolutionary process, appreciating and establishing humanity itself—the integrated and harmonious human personality—as the culmination of an evolutionary process. This special place of humanity not only invites unconstrained self-fulfillment but also renders social and ecological stewardship a duty.[6]

In reference to these ethical, humanistic, and spiritual questions, Huxley was hardly alone, even if his own views of human difference and teleology seemed to some to be holdovers from the previous century. Married to these visions are those views associated with the figures who played a role in the "evolutionary synthesis" in the rise of modern biology, and/or its impact on the definition of the human—figures who kept company with each other and debated ethics and future fates. Among these people were J. B. S. Haldane, J. D. Bernal, C. H. (Conrad Hal) Waddington, Joshua

5 Note that even here, Huxley also emphasizes that his transhumanism means "man remaining man, but transcending himself, by realizing new possibilities of and for his human nature" (1957, p. 17).

6 For a characterization of Huxley's views, see Tirosh-Samuelson (2012, esp. pp. 719–720).

Lederberg, and Jacob Bronowski.[7] Apart from Huxley, Bronowski was perhaps the most prominent to invoke humanism explicitly, defending science as an essentially humanist enterprise and seeing the emergence of a specifically Western ethics as the heritage of both the humanist Renaissance and the empiricist Scientific Revolution. At the same time, Lederberg adds to this constellation by advocating for research on artificial intelligence and space exploration, partly justifying them as studies of the human.[8] Lederberg shared with Huxley a vision of psychosocial evolution that, he argued, propelled humanity into space for the purpose of understanding better the nature of life itself and in acceptance of a duty ambitious and embracing enough to constitute a collective orienting mission.

The attempt to find some compromise between or synthesis subsuming humanistic and scientific understandings of humanity was, thus, pervasive and varied, even among Huxley's immediate acquaintances. And the effort embraced the widespread vocabulary and debate emerging in the late 1950s over C. P. (Charles Percy) Snow's "two cultures" at a time when many of the central examples of scientific impact and worldview for Bernal, Bronowski, Snow, and others were technological artifacts.

What science and technology were doing to and for people—and therefore how they were redefining not only humanism but humanity—was and is an extensive contestation, deeply inflecting the varied possible meanings of Huxley's transhumanism, for himself and others. Read against Bronowski, Lederberg, and Snow, Huxley's own earlier writings, including his defense of eugenics and his writings on "transhumanism" itself, the "trans" that Huxley had in mind did not tend to signal a desire to overthrow entirely humanist understandings of the human or simply to discard conventional human materiality. Instead, his "trans" tended to signal transcendence of national, cultural, and ethical boundaries of disciplines of knowledge and specialization, of unnecessary limits to what Huxley often presented as a fixed store of human potential. This was a transhumanism connecting all these modes of being human and, Huxley argued, reinforcing and clarifying human meaning as such, transcending stymieing constraints, though without celebrating endless human plasticity or the death of individual death.

Nevertheless, as with any complex set of texts and collaborations, different moments stand as resources for transcending humanism in other senses. In the early 1960s, humanity could be defined at least genotypically. Lederberg's encouragement of euphenics tended less toward understanding human subjectivity as evolutionary

---

7 In his 1951 lecture (p. 128), Huxley names Bronowski's *The Common Sense of Science* as one of the texts that contributed to his own thought.

8 See, for example, Wolfe (2002), who links Lederberg's exobiology to research into the possibility of biological warfare.

outcome and more toward understanding it as potential and process. Connecting to a number of debates over humanism and utopianism, even while attempting to transcend them, Lederberg warned, “Eugenics and euphenics are the biological counterparts of education, a panacea that has a longer but equally contentious tradition. The troubled history of Utopian education warns us to take care in re-building human personality on infirm philosophy” (Lederberg, 1963, pp. 269–270). For such a vision of scientific humanism, humanity was its own working project, tasked with improving its own material body, individual and collective, toward a more utopic condition. And Peking man might seem less-than-human enough to render Huxley’s analogy an endorsement of an attempt to become superhuman, however that superhumanity might be understood.

If, then, we add to this specific configuration and representation of the human subject in actuality and potentiality wider speculation on computing and artificial intelligence, on cybernetics and cyborgism—as scholars such as Paul Edwards (1996) have examined and root in Cold War-era discourses—and on the perpetuation of life or the forestalling of death—as Asif Siddiqi (2010), among others, observed has long been a part of the imaginary circling around space travel—we reapproach some of the technological planks and the ideological fabric of common, contemporary transhumanist tropes.

From the perspective of scientific humanisms, themselves partly rooted in related Cold War views, transhumanism appears less as a culmination and more as a concentrated configuration of the glamorous individual technologies and the many active theorizations, hopes, and anxieties associated with those technologies in a wider cultural context. It stands as one specific thought experiment made culturally actual—actual in terms of a narrow band of institutions promoting a technologically charged and determined vision of the human present and future, a vision that is more or less utopian depending on the extent of collective benefit imagined, or imagined as beside the point.

Original Mysteries, Transhumanist Futures, and the Limits of Grammar

But is even a utopian version of transhumanism itself one of Taylor’s “exclusive humanisms,” providing little hope of future fuller experience? The sublimes of Nye or the deep spaces and times Taylor almost laments are still more suggestive of the broader, transcendentally minded patterns in which both scientific humanisms and later transhumanisms participate.

As with the technological sublime, these scientific worldviews provoked spiritual sentiments. The religious feeling of Julian Huxley might appear to be distinctive among a number of his more technologically minded colleagues. But his views were consonant with a pantheistic scientific naturalism apparent in the late 19th century and thereafter.[9] Such pantheistic naturalism finds inaccessible mystery in the phenomenon of the immanent flowering of consciousness from inorganic materials; in the expanse of earth and heavens; and in the conclusion that the existence of expressible and discoverable natural laws cannot be explained. To some, such naturalism, by reinforcing the limits of reason, re-enchanted or kept the world's enchantment alive, both well before and while Weber was announcing its science-induced disenchantment.

As an example, Ruth Barton (1987) expands on the religious feeling of one of the apparent apostles of 19th-century materialism, John Tyndall.[10] In Tyndall's vision, which was shared in central ways with X-Club colleagues such as Herbert Spencer and Julian Huxley's grandfather Thomas Henry Huxley, matter itself is filled with the mysterious promise of life, and physical laws present the ultimate, cloudy mystery that they establish and express themselves. The task of science is little more than an attempt to describe these laws, the task of technology to manipulate them, without in either case the ability to get behind those laws: "under all things there lies an impenetrable mystery" (Spencer, 1857, p. 485).

Some of the central grounds for that scientific naturalism were deepened in the 20th century by the extension of time and space and in the research questions provoked by Alexander Oparin and J. B. S. Haldane, and later by Harold Urey and Stanley Miller, even if they were prone to the charge of materialism. This 20th-century scientific naturalism was also broadly consistent with aspects of the "cosmic religion" associated with figures such as Albert Einstein. The attendant mysteries of such a view also derive from the inexplicable fact of an explicable, sublimely scaled universe, all the more mysterious in the face of scales that further marginalized the physical position of humanity. To other scientists engaged in religious encounters and conferences (such as the American astronomer Harlow Shapley), the newer views of the heavens reinforced humanity's metaphysical "suburban-ness"—its spiritual "anthroperiphery."[11]

---

9 Julian Huxley himself invokes naturalism and mystery in his 1951 lecture, pp. 147–148.

10 See also Turner 1974 and 1981.

11 This is a characterization of C. S. (Clive Staples) Lewis in (1994, p. 58). As for the suburban, Lewis notes, "I think that there remained throughout the Middle Ages an unresolved discord between those elements in their religion which tended to an anthropocentric view

These older scientific naturalisms and cosmic religions might also point to contrasts with prominent putative theological configurations within transhumanism. These older views share an emphasis on the humble place of humanity in the natural order (whether physical or spiritual), with an enduring promise to understand the world up to the limits of understanding. Even in the hands of a technoscientific optimist such as Carl Sagan, humanity might be promoted to no more than the voice of the local cosmos—though its knowledge of its limits might have no qualification, just as its connection to everything (as with Julian Huxley) might admit no limits.[12]

Nevertheless, such views also often intimated the assertiveness Taylor finds in the buffered self, at the moments when the universe is not presumed quite so sublime or opaque. The nature of disagreement among positions of scientific naturalism, cosmic religion, and transhumanism on the score of the marginality of humanity is elusive and certainly arguable, much as the contrasts between modern and medieval human beings are debatable by the same measure.

The differences as such appear less in the evaluation of the current ontological status of humanity and more in depth of their futurisms. As Hava Tirosh-Samuelson (2012) has analyzed in this regard, there is a strong eschatology running through transhumanist literature. In contrast to scientific naturalism/cosmic religion, however, there is comparatively less cosmogony, as the latter is so much the focus of the current scientific epic or cosmic evolution.

Technological transhumanist extrapolations flowing forward can be reversed in their course, and many of the conventional secular dimensions of transhumanism presumably rely on the account of cosmic and organic evolution. Whatever the case, where other scientific humanisms and naturalisms might offer crisper accounts of the past, contemporary transhumanism is instead more thoroughly and coherently futurist—even when disputing the fact of singular or omega points. Unlike postlapsarian accounts of human decline or even progressive accounts of human evolution and potential consummation, much of the essence of what is human in this imagination is projected out across a future, unrealized history.

In this transhumanist future, what is conventionally theological is often construed through the superhuman. To be fuller is to live longer and with greater human flourishing—to equate fullness and flourishing. This is not generally a transcendental understanding in terms of the inapproachable, the noumenal, the limits of

---

and those in the Model which made man a marginal—almost… a suburban—creature" (1994, p. 51).

12 To an often more earthy Haldane, the universe was, in much-quoted words, "not only queerer than we suppose, but queerer than we can suppose" (Haldane, 1940, p. 263).

human subjectivity. There may not be and perhaps cannot be the multidimensional transcendence invoked by Taylor's analysis.

Thus, a belief that transhumanism in some way undermines (or opposes or replaces) *conventional* religion projects onto any such putative theological opposition a stricter vision of either the gratification of individual desire or the solution of collective problems, as suggested by Segal (2005). A transhumanist future might further offer a more generous vision of human unity and reconciliation, a robust technological utopia. But it tends to preclude any theological dimension that, consonant with Taylor's analysis, argues against the possibility of an anthropocentric vision as thoroughly or sufficiently meaningful—that is, a dimension with space for the noumenal, the uncognizable, or even the transvalued, leaving hope for meaning and fullness in some transcendent state that cannot be extrapolated, much less determined, from the present. It rejects, then, any Western or earlier tradition that might claim that, even if humanity lived without fear of death, want, loss, and other ills, true religion would essentially be uncompromised—that there would still be a place for God or the transcendental.

The utopic impulse in transhumanism, thus, appears to be radically different from older utopian traditions analyzed by religious scholars such as J. Z. Smith, its visions of transcendence as different as the several forms transcendence can take. By these lights, any current contestation concerning transhumanism—whether it is linked or contrary to other technological canons, scientific humanisms, and religions, or whether it strongly participates in secular or theological conceptions and practices—appears to underscore the work necessary to demarcate these domains or grammars. It also suggests the ease with which the putative subjects of transhumanism (the nature of the human above all), though perhaps not yet coherently institutionalized, might easily enough be appropriated to different, even nominally opposing imaginations. The intersection of differing imaginative modes has itself been constitutive of imagined technological futures: for example, Lederberg easily rendered Peter Medawar's Nobel Prize-winning work on the potential of organ transplantation into a futurist vision of a "spare-part" humanity. More recently, and with explicit religious inflection, Ray Kurzweil and others have rendered technologies, such as Kurzweil's own scanning products, as capable of miracles, allowing the blind to see or the deaf to hear.

In this sense, even a future singularity as defended by some transhumanists seems less to declare the limits of understanding in the present and more to point to a utopia that, if not visualizable, is nevertheless an outcome of a continuous trajectory from the present. In that future, there appears to be little possibility for the persistence of feelings of diminishment, boredom, and waste. But these improvements need not be counted as essential alterations to humanity. From this point of view, the conventionally spiritual does appear to be the more humbling perspective. Adam needed God even in Eden; to be human in this light is to be forever incomplete, where nothing immanent can ever supply full consummation.

But, if upheld, such a contrast does not pronounce transhumanism as somehow comfortably, conventionally secular by present-day lights. To do so would be to side with one kind of transcendence as more secular than another. Instead, to attempt to capture the nature of transhumanistic visions, their "hybridization of the religious and the secular" as Tirosh-Samuelson refers to it (2012, p. 729) we shift back and forth between contemporary technological and secular frames. Any necessity to switch frames can, if with some overreaching of its own, be argued to add to problematizations of secularization theses construed as clean historical trajectories. It provokes questions of whether the world ever experienced its own enchantment and fullness or whether enchantment is a value that itself requires a comparative, external vantage point. And could humanity's enduring anthroperiphery often have provoked the experiencing of oneself as a worthless creation?

But the need for multiple frames might itself be a function of the further elements of analysis and institutionalization needed to ground the theological. If, for example, past utopian religious and transcendent-mind movements were at times linked to religions in diaspora (as Smith has noted), what more specific political and social configurations provoke the particular forms of transcendence at play in transhumanism? If Smith detects a recurrent impulse in religious utopias, and transhumanisms participate in such logics (transhistorical in reflex, yet schematized through specific historical prisms), from and out of what sociopolitical conditions has contemporary transhumanism taken flight?

The question appears immediately problematic. Can we speak of fairly stable and self-disciplining transhumanist institutions, training, and rites? Is there a religious discipline and moral discourse that is as much the site of subject construction as the projections of a futurist imagination? And would such practices, if they emerge enduringly, dictate more clearly the ways in which a social mechanism will decide where and how to place transhumanism and what transhumanism hopes to escape or to reclaim?

The ability to understand aspects of transhumanism in contemporary paradigmatic religious terms has implications for the breadth of religious understanding. It becomes an added measure of how much current religious thought need not involve itself in cosmogonies, in absolute as opposed to consumerist transcendence, or even perhaps in shared stable practice. Present-day religious imagination can, in turn, be constructed according to a principle of extrapolation that may or may not lead to disjunction with the past and that may or may not turn on individual gratification.

Likewise, the extent to which transhumanism can be understood through conventional *secular* notions suggests how much the varieties of *humanism*—like those of natural science—have enfolded comprehensive imaginations outstripping the apparently limited license of reason and have asserted reason's own authority to prove the inexplicability of its own reach. Spencer's position, or Tyndall's, might in today's context as easily be read as irrational mystification as requisite scientific modesty.

What any current contestation over the status of transhumanism suggests more widely, apart from the weighty question of institutionalization and practice, is that secular and religious frames are themselves, if conventional, at least to a degree perspectival. What by the lights of one genealogy or cultural embedding can be judged as secular can by the lights of another genealogy or embedding be rendered theological. Medical interventions become transfigurations, and products become miracles. This oscillation parallels the dualities of experiential contexts: where some might see the progression of natural law or the intrusion of historical coincidence, others might see the signs of providence. But, put in terms of varying theorizations of the secular, any such oscillation might suggest the extent to which the supposed retreats of Gods are and perhaps always were more complicated maneuvers and might suggest how foggy the political implications of the designation "religion" in the context of transhumanism may remain.

Anticipation versus Salvation

The specific soteriology of contemporary transhumanism links to these deliberations since it suggests what in the present condition requires redemption—from what must we be saved and to what must we return? The answers to such questions determine the extent to which transhumanism itself is a measure of the soundness of any firm, authorized distinction between secular and sacred.

In this regard, we might invoke the more recent work of Asad, who, in his discussions of the genealogies of religion, attempts to reconfigure these contrasts in terms of intricate and extended genealogies of grammars of the secular. His work as a whole suggests that societies tend to establish broad distinctions between "secular" and "religious" activities and modes, with different if intersecting orders of institutionalization, authority, and groupings.[13] Though not beside the point, the

13 Such a theorization itself might appear as a reconfiguration of a nominalist Durkheimian distinction between sacred and profane, even though it is part of the grammar of the

content of belief is less the hinge between the secular and the religious than are the sociopolitical forces at a particular place and time, embedded within frames of tradition and culture.

In recent discussions, partly through the critique of Taylor's genealogies of the secular and his emphasis on the "buffered self," Asad has focused on body disciplines and practices. These are reminders of the extent to which religious activity, self-construction, and secular demarcation are not primarily discursive, not primarily centered on quotable belief. This is a point with which Taylor would hardly disagree, though Asad (and Gordon 2008) might find his emphasis on a Western history of ideas problematic nevertheless.

A more pressing distinction between Taylor and Asad for our purposes is made clear in their discussions of the nature of the secular. Taylor depicts the secular as a function of the spaces opened by the retreat of God or religion. As such, secularism can provide the conditions for consensus and for political (rather than religious) self-conception, whether on an individual or communal level. Asad agrees with Taylor on this score to the extent that the secular is such a "transcendent mediation" across social distinctions. But by contrast, for Asad,

> Secularism is not simply an intellectual answer to a question about enduring social peace and toleration. It is an enactment by which a *political medium* (representation of citizenship) redefines and transcends particular and differentiating practices of the self that are articulated through class, gender and religion. (Asad, 2003, p. 5)

In each, transcendence is key, but, for Taylor, the political consequences of the secular appear to be the result of the deuniversalization of the religious. For Asad—whose genealogies of self-conscious (that is, anthropological) religious categories themselves depend on universalizations associated with empire and the determinations of Enlightenment liberalism—the political forms and deploys the secular.[14]

---

religious and secular to establish and maintain that distinction. Also, earlier in the 1980s, Asad's work had attempted to destabilize straightforward accounts of secularization, already married to rising political anxieties about the role of religion. The question of the theological in the context of Western secularism predated the events of the new millennium, but Asad, nevertheless, appears to agree that the questioning of the secular has since intensified.

14 "Thus, what appears to anthropologists today to be self-evident, namely that religion is essentially a matter of symbolic meanings linked to ideas of general order (expressed through either or both rite and doctrine), that it has generic functions/features, and that it must not be confused with any of its particular historical or cultural forms, is in fact a view that has a specific Christian history. From being a concrete set of practical rules attached to specific processes of power and knowledge, religion has come to be

Asad's work, and that of other anthropologists and sociologists of religion, therefore points to the necessity of examining not only eschatology, soteriology, and imagination but also the practice and relationships of the religious to other cultural grammars. This task is made all the more difficult when such practices have not and may not take stable and enduring forms. Moreover, the analysis of practices with and beyond imaginations may also involve the analyst in the legislative work of nominating certain activities as religious, rather than inheriting such a designation from a wider social context. To argue too strongly for the secular or religious dimensions of transhumanism might itself be a practice of seeing faces in clouds or rocks. More justly, these classifications might function more as predictions as to how transhumanism will be further institutionalized and judged.

Such cases further suggest that the nature of any theological dimension of transhumanism depends not only on potential soteriologies but also (judged according to conventional religious grammars) on whether there are transhumanist rituals that are the sites of embodied self-disciplining and self-construction. How pertinent or beside-the-point is it to ask whether the dietary recommendations of a Kurzweil or the many eugenic-euphenic "human enhancement" arguments or the funerary practices of freezing the body are ritualized among a community and given the status of the religious or cultic by cultural or political bodies? How much are analysts involved in the collective practice of determining the religious status of transhumanism in the very attempt to diagnose its participation in wider religious and secular imaginaries?

At least in the present, Nye, Segal, and Taylor all point to factors that are revealing of the problems common transhumanist tropes imagine themselves to address. In the West, with its own negotiated and porous boundaries, in which neither secular nor religious grammars signal the possibility of a future life as strongly as they once did and in which meaning is forcibly predicated on both individual and species mortality, the perpetuity of individual life may indeed become a more tantalizing object—provided there was enough collective faith in the past to function as a guarantor for life. Furthermore, in the persistent shadow of Cold War logic, with its optimisms for technological empowerment and its pessimisms about its use, with its prying apart of human dominance and responsible stewardship, transhumanism holds out the possibility of transcending (individually or collectively) the quandary

---

abstracted and universalized" (Asad, 1993, p. 42). It might be noted that J. Z. (Jonathan Zittell) Smith has, from a very different perspective, reinforced such a picture, having observed that the contemporary study of religion and its enabling idea of a "generic religion" is itself a product and project of the Enlightenment, a project that, in embracing the aphorism "nothing human is foreign to me" (Smith, 1982, pp. 104–105), Smith at least partly endorses.

of not being in control of technological systems and powers that are apparently human. The specifics of that transcendence can celebrate the fusion of man and machine, the escape from the body or from earth, or the production of human solutions. This last, in its secular key, appears to be little more than a technological optimism decried by those who object to faith in technological solutions.

Put another way, part of what supports the transhumanist vision both within itself and in cultural circulation is how much it is predicated on secular and mainstream discourses persistently declaring distrust, finitude, and loss and asking for direct compensation for these ills. In this context, transhumanism presents itself not only as a kind of intensification of technological optimism in the present but also as one that can be disseminated precisely because of the pervasiveness of its claims—for the hopes it promotes that technology can provide the solution after all, rendering government either irrelevant or more just, overcoming mortality and alienation, and resurrecting the past.

Transhumanism is neither singular in its raising of technology to such a status nor the most gloried bearer of this technological standard—as indeed the multiple targets of the multiple critiques of technological determinism demonstrate. In the present, contemporary institutions of and debates over anticipatory governance and the status of emerging technologies as a category also treat technology as a primary site for solutions to problems older technologies were often conceived as having created. As Alfred Nordmann recently observed,

> History—and the history of technology, in particular—offers another reason for rejecting the idea that a non-trivial future is already in the present, like a seed perhaps that only needs to germinate. History first requires us to refrain from the technological determinism that is inherent in the image of the rising tide [which Nordmann associates with the metaphor of "emerging technology"] or of the seed and what it harbors. Such imagery renders the inhabitants of the present world mere recipients of what is to come, allowing them only to prepare for the future, at best to modulate the course of events. (Nordmann, 2014, p. 93)

Whether various academics, scholars, and consultants involved in technological assessments of the future actively trade in "trivial futures" has provoked debate. But to the extent that futurist expertise—linked to managerial, pedagogical, and entrepreneurial idioms—is expertise about the future of technology, Nordmann's critique is significant as a reminder that even the conception of nontrivial futures can assume the predominance of technological orders: that we live or die by technological feats. Even the fear of technological hubris might, if emphasizing different terms than Nordmann's, reinforce technological determinism—if that hubris is formulated more in relation to technology than to the future, as overstepping in relation to the power and dominance of technological development. How dare we attempt to forecast the future of technology?

The wider tendency in this and other literature depicting technology as prime mover—the future of technology as *the* future—underpins the transhumanist imaginary and makes more ordinary the fact, if not the content, of its own projections. The specific salvations it holds out need not be shared or endorsed by those who, in a more politically and conventionally secular key, desire and promote (governmentally or otherwise) technological innovation as the key to future welfare. If technological innovations are providing goods and securing against harms, including technologically induced harms, then, on the level of the governmental or the entrepreneurial, soteriology might be at best a secondary concern. That technology is redemptive is a largely secular belief, just as it is the way and the degree to which it might be understood as transcendent that determines its religious character—not the fact that technology is construed as transcendent or sublime in itself.

Conclusion

Smith has distinguished between locative and utopian religions, the former concerned primarily with maintenance of the purity of an immanent ordered state that might slip into chaos and the latter directed toward a utopic, transcendent state that stands as redemption for and beyond the present.[15] Likewise, Nordmann

15 Smith articulates this distinction in relation to "Mediterranean religions" of antiquity. The locative is "concerned primarily with cosmic and social issues of keeping one's place and reinforcing boundaries. The vision is one of stability and confidence with respect to an essentially fragile cosmos, one that has been reorganized.... We may term such locative traditions, religions of sanctification" (Smith, 1990, p. 121). Note that, in an earlier article, in discussing religions in diaspora, Smith draws the following contrasts: "To the native religionist, homeplace, the place to which one belongs, was an important religious category. One's self-definition, one's reality, was the place into which one had been born geographically and socially. To the new immigrant in the diaspora, nostalgia for homeplace and cultic substitutes for the sacred center became the important religious categories (e.g., Jewish diasporic synagogues were oriented toward Jerusalem; all water used in the cultus of the Roman Isis temples was considered to come from the Nile). To the thoroughly diasporic member, who may not even have belonged to the deity's ethnic group, freedom from place became the major religious category. For such an individual the values of the native religion appeared reversed. For the native religionist to be in place was to truly live. His freedom found fulfillment in the service of the national deity. For the diasporic religionist to be in place was death. It was to be shackled to that which was confining and hence untrue. Diasporic religion was 'utopian' in the strictest sense

recently has attempted to distinguish between science and technoscience on the basis of concerns with purity—a desire to provide a comprehensive representation of the world as it is—and transcendence—bracketing purity in order to refashion the world (Nordmann, 2010). These are not in themselves "ontological" distinctions, as both scholars observe; in neither case are there examples in which these categories exist in distillated form. Purity is entangled with transcendence, science with technoscience. But, as cultural commitments, they summarize something of the location of the transhumanist imagination and the ambitions it represents. In its participation in a religious grammar, that imagination is more overtly utopian than locative in Smith's senses, if not transcendent by Taylor's lights or utopian by Segal's. In its participation in more secular grammars, it is more technoscientific than scientific in Nordmann's senses, though neither humanist nor scientific humanist in Bronowski or Huxley's terms. The sublime it projects is invested in the wonder of actualized technologies, now often naturalized (optical scanners, musical synthesizers, humanoid robotics, and so on), participating in the pluralist celebration Nye documents. Crisscrossed by overlapping and transnational traditions of technological utopianism, sublimity, and scientific humanisms, that imagination alternately resonates with secular and religious understandings, suggesting not only the breadth of those frames, and the comparative narrowness of contemporary transhumanism, but the ways in which the frames themselves yield to each other and to the hands that do the enframing.

References

Asad, T. (1992). Religion and politics: An introduction. *Social Research, 59* (1), 3<N>16.

Asad, T. (1993) *Genealogies of religion: Discipline and reasons of power in Christianity and Islam*. Baltimore, MD: The Johns Hopkins University Press.

Asad, T. (2003). *Formations of the secular: Christianity, Islam, modernity*. Stanford, CA: Stanford University Press.

Barton, R. (1987). John Tyndall, pantheist: A rereading of the Belfast address. *Osiris*, 3, 2nd ser., 111–134.

of the word, it was a religion of 'nowhere,' a religion of transcendence" (Smith, 1971, p. 238). Note further that, in discussing the consequences of the move away from "cult centers," Smith observes a "change from concern for a religion of national prosperity to one of individual salvation (or, in some traditions, from focus on a particular ethnic group to concern for every man). The chief religious figure was no longer the priest and king but rather the prophet or savior" (Smith, 1971, p. 237).

Bimber, B. (1994). Three faces of technological determinism. In M. R. Smith & L. Marx (Eds.), *Does technology drive history? The dilemma of technological determinism* (pp. 79–100). Cambridge, MA: MIT Press.
Edwards, P. N. (1996). *The closed world: Computers and politics of discourse in cold war America*. Cambridge, MA: MIT Press.
Gordon, P. E. (2008). The place of the sacred in the absence of God: Charles Taylor's "a secular age." *Journal of the History of Ideas, 69* (4), 647–673.
Haldane, J. B. S. (1940). Possible worlds. In *Possible worlds* (pp. 241–263). London, England: Evergreen Books.
Huxley, J. (1951). Knowledge, morality, and destiny—The William Alanson White Memorial Lectures, third series. *Psychiatry, 14* (2), 127–151.
Huxley, J. (1957). Transhumanism. In *new bottles for new wines: Essays* (pp. 13–17). London, England: Chatto & Windus.
Lederberg, J. (1963). Biological future of man. In G. Wolstenholme (Ed.), *Man and his future: A Ciba Foundation volume* (pp. 263–273). Boston, MA: Little, Brown, and Co.
Lewis, C. S. (1994). *The discarded image: An introduction to medieval and renaissance literature*. Cambridge, England: Cambridge University Press.
Nordmann, A. (2010). A forensics of wishing: Technology assessment in the age of technoscience. *Poiesis & Praxis, 7*, 5–15.
Nordmann, A. (2014). Responsible innovation: The art and craft of anticipation. *Journal of Responsible Innovation, 1* (1), 87–98.
Nye, D. (1994). *American technological sublime*. Cambridge, MA: MIT Press.
Segal, H. P. (2005). *Technological utopianism in American culture*. Syracuse, NY: Syracuse University Press.
Sidiqqi, A. (2010). Competing technologies, national(ist) narratives, and universal claims: Toward a global history of space exploration. *Technology and Culture, 51* (2), 425–443.
Smith, J. Z. (1971). Native cults in the Hellenistic period. *History of Religions, 11* (2), 236–249.
Smith, J. Z. (1982) *Imagining religion: From Babylon to Jonestown*. Chicago, IL: The University of Chicago Press.
Smith, J. Z. (1990). *Drudgery divine: On the comparison of early Christianities and the religions of late antiquity*. Chicago, IL: University of Chicago Press.
Spencer, H.. (1857). Progress: Its law and cause. *Westminster Review, 67* (132), 445–485.
Taylor, C. (2007). *A secular age*. Cambridge, MA: Belknap Press.
Tirosh-Samuelson, H. (2012). Transhumanism as a Secularist Faith. *Zygon, 7* (4), 710–734.
Turner, F. M. (1974). *Between science and religion: The reaction to scientific naturalism in late Victorian England*. New Haven, CT: Yale University Press.
Turner, F. M. (1981). John Tyndall and Victorian scientific naturalism. In W. H. Brock, N. D. McMillan, & R. C. Mollan (Eds.), *John Tyndall: Essays on a natural philosopher* (pp. 169–180). Dublin, Ireland: The Royal Dublin Society.
Wolfe, A. J. (2002). Germs in space: Joshua Lederberg, exobiology, and the public imagination, 1958–1964. *Isis, 93* (2), 183–205.

Future Vision in Transhumanist Writings and the Religious Public

13

John H. Evans

Introduction

According to Hava Tirosh-Samuelson and J. Benjamin Hurlbut (cf. the introduction to this volume), transhumanism (TH) "anticipates the application of science and technology to the amelioration of the human condition through genetic engineering, robotics, informatics, and nanotechnology." TH represents a particular type of amelioration that will be important for this paper: that TH is "the belief that the human species can and should *transcend itself* 'by realizing new possibilities' of and for human nature" (Hauskeller, 2012, p. 39, *my emphasis*). A typical sentiment of transhumanists can be found in the words of British philosopher John Harris, who writes that "'taking control of evolution and our future development to the point, and indeed beyond the point, where we humans will have changed, perhaps into a new and certainly into a better species altogether,' is 'nothing short of a clear imperative to make the world a better place'" (quoted in Hauskeller, 2012, p. 40).

TH is not a religion (Hopkins, 2005). However, it has many cultural similarities to, at least, Western religion, such as narratives of progress, teleology, and salvation. For example, Harris's claim above is that there is a "clear imperative to make the world a better place," an impulse shared by most Christians. Similarly, theologian Brent Waters considers TH and Christianity to be "contending salvific religions." His summary of the similarities is worth quoting at length:

> Transhumanism represents a late modern religious response to the finite and moral constraints of human existence. It is not a religion in a formal sense, but as Martin Luther suggests, wherever one places one's confidence is necessarily one's god—or, more broadly, one's object of faith or ultimate concern. In this respect, transhumanism and Christianity appear to have a number of similarities, particularly with regard to soteriology and eschatology. Transhumanists and Christians agree, for instance, that the finite and mortal human condition is far from ideal. For transhumanists humans

> have fallen short of achieving their true potential, whereas for Christians humans have not yet become the kinds of creatures God intends them to be. In response both agree that humans require release from their current condition. . . . Both agree that death is the final enemy; transhumanists conquer this foe by achieving the immortality of endless time, whereas Christians are resurrected into eternity, where there is no time. (Waters, 2011, p. 164)

However, TH is an explicitly secular project and often posits itself as an enemy of traditional religion (Cady, 2012, p. 83). While there are real differences between traditional religion and TH, one of the reasons for the ire thrown at religion by TH is the scientific visionaries' ignorance about contemporary Western religion. These scientific advocates assume that opposition to human modification is the result of religious people not wanting to modify a sacred and unchangeable human. A more subtle approach to Western religion would presume the opposition is to the goals of the modification.

Portraying religious opposition as a defense of the sacred human body that cannot be modified except by God produces a caricature that is easy to vilify. For example, biologist Lee Silver implies that the religious people who are opposed to cloning are so because "cloning leaves God out of the process of human creation and that man is venturing into places he does not belong" (2000, p. 63). Similarly, transhumanist John Harris says religious people oppose genetic engineering of humans because this is "usurping God's prerogative and creating new life forms or perhaps disturbing the course of nature." Harris continues by arguing against a straw figure of a hypothetical religious believer who believes the natural condition of the body is totally sacred and cannot be changed at all: "even believers must believe it can be right to disturb and redirect the course of nature; otherwise the practice of medicine itself would be wicked. . . . The practice of medicine . . . must be seen, if anything, as the comprehensive attempt to frustrate the course of nature" (Harris, 1998, pp. 177–178). Philosopher Gregory Pence writes that "some people believe that humans have no right to change the way that humans are created because sexual reproduction was ordained by God and that to try to change this way is sinful" (1998, p. 119).

There are undoubtedly some religious people who fit this description. However, I will argue that they are not the largest group. In actuality, religious people and transhumanists have a lot in common. These commonalities are evident in how TH clearly uses at least secularized Christian narratives as part of its appeal. As Hurlbut and Tirosh-Samuelson argue in the introduction to this volume, technological transhumanism can be understood as a unique hybridization of religious and secular ideas.

Nearly all of the scholarly analyses of TH concern its prophetic writings because, by and large, it is not as of yet a social movement with actual followers (Hughes, 2004; Kurzweil, 2005; Young, 2006). It is a strain of thought popular among some philosophers and a number of influential elite scientists, and, to the extent those elites can determine what happens in the world, it is important to analyze. Analyses of the religious, utopian, or teleological themes in TH thought have also been based largely on elite interpretations of religious traditions, typically produced by theologians. This sort of humanistic debate is important so that scholars can sharpen their ideas.

However, if technology progresses, TH visions may move toward actual implementation. And there are people working on these technologies. For example, human genome-mapping pioneer J. Craig Venter recently announced that he has formed a company called Human Longevity to, in the words of one newspaper writer, "cheat aging and death" (Pollack, 2014). The cofounder is Peter Diamandis, who is cofounder and executive chairman of Singularity University, the TH-inflected institution discussed by Margarita Boenig-Lipstin and Ben Hurlbut in chapter 11 of this volume.

If the TH vision moves toward implementation, then we can be assured that TH will become more of a public debate than it is now. At that point, the views of ordinary religious people become at least as important as the views of the theologians. Therefore, in this paper, to gain leverage on what ordinary religious people's reaction to TH would be, I examine the teleological narratives of progress through genetic transformation of the human body articulated by ordinary religious Americans.

Existing Research on Religion and Faith in Transhumanism

In this paper, I emphasize some similarities within the visions of transhumanists and religious people in the United States. However, previous research has identified a difference that should be kept in mind, which is that religious people have less faith in science than do the nonreligious. It is not that religious people do not believe in science as a means of explaining the physical world, which they largely do (Evans, 2011). Rather, they lack faith, which is the "firm belief in something for which there is no proof" (Merriam-Webster online). Transhumanism relies on faith that scientific technologies will be able safely to conduct all sorts of improvements to the human species. It also relies on a sort of teleological faith in science as salvation.

In an earlier paper, I examined the public's faith in science, using public opinion data across Western nations from 1993 to 2010 (Evans, 2014). I conceptualize three types of faith in science that are prerequisites for advocating transhumanism. The first is faith in science to provide meaning for society, akin to how religions have traditionally provided meaning. The second is faith that science can effectively solve any problem. The third, more limited type of faith is that science can solve problems in the physical world with technology. Focusing on the U.S. results, compared to the nonreligious, members of each Christian group measured have less faith in science producing meaning than do the nonreligious. This is also generally true across other Western countries.

American religious people also have less faith in science solving general problems than do the nonreligious. In this instance, American religious people are largely distinct in that this difference is generally not found in other countries of the West. Finally, in the United States, as well as in the other countries examined in Europe, religion does not lead to a lack of faith in science's solving specific problems in the physical world, again compared to the nonreligious. Put simply, American religious people have faith in sciences solving concrete problems, such as in medicine.

If transhumanism were portrayed as solving concrete problems in the world, there would be less religious opposition. However, if transhumanism is portrayed as a meaning system of its own, as a sort of substitute religion to believe in, there will be a strong negative reaction among U.S. religious people. These findings are consistent with what I will conclude below, which is that religious people are not opposed to technology or science per se, only when technology or science becomes a meaning system that sets its own goals.

Comparing TH Elites and the Religious

The interview data used in this paper can be used to make more fine-grained analyses than can be had with survey questions. I will make the following points below. First, TH elites and religious Americans both see technology as a source of hope and an engine of human progress. TH elites and religious people diverge on the end goal of this progress. For TH elites, technology leads to the goal of individual human freedom. If someone wants to live for 300 years and grow wings, technology will provide these wonderful choices. The religious people I spoke with are fundamentally hopeful about the genetic future. For the religious, the goal for these technologies is that they will restore the nonsuffering human, not create a different superhuman with qualities beyond the relief of suffering.

Second, TH elites believe in genetically modifying the human genome to improve the species. Common wisdom and TH writings suggest that religious people are opposed to this modification because God created humans as they should be. I will argue that this is not correct and that the majority of religious people are not opposed to genetically modifying the human species per se and agree with transhumanists that the current human body is flawed. They do not view the current human genome as somehow sacred and not to be "tampered with."

Third, TH elites believe that humans are the agents who work on transcending our own bodily limitations. Religious Americans agree but add a critical caveat. While transhumanists say that humans should come up with the design of humanity 2.0, religious people say that we should create humanity 2.0 but use a design that God would approve of. That is, the two sides agree that humans should fix the body, but they disagree about the source of the plans for the redesign.

Method

To examine these narratives of progress, I conduct an inductive discourse analysis of responses from an in-depth interview study concerning what religious people in the United States think about reproductive genetic technology (Evans, 2010). In 2003 and 2004, my research team and I interviewed 145 members of religious congregations in the United States and 35 nonreligious people. The number of religious respondents from each of the large religious traditions in the United States was set to match the percentage of congregational members from each of those groups in the general population. The groups were Catholics, evangelicals, and mainline Protestants, with an oversample of Jews.

The interviews began with asking the respondent to identify her religion and to describe her religious practices. The interview then turned to a series of hypothetical scenarios about couples who are planning on having children and asking the respondent for what advice she would give the couple. The first scenario was about a couple who have found that they are both carriers for cystic fibrosis. "What should they do?" the respondent was asked. In the next scenario, the woman is already pregnant, and the couple is offered amniocentesis possibly followed by abortion. "Should they do this?" the respondents were asked. In the next scenario, a couple is offered preimplantation genetic diagnosis to avoid cystic fibrosis. Another couple is offered preimplantation genetic diagnosis for early onset Altzheimer's disease, and another for deafness; yet another is offered preimplantation genetic diagnosis to avoid having an obese child. Finally, a hypothetical couple is offered preimplantation genetic diagnosis to obtain the smartest child possible. Another scenario involves sex selection, and another sperm sorting to determine the gender of the baby. The scenarios then turn to germline human genetic engineering to engineer in or out traits such as cystic fibrosis, obesity, and intelligence. A final scenario asks about cloning. Questions then turn to what the respondent thinks the effect of these technologies will be on society, some questions about religion, and finally how the respondent thinks our society should have a debate about reproductive genetic techologies.[1]

It is important for my claims to note that the questions I primarily focus on for this chapter were near the very end of the interview after an hour or more of conversation about religion and reproductive genetics. By the end of the interview, the respondent has had an education in the types of reproductive genetic technologies that would be used to enact the transhumanist vision, with discussions of the wisdom of improving both individual humans, as well as the species. The conversation has been, at least for the religious respondents, interwoven with religious discourse, and, due to the priming early in the interview, many responded with religious reasons for and against the use of these reproductive genetic technologies.

The first of the questions I examine occurs at the very end of the interview. It was different from the previous questions in that the interviewer handed the interviewee a card with ten words listed in a column, and the interviewee was asked "When you think about all the issues that we have talked about today, which one of these words best summarizes your feelings. Or, you can pick a word that is not on the list. Or, you can talk about more than one if you want." When the respondent would pick a word, she was asked why that word. In their reasons for picking that word, we can determine their notions of hope, redemption, and progress related to human improvement. The words on the card were "humility," "worry," "fear," "hopefulness," "happiness," "hopelessness," "anger," "helplessness," "reverence," and "awe." The other question I focus on for this paper asks "Should the ability to change the genes of the human species be reserved for God?" This is designed to get to the question of who should modify and/or provide the design for future humans.

Results

Vision of the Future

Let me start with the data generated from the word selection exercise, which addresses my first point: that religious people in the United States generally see reproductive genetic technologies through a lens of hope. The question comes after extended

1 For a copy of the interview guide, see Appendix B in Evans, 2010.

discussions of technologies, some of which nearly all respondents are opposed to. For example, almost nobody was in favor of human cloning, with most people finding it repellant. The vast majority of respondents had some technologies and applications that they approved of, and some that they did not. Overall, what was the conclusion about reproductive genetics? Were they hopeful and, thus, tending to ignore the "threat" of technologies like cloning and creating superintelligent babies? Or did they focus on the negative and see human bodily modification through genetics as a foreboding picture of our future?

I deductively coded the responses by the word selected and then inductively coded the reasons given for selecting that word. There were certainly people who fit the stereotyped depiction of religious people seeing *dread*, with some selecting "worry," "fear," "worry and fear," "anger and fear," and "hopelessness." A few suggested their own word to summarize their thoughts about reproductive genetic technologies—"concern." The response of one Catholic respondent was typical. His word was "worry"; when asked why, he said, "I would hate to see society come to the place where we can manufacture a human being the way we want it to be. That would be very worrisome to me."

Similarly, a mainline Protestant who approved of the individual use of many of the technologies talked through the other possibilities on the card. "Well, not 'happiness' or 'anger,'" he said. "I think probably 'fear' and 'worry' more than the 'hopefulness' reference and 'awe,' which probably reflects my cynicism more than anything else. But . . . I mean it could potentially be a very good thing, but human history being what it is, you know, when we meddle in these things, we tend to end up doing more harm than good. So, I think 'fear' and 'worry.'"

However, these negative visions of the future of genetic modifications of humans were not the dominant response. The majority of words selected by conservative Protestants, mainline Protestants, and Catholics could be categorized as either "all good" or "good and bad." Mainliners were the most positive, with almost all selecting "all good" words or "good and bad" words. Conservative Protestants were a bit less so, followed by Catholics.

By far, the "all good" word most chosen by respondents was "hopefulness," and nearly everyone said the reason he or she selected that word was that we would soon be able to relieve the suffering of disease by using reproductive genetic technologies. As one mainline Protestant respondent put it,

> I think "hopefulness" would be the selection that I would make. I get to thinking about it and think that we can contribute significantly through the development of these technologies to reduce human suffering. I can't imagine anyone not being hopeful about that. It's not going to be easy. It's going to be very complicated. We

> need to exercise humility in the process, but I don't find it hopeless or helpless. I'm hopeful.

Earlier, she was asked about a scenario where a doctor could "fix the genes in a fertilized egg to remove the chance that any baby would have cystic fibrosis," so that "not only would this child be free of this gene, but so would the child's children, the child's grandchildren, and so on." She responded that she would be supportive of the use of this technology to change "the lineage," in her words, because "I think it's a great technique, if that exists, and I think it would be wonderful to eliminate cystic fibrosis from our world, and that will take some time to do, I'm sure."

A fundamentalist Protestant said he picked "hopefulness" because "I think we're on the verge of some good things for people for our society and hopefully we'll be able to handle these new technologies with wisdom. And if we're wise in what we do, we only improve society. If we're not wise, they're very dangerous. So I guess I'm hopeful that we'll be wise in the way we can handle them."

Earlier, in a discussion about using germline genetic engineering to remove cystic fibrosis from the family line, this fundamentalist Protestant was not concerned about the technology itself, only that it did not involve killing embryos or fetuses, a not uncommon response for conservative Protestants. "Sure if that technology is available, certainly," they should use it, he said. When asked, "Do you have any concerns about this technology at all?" he said no, as long as "it wasn't done with—by using—as long as it's not done by replacing a gene with fetuses or that stem cell stuff."

A Catholic man said the choice of words was easy, and his choice was "hopefulness." The reason was "because we have within our grasp, like no other time in history, the chance to eradicate so much human suffering. We're on the cusp of a great era of discovery and, you know, my . . . hopefully my children will live to see it. . . . I remember getting . . . polio shots and, you know, polio and small pox [have] been eradicated. Amazing things have happened in my lifetime and my kids are going to see remarkable things."

Perhaps more interesting was the very common impulse to select a positive and negative word to indicate the good and the bad. The positive is always a word that the respondent chooses to represent the hope for relief of suffering through genetic technology. The negative is a word to represent either fear of misuse (typically radical enhancements) or of forgetting that we humans are not God. The most common words to select are the combination of "hopefulness" and "worry." One respondent stated, "Actually, two things kind of pop out. The hopefulness—because I think that a lot of good could come from being able to control some really major medical problems—and worry. I'm a worrier anyway. But, a little worry about just how far the scientific community would maybe want to go with the technology that they are developing or that they have."

Similarly, a mainline Protestant stated that "I would put 'hopefulness' in there and I think I have to add 'worry.'" "Hopefulness" is selected because of "the good things that can happen with the technology that you have described," and "worry" is selected because "the technology could be . . . misused for frivolous reasons." She was typical in her support for revising the human species to eliminate disease but not for changes that she thought were not diseases. She was in favor of germline genetic engineering for cystic fibrosis as long as there are no "unforeseen consequences." The interviewer then began, "What about for . . ." and she interrupted with, "For blue eyes…NO!"

Religious respondents often also selected terms of "reverence" and "awe," signifying caution, that we are interfering, perhaps with good reason, in something that is far above our human perspective. Similarly, it was common for respondents to select "hopefulness and humility," with "humility" a reminder that we are not God. As one conservative Protestant put it, he selected the term "humility" "because it helps . . . you know, helps me realize that I'm not God. I'm not able to make perfectly correct decisions. I mean, I can't even decide which website to visit, I mean, let alone which genetic illnesses are really worth selecting against, you know?" He was hopeful "because I think that this is a technology [that] will help people."

A smaller group of people selected "hopefulness" and "awe," with the hopefulness being to cure disease and "awe" meaning "at awe with our awesome technological abilities." A fundamentalist Protestant said he picked these words because

> I believe, of course, there is, you know, as we progress along as a society, there are things to be hopeful for that diseases will be eradicated and that people will live comfortable lives, you know, and an ultimate hopefulness of someday getting to heaven and sitting face-to-face with the creator. It is kind of awesome to think that we have this technology to be able to do some of these things, but I believe, of course, that credit should be given where credit is due. It's not necessarily us that are actually doing this. We're the clay in the potter's hands, even those that may not necessarily be someone who believes in Christ, because God certainly has the capability and power to do that. So I would say probably "hopefulness" and "awe." Some of these more negative feelings, you know, I just don't see that. I for one, I'm certainly not worried or fearful.

Roughly equal numbers of religious people in each religious tradition selected "all good," "good and bad," and "dread" terms. There was a slight tendency for mainliners to be more "glass is half full" and Catholics to be more "glass is half empty." For the "all good" category, these were people who, despite being opposed to some aspects of reproductive genetic technology, had an overall exclusively positive, hopeful vision for genetic intervention going forward.

In terms of vision of the future vis-à-vis genetic technology, a transhumanist would be all positive. Theirs is, after all, a utopian quest. Religious people are largely positive as well. They are not antiscience or antitechnology. By and large, they do not see the human genome as sacred and, thus, inviolable. They believe in what they perceive as the relief of suffering, most notably through "medical" interventions. However, religious people in the United States seem to lack a positive eugenic vision, where we intervene in ourselves to "improve" ourselves. Rather, the present-day healthy, "walking-around" sort of human is the ideal, and the goal is to get everyone to that pristine state. However, their positive vision going forward is largely tempered by a caution at what humans might do wrong. In general, one could imagine religious people in the United States in favor of creating a humanity 2.0, as long as the blueprint were limited to healing disease. They would probably be opposed to the open-ended humanity 2.0 of the transhumanists.

Creating Humanity 2.0 for the Glory of God

The transhumanist vision is to transcend the existing human to create a human that maximizes our present desires. If the current culture says that an ideal human life should be free of disease, then that is what humanity 2.0 should have. If the current culture values superintelligence, then that is what humanity 2.0 should have. These are all framed as "improvements" on the human. In this section, I make my second and third points. The second point is that religious people are not opposed to modification per se, and the third is that humans are to enact this modification, not God.

Religious respondents, similarly to transhumanists, also believe in "improving" the current human. They simply reject the open-ended horizon of the transhumanists but want to constrain improvements to those that would please God. Of course, an academic analyst could say that these people should recognize that what is thought to be "pleasing to God" is also a matter of social consensus, ultimately no different than simply polling a society. However, this is not how religious people see it.

The limitations on human improvement in the Western religious vision can be found in the elite theological discourse about the relationship between God and humans. In at least Western religious discourse, there is a strong dichotomy between God and humans. Humans are not God but were created by God, along with all of "nature." A long-running part of Christian theology concerns what actions in the physical world are then the responsibility of humans and which are to remain in the province of God. This controversy is well represented by the theological debate concerning whether we humans are simply a created creature of God, cocreators with God, or created co-creators with God (Cole-Turner, 1993). Cole-Turner writes about cocreation: "to what extent may we human beings, throughout our scientific understanding and our technological ability, serve God the Creator in this ongoing creativity? Human work, especially our technology, may be seen as a partnership with God in the continuing work of creation (Cole-Turner, 1993, pp. 100).

With a technology like the genetic engineering of humans, if we are creatures, God is entirely responsible for our human bodily form, and we are not to interfere. To take this to its logical conclusion, we should not heal diseases humans are born with because God created them. However, if we are cocreators with God, we participate in fulfilling God's desires for the human bodily form. In this vision, God wants us to heal disease and relieve the suffering of an imperfect world and an imperfect human. Being a created cocreator means something similar but emphasizes that humans are ultimately not God.

In the interviews we conducted, well after we asked all of the questions about what people should do in the face of various decisions concerning genetics, we asked the respondent the following: "Do you think that the ability to change the genes of the human species should be reserved for God?" The responses here point directly to the respondent's notion of whether we are cocreators with God and whether something like human genetic improvement should be something "left to God." We can also see what is left to God and what is left to humans. To anticipate the conclusion, it is God's job to come up with the plan, and it is humanity's job to implement it.

Similar to the findings above where a minority of the respondents reacted with pure foreboding to these technologies, a minority of religious people do say that, yes, the ability to change the genes of the human species should be left to God. These are typically people who are opposed to all technologies and their applications. It is hard to imagine that they do not believe in the human improvement of God's creation—as this would be put in the conservative Protestant tradition—because they presumably all believe in medicine, mechanized agriculture, and so on. I suspect that it is the intervention into the human that is particularly problematic for this group.

It is hard to tell what these outright opponents were thinking because they rarely gave their reasons and just implied that it was obvious from their previous comments. The few who really articulated their reasons sound like the stereotypes of religious people as theological Luddites. The common wisdom would be that religious people would say that we "should not play God" with technologies, and one Pentacostal respondent seemed to agree, saying "yes," we should leave it to God

> because He created us... I don't think someone should change you just because they want you to be a certain way.... If God wants you to be better then He will make you better .... No, I think it's up to God to do it, not to wait for scientists to put you out

> in a lab and you're sort of like a guinea pig or something. So I think it's up to Him. I mean, we're here. He made us a certain way for a reason.

This summary was reflected in this respondent's earlier statements about specific technologies, which reflected the radical immanence of Pentecostals. For example, for preimplantation genetic diagnosis to remove the possibility that a child would have cystic fibrosis, she said that "personally, I think you should get what—there's a reason for everything. . . . God gave that baby to you for a reason if you become pregnant, if you have that disease, there is a reason for everything. . . . God gives you, I think in life itself, we are given things to see if we can handle. I mean, God gives us things. He wouldn't give us things if we couldn't handle it. I think it's just another test."

However, the strong majority of respondents said "no," the ability to change the genes of the human species should *not* be left to God. There were three major reasons: that God works through humans; that we should transform the human if limited to good, God-given reasons; and that we have God-given free will to do both good and bad.

The first reason is well articulated by one Catholic respondent who put it succinctly, stating that "He created the people and He helps people to create technology to change these things. I don't think it should be just God. . . . No, I think He's out there saying, you know, 'I hope you find a cure for cancer' or whatever." A mainline Protestant said that "we are God's hands on earth. So it's up to us to use it, not—God's not going to do it. [Normally], He's going to have us do it. So, it's our responsibility to use those techniques." Similarly, an evangelical responded that

> it goes back to the sense that God has given us these abilities and this intelligence to create penicillin to wipe out, not wipe out, but to able to fight bacterial diseases, to create drugs that combat cancer and other disease. It gives me medicine to take for my thyroid to make it work right, and I think that is all good stuff. I think there can be good stuff coming out of genetics and genetic technology. But there is always the flip side of how that technology is going to be used. I mean, we create all these drugs, and we also create dangerous viruses that can be used as weapons. That is the flip side of everything. God has given us the ability, so I can't....I don't believe you can say, "Okay, God alone can do this," but, on the other hand, I believe you have to be willing to look at things as not just as a "What can I do?" but "What should I do? What is the right thing to do?" Unfortunately, too many people don't care about what is right. They just want to do whatever they want to do because they can do it.

The second common response is that genetic transformation should not be reserved for God, as long as we are doing God's will, which is obviously related to the first reason. To cite the phrase of the late conservative Protestant theologian Paul

Ramsey, we should "play God as God plays God" (Evans, 2002, p. 130). A mainline Protestant respondent said,

> If it's something like medical science would determine that, if they removed a gene and prevent someone from being born with the disease or could prevent disease from occurring in that child, yeah, I think that's great. Again I look at that as medical science, that knowledge . . . a God-given knowledge. It's not creating life; ...it's improving the quality of life. It's human life, but I think that's as far as I can go. But altering genes to make a superhuman being or, you know, making someone... making a life, cloning, cloning a life to be something that wouldn't otherwise be naturally, I don't think that's within our purpose as human beings. I don't think it's our right.

An evangelical responded that "some of this genetic engineering is good and some of it's bad, but overall, I think He's given us the intelligence that I think our soul and our spirit—if the soul and the spirit's not lighted up with God, then that's where you get into trouble." We have to be doing it "from the right perspective... and from God's perspective." Finally, a mainliner said that

> it goes back again to, what are you going to do with it? Like, that's the big question in all of this. It's not so much that should you or shouldn't you do it. It's like, once you learn how to do it, the genie is out of the bottle, right? You're never going to put him back in. What are you going to do with it, now that you're able to do this? Are you going to do things good for society and for humankind or use it for trivial things like picking kids' eye color. That's the question.

The third reason is not so much that God's plan should be in control but that we humans are on our own to select good and evil. We have free will. There is still good, and the respondent seems to know what it is, but this choice is part of the human condition. God is not a micromanager. As one Catholic man said,

> We're always going to be able to do miraculous things. We're always going to be able to go to the moon, things that just, you know, that are going to astound me whenever I'm really old that I had no clue. It's what we do with it and how we use it is where we stay connected to God. I think we're capable of dickering with just about anything we want to. That's just our nature because we're that smart or that intelligent and we can. We have that free will.

Similarly, a conservative Protestant woman told us that

> my belief is that the Lord has these things at our "disposal" or at our use, if needed, if necessary, and we make the choice of how we want to move with that. Do we want to use it—do we not want to use it? Why do we want to use it? I don't think it's an issue of are we going to upset God if we do it. Are we going to upset Him if we don't do it?

> No. I think these are the different options we have. What are we going to choose and what are we doing in making our choice?

Conclusion

It is easy to imagine that, if and when people start explicitly framing concrete scientific and medical decisions with reference to TH ideas, they will face opposition from religious persons. It is important that this opposition be more nuanced than our stereotypes about religious people's orientation to the human body would suggest.

First, both TH thought and ordinary citizens who are members of the Western religious tradition see technology that can modify the human genome as a source of hope and an engine of human progress. They do differ on the question of what we should be progressing toward. The religious people in this study generally saw hope in technology as leading to a future free of human disease. TH thought sees hope in a freeing of the human from any constraint on one's desires. If someone wants to live to 300 years old or have superintelligence, he should have the freedom to do so, according to TH thought.

Second, religious people do not hold the body to be sacred and unchangeable. There are multiple theological discourses for dealing with the issue of a transcendent God and a humanity that needs to be repaired (such as the doctrine of the Fall in Protestantism). While ordinary believers cannot articulate this specific theology, its general conclusions are well represented in what they say. Religious people agree with the transhumanists that humanity 1.0 is flawed, and both advocate for a humanity 2.0.

Third, a true difference between religious Americans and transhumanists is much more subtle than the idea that religious people want an unchanging body and transhumanists want to change it. It is not a question of changing the body, but where the design plans come from. Transhumanists want either individual human desire or at least collective human opinion to come up with the redesign. Religious people want the redesign to articulate with God's desire for the human species. Of course, how religious people are to know what God's desires are is *the* problem of theology over the past few thousand years, but this does not stop individual believers from having reached their own conclusions.

Transhumanism has fully adopted a secularized notion of religious progress that originated in Christianity. The two systems of ideas have more in common with each than is traditionally assumed. When a debate between the two begins in earnest, I hope that they will start with their true differences instead of stereotypes about the other. That may be soon, as the cultural conditions for transhumanism to become more than a set of ideas and become an actual social movement are coming into place. In the earlier paper analyzing survey data across the West over time, I found that, in general, faith in science as proving meaning—as a sort of substitute religion—is on the rise over time. It is on the rise in the United Kingdom, the United States, Norway, Czech Republic, Slovenia, Bulgaria, New Zealand, Canada and Spain—but not Germany, Russia and Israel (Evans, 2014, pp. 824). The conditions are becoming more favorable in the United States for belief in the religion-like qualities of transhumanism Moreover, the type of faith in science that is growing the fastest is exactly that which will put it on a collision course with organized religion in the United States and the West more generally.

References

Cady, L. E. (2012). Religion and the technowonderland of transhumanism. In H. Tirosh-Samuelson & K. L. Mossman (Eds.), *Building better humans? Refocusing the debate on transhumanism* (pp. 83–104). Frankfurt am Main, Germany: Peter Lang.

Cole-Turner, R. (1993). *The new Genesis: Theology and the genetic revolution.* Louisville, KY: Westminster/John Knox Press.

Evans, J. H. (2002). *Playing God? Human genetic engineering and the rationalization of public bioethical debate.* Chicago, IL: University of Chicago Press.

Evans, J. H. (2010). *Contested reproduction: Genetic technologies, religion, and public debate.* Chicago, IL: University of Chicago Press.

Evans, J. H. (2011). Epistemological and moral conflict between religion and science. *Journal for the Scientific Study of Religion, 50* (4), 707–727.

Evans, J. H. (2014). Faith in science in global perspective: Implications for transhumanism. *Public Understanding of Science, 23* (7), 814–832.

Hauskeller, M. (2012). Reinventing Cockaigne: Utopian themes in transhumanist thought. *The Hastings Center Report, 42* (2), 39–47.

Harris, J. (1998). *Clones, genes and immortality: Ethics and the genetic revolution.* New York, NY: Oxford University Press.

Hopkins, P. D. (2005). Transcending the animal: How transhumanism and religion are and are not alike. *Journal of Evolution and Technology, 14* (2),13–28.

Hughes, J. H. (2004). *Citizen cyborg: Why democratic societies must respond to the redesigned human of the future.* Boulder, CO: Westview Press.

Kurzweil, R. (2005). *The singularity is near: When humans transcend biology.* New York, NY: Penguin.

Pence, G. E. (1998). *Who's afraid of human cloning?* Lanham, MD: Rowman and Littlefield.

Pollack, A. (2014, March 5). A genetic entrepreneur sets his sights on aging and death. *New York Times*, B1.

Silver, L. M. (2000). Reprogenetics: How reproductive and genetic technologies will be combined to provide new opportunities for people to reach their reproductive goals. In G. Stock & J. Campbell (Eds.), *Engineering the human germline* (pp. 57–71). New York, NY: Oxford University Press.

Waters, B. (2011). Whose salvation? Which eschatology? Transhumanism and Christianity as contending salvific religions. In R. Cole-Turner (Ed.), *Transhumanism and transcendence: Christian hope in an age of technological enhancement* (pp. 163–175). Washington, D.C.: Georgetown University Press.

Young, S. (2006). *Designer evolution: A transhumanist manifesto*. Amherst, NY: Prometheus Books.

The Quest for Enhanced Consciousness in Everyday Life 14

G. Pascal Zachary

The first time I took LSD came after a long period of study. I was an undergraduate philosophy student, reading eight to ten hours a day, for years, mostly in various places in the university's spacious library. Questions about mind, consciousness, and cosmos arose naturally. I would make an early conversion to Kantianism, embracing the appealing notion that humans were hard-wired to make sense of reality in certain ways and not others. Kant's categories and imperatives sometimes struck me as arbitrary, which is different than "wrong," yet his insistence that we could never organize our coherent sense of reality purely by improvisation, on the fly, bereft of a master plan, a flexible program of sorts, persuaded me completely. What William James described in his *Principles of Psychology* as a "stream" of consciousness no doubt required an inner whip; to set the cacophony into rough order, to give rise to the light of reason, I surmised, required the imposition of structure on the stream (James, 1890).

I often wondered what was beyond the categories of reason, and I suspected some combination of art and madness. Yet, even as I preferred to think of myself as an artist but feared I was actually mad, I refused to take complete refuge in reason. It was the mid-1970s, and I harbored doubts and anxieties about the entire rationalist project. The democratization of computing and digital communications loomed on the horizon, impersonal forces not yet enmeshed in our daily lives. Yet digital science and technology were pregnant with the possibilities for "creative destruction" of our tidy mental boxes. Empiricism and logic ruled the day. Rudolf Carnap's Vienna Circle advanced the perspective that the entire universe, and every living thing inhabiting it, could be explained by a set of logical mathematical principles. The mind was only a kind of programmable computer; the brain, if forever alienated from mind by the Cartesian dualism, was also a machine. Mechanistic readings of reality easily seduced people who delight in a single capacious explanation, or what the historian of ideas Arthur Lovejoy neatly described as "the great chain of being" (Lovejoy, 1936).

When I dropped acid in the presence of two close friends, I thought I was prepared. After all, I had read Aldous Huxley's *Doors of Perception* (1954). I had joined the philosopher Robert Garvin's semester-long tour of existential thought in literature and religion. I had become, improbably for an avowed atheist, enamored of the medieval mystics, especially the contemplative text *The Cloud of Unknowing*. For a time, Meister Eckhart was my cat's meow. While the rationalist project nudged me toward the realm of verifiable knowledge, deduction, and a kind of certainty, my persistent anxieties helped amplify the existential invitation to irrationality.

So, we dropped acid late one cloudless night, Margarita and I. She was the supplier and, she claimed, a veteran freak. She was an art student, possessing wild, dark hair, olive-colored skin, deep, penetrating eyes, and a soft, handsome face. For a brief time, she was my best friend's lover and always around. A potent force, she entered the stage like a thunderstorm, upsetting everything and halting, immediately, my long bouts of reading. When she finally left the university, some substantial time before graduation, she vanished without saying goodbye. Her lover, my best friend Mark, never saw her again. Neither did I.

The night of our psychedelic interlude, Margarita never left my side. Steadfast and devoted, she endured my freak-out, insisting only that my best friend, who did not drop acid with us, stick close by, acting as what we would later call *the designated driver*. For hours in the dark of night, we walked in circles around the university's main campus. Time and again, we sat to rest under trees. Much to Margarita's dismay, and against her explicit and insistent counsel, I fought the drug. She pleaded with me to surrender. I would not. I fiercely held on to my rational Kantian categories. I refused to let go.

We walked for hours that night, the three of us, circling the campus so many times that, by the end, when we returned to our dorm room, we collapsed in exhaustion. Until I finally slept, I kept babbling about the war inside me between logic and experience, analysis and letting go. In the end, I kept trying to categorize and explain, to order and justify. My trip, in short, failed.

***

Fifteen years later, in the early 1990s, I recalled my botched acid-dropping adventure when I met Timothy Leary in northern California. Leary, the troubadour of LSD, had grown passionate in his dotage about the possibility for digital computers and electronics to provide the very sort of mind-expanding experiences he once believed only pharmaceuticals could provide (Roszak, 1969, pp. 166–169). He was older now, much older, and some skeptics suspected he had come unhinged from all those acid trips. His embrace of computing revived his celebrity, by introducing him to a new generation of youth. Leary, moreover, supplied some desired legitimacy to passionate prophets of digitally mediated experiences—computer simulations and other forms of electronic games and media that struck me and others as superficial and escapist. Leary's blessing seemed to connect the misfits and nerds of Silicon Valley to a deeper, broader tradition of technological populism and even revolt. A godfather of the counterculture, Leary had come under the sway of young Silicon Valley artist and engineer Jaron Lanier. An autodidact and a visionary advocate of emerging forms of immersive experiences in computer simulations, Lanier wore his blonde hair in dreadlocks and sported a dreamy look on his face that suggested either he had achieved spiritual contentment or possessed a reliable supplier of high-quality marijuana (Zachary, 1990).

In one short period, I met Lanier about a dozen times in northern California, and Leary joined us on one occasion. He showed no embarrassment over promoting electronic devices as the very path to enlightenment that he once reserved for the lawless embrace of rogue drugs. Animated and prone to moments of nonsensical mania, Leary seemed to have abandoned the sensibility that led him famously to advise youth to "turn on, tune in and drop out." Instead, he announced the supremacy of the digital. Along with Lanier, he insisted on the inevitability of what he called *virtual* reality—or vivid computer-mediated experiences that he insisted resembled an actual hallucination. This new state occurred when a person strapped on a variety of wearable electronic gear, which covered their eyes, ears, arms and feet. With Lanier in his futuristic digital suit, Leary announced that someday Huxley's "doors of perception" would be opened by computers.

The claim was ironic, I would later learn. The improbable story goes like this. In the 1960s, one of the world's leading computer designers, Doug Engelbart, led a laboratory in Menlo Park, California, a few blocks from where I lived during my Lanier-Leary phase. Engelbart's lab became the stuff of legend. The essential elements of the desktop computing experience and the Internet were born in Engelbart's lab. His scientists and engineers wrote some of the first emails ever written. Engelbart himself is credited with inventing the ubiquitous computer mouse. For many years, a dark secret lay at the heart of Engelbart's staggering inventiveness (Rheingold, 1985, pp. 174–204). He and many on his team dropped acid, and those mind-expanding experiences some credited with helping them envision the contours of contemporary computing (Markoff, 2005, preface).

There's more. The acid, in one of those perfectly logical twists in Cold-War American life, came indirectly from the Central Intelligence Agency, whose operatives seeded elites in various sectors of society with the drug (Lee & Schlain,

1992, pp. 3–7). Engelbart's lab, as it happened, was a few miles from the birthplace of the *Grateful Dead*, and a sound-and-light engineer for local rock concerts also played the same role at Engelbart's most important public presentation, a legendary event in December 1968 when he unveiled the computer mouse, hypertext, object addressing, and dynamic file linking, as well as shared-screen collaboration involving two persons at different sites communicating over a network with audio and video interface (Brand, 1995). The saga of Engelbart, acid, and the counterculture is documented in the superbly researched *What the Dormouse Said*. Writes author John Markoff, "Computing went from being dismissed as a tool of bureaucratic control to being the symbol of individual expression and liberation" (2005, pp. vii).

The curious role of mind-expanding pharmaceuticals in the rise of personal computing was carried on by Steve Jobs, cofounder of Apple and champion of the Macintosh, iPod, iPhone, and Mac Air families before his death in 2011. In his youth in the 1970s, Jobs took LSD some 15 times, and it was a "life-changing experience," he revealed on a 1988 security-clearance application he filled out for the Department of Defense. "It was a positive life changing experience for me, and I am glad I went through that experience," Jobs said of his acid trips. "I would ingest the LSD on a sugar cube or in a hard form of gelatin," he said on the application obtained by *Wired*. "I would usually take the LSD when I was by myself." Jobs later insisted, to his biographer Walter Isaacson, that his approach to digital design could only be understood by people who had also *tripped*. He even went as far as to insist that taking LSD was one of the two or three most important things he had ever done. He said there were "things about him that people who had not tried psychedelics—even people who knew him well, including his wife—could never understand" (Isaacson, 2011, p. xxx).

I met Engelbart and Jobs long after all three of us had taken LSD. Judging by their accounts, their experience with acid had been more satisfying than mine! So, to me, there was always a paradox in the possibility that computers were themselves a pathway to mind expansion because I saw a rivalry between digital and biopharmaceutical inventions. I also could not shake the sense that digital electronics perhaps offered a more effective, controllable pathway to cognitive enhancement than LSD. As Alan Turing and John Von Neumann, the twin geniuses who launched digital computing often noted, computers were *universal* machines, plastic in the sense that they were subject to an infinite variety of encoded instructions. In short, computers did what they were told; so there was hope that they could be programmed to enhance or improve the *machinery of the mind* in intended ways, thus avoiding what appeared to be the *unintended* consequences of ingesting drugs (Rheingold, 1985, pp. 67–98).

For Engelbart, the idea of mind expansion through computing arose long before the 1960s. He could recall distinctly, while stationed as an American soldier in the jungles of the Philippines in the summer of 1945, reading an article by the mathematician and computer designer Vannevar Bush in the July issue of the popular monthly magazine *The Atlantic*. In the article, "As We May Think," Bush envisioned a novel information-processing machine that sat on a desk and could be customized to the needs and tastes of its owner. Bush described the memex as a device, embedded in an ordinary desk, in which an individual stores all his books, records, and communications and which is controlled by a typewriter keyboard and depends on automated retrieval so that information can be retrieved quickly (Zachary, 1997, pp. 267–268).

In 1967, when Engelbart's research on personal computers and digital networks was bearing fruit, he wrote Bush about his successes (Zachary, 1997, p. 399). Bush, in turn, reflected on his concept of the memex, suggesting that digital technology offered a route to a kind of immortality, if not for everyone, then at least for scientists and scholars, artists, and writers—in short, for people who left a written or visual record of their ideas and actions. On death, the creative person would now bequeath to friends, family, and disciples the electronic record of his cognitive trails:

> The inheritance from the master will become, not only his additions to the world's record, but for his disciples the entire scaffolding by which they were erected. Each generation will receive from its predecessor, not a conglomerate mass of discrete facts and theories, but an interconnected web which covers all that the [human] race has thus far attained. (Zachary, 1997, p. 401)

Engelbart's fascination with Vannevar Bush helped ignite my own curiosity. When I set out to write a full-scale biography of Bush 25 years ago, I did so partly because of Engelbart's insistence that Bush would be a major figure in some future intellectual history of humanity that charted the rise of superconsciousness, a kind of individual and collective omniscience that Engelbart saw arising out of still barely imaginable computer networks. After one especially long session with Engelbart in his office, reviewing his ample memories and documentary archives, I came away fired with enthusiasm for putting Bush's intellectual achievements in the context of his unusually rich life as a scientific statesman and charismatic engineering leader.

Bush's notions about electromechanical techniques of mind expansion were rooted firmly in the field of library management and in an emerging new paradigm that came to be known as *information science*. Born in 1890, Bush designed and built the world's most powerful mechanical computers in the 1930s; the machine age deeply shaped his conceptions of the possible. Entranced by the potential of microfilm to store vast amounts of documents, during the 1940s Bush oversaw the design of a series of *rapid selector* devices that could quickly retrieve specific images from large amounts of microfilm. Having so much information, so well organized and available nearly immediately, Bush believed, would dramatically improve the quality of technoscientific analysis (Zachary, 1997, pp. 75–78).

Bush, who during World War II organized the Manhattan Project and managed all wartime science directed to military ends, believed that cognitive enhancements from personal-information technologies would chiefly help scholars, scientists, and professionals cope with "information overload." He did not expect ordinary people to benefit especially and never anticipated the explosion of entertainment activities delivered by digital networks. But he also foresaw that a consequence of empowering individual elites would be the establishment of societal-wide capacities for storing knowledge and making sense of increasingly complex data and information. Bush thus anticipated the communal nature of mind expansion so that the proposition that the *collective intelligence* of the species might also grow through sharing between strangers made sense to him. Through "an interconnected web," Bush argued, the human mind would become liberated from mental drudgery and, thus, civilization's advance would accelerate. The happy outcome depended, he reasoned, on "the task of selection," which today is known as "search," the basic tool for managing the "web," a term Bush may well have first used to refer to massively shared stored information (Zachary, 1997, pp. 261–276).

Bush's notion of the memex, which animated Engelbart and his team and indirectly infused Steve Jobs and Apple with a positive vision for cognitive enhancement, did not spring entirely from Bush's imagination. Bush borrowed more than a bit from H. G. (Herbert George) Wells, the English futurist and writer of speculative fiction. In the fall of 1937, Wells toured the United States, even lunching with President Franklin Roosevelt. His message: technology was emerging that would produce a device akin to "a cerebral cortex which (when it is fully developed) will constitute a memory and a perception of current reality for the entire human race" (Wells, 1994[1937], p. 123).

At a meeting in Paris, before his American trip in 1937, Wells called his outline for an information network, "the beginning of a world brain." He also predicted that "a replica" of the "brain of mankind" might exist in many places in the world, so that, if some were destroyed by war or catastrophe, the collective knowledge of humanity would survive. To skeptics, Wells countered,

> Please do not imagine that I am indulging in any fantasy when I talk of … a world brain. I am speaking of a process of mental organization throughout the world which I believe to be as inevitable as anything can be in human affairs. The world has to pull its mind together, and this is the beginning of its effort. The world is a Phoenix. It perishes in flames and even as it dies it is born again. (Wells, 1994[1937], pp. 123–124)

The imaginations of Bush and Wells were not technoscientific plans in themselves, but, by delivering appealing stories about the future, they nudged engineers and scientists on to a sense-making trajectory. As the historian of technology Rosalind Williams wrote in a recent issue of the journal *Technology and Culture*, "Both individuals and history need plot, characters and settings to make sense of who we are and where we are headed." By imagining alternative futures, thinkers and writers such as Bush, Wells, and other contemporaries "have been enormously influential in defining our technological age," Williams writes, "by giving it a story" (Williams, 2014, p. 473).

***

Hindsight is 20–20. In retrospect, the Internet seems inevitable; from the perspective of today, the Wellsian "world brain" is a no-brainer, the outcome of a brute-force incremental accretion of computing and communications power. The Internet had to be. The declining cost of storage and processing speed powered the enormous reach of what today is known, reverentially, as *the cloud*. Perhaps our minds cannot all be stored in the cloud, but what about our photographs? The digital dreck of our lives increasingly looks likely to live on forever in the cloud, and already some Twitter account holders find a brittle form of immortality after death when they—or whoever inherited their Twitter account—keeps *tweeting in their name*. Programs exist today that even continue tweeting after someone's gone by imitating the dead person's tweeting style (CNN, 2013). A new box to check in your will? Who inherits the responsibility for managing my tweets *from the grave*? I must confess that I decided to construct a purely online literature class, studded with a few dozen videos of me passionately lecturing, partly because my university's online educational arm assured me that my class could be offered *after my death*. So scholars now face the Oz-like possibility that the tenure system may die—and they may even die—but their teaching will live on forever (perhaps with the help of a "postdoc" hidden behind the screen).

While Bill Gates famously failed to mention the Internet in the original edition of his 1995 book *The Road Ahead,* he perspicaciously perceived that everyone *wired* would possess "information at your fingertips." For anyone old enough to remember the time and toil required to visit a library or any archives to inspect a single book or document, the instantaneous access of the vast trove of publications available with a keystroke positively relieve humans of what Bush once promised of his mythical memex: an end to "mental drudgery."

But relief from drudgery, however welcome, is not the same as clearing a path toward enhanced creativity or problem solving. The automation of the mind does not instantly deliver new and improved thoughts or even greater quantity of them. Indeed, the digital path to superconsciousness is contested, facing a chorus of criticisms, including the bold counterclaim that dependence on digital devices hampers and perhaps even cripples the human mind (by rewarding laziness).

These complaints are worth examining more closely, by considering the interesting positions of two prominent critics of digital life and culture: Alex Soojung-Kim Pang and Nicholas Carr. Let's start with Pang, who holds a doctorate in history and studied at University of Pennsylvania under the influential historian of technology Thomas Hughes. Pang identifies the characteristic problem of our era as *distraction*—caused by the proliferating, relentless flows of digital information, from all manner of sources, that rob us of our ability to stay on task, to stick to our priorities. The torrent of data, Pang argues, steals concentration. Distraction rules. We battle to pay attention while immersed in our (shared and private) digital worlds. We are the prisoners of our digital distractions (Pang, 2013).

The Internet and our digital devices—our laptops and smart phones especially—are often blamed for our attention deficits. Is there no remedy against endless interruptions save for getting off the grid entirely? Pang carefully constructs a remedy he calls "contemplative computing." The need for an alternative to the tyranny of the Web is clear to him: "Living with the Internet has created a reflexive need to be connected, even when we think we would benefit from being offline," Pang writes (2013, p. 53).

To kick our distraction addiction, Pang looks for tips and insights in the latest research on neuroscience and psychology, as well as in techniques of meditation and Eastern religions. He embraces a paradigm for improved thinking that depends on an elusive but intriguing concept of *flow*, popularized by the Hungarian psychologist Mihaly Csikszentmihalyi. A state of digital grace, which Pang defines as "calm," is the ultimate goal because, he writes, "my calm mind is able to do things that my ordinary mind never can" (2013, p. 87).

Achieving heightened consciousness has been sought for centuries by mystics and seers, poets and artists. Pang insists that, because of our immersion in the Web, all of us must find ways to raise our level of thought. Many software programs, he reports, can cleverly shield us from distractions. He finds so many worthwhile calming programs—many available for free—he even coins a clever term, *zenware*. Conceding "there's no perfect technological fix" to our distraction addiction, Pang embraces the very large irony that we digital addicts must rely for solutions on the very computing tools that shatter our attention in the first place (2013, pp. 72–74).

While he celebrates digitally mediated experience— and concedes that computers and the Web improve lives in manifold ways—Pang favors taking a regular "digital Sabbath." He is optimistic that waging a lonely battle against distraction will reap rewards even in the face of emerging technologies—driverless cars and Google Glasses, for instance—that promise to increase them, perhaps even dramatically. "We needn't accept the idea that a future in which computers ... think with and for us is inevitable," he writes. "Don't resign," he insists. "Redesign" (Pang, 2013, p. 138)

That is also Pang's advice to computer and software companies that, he argues, are often guilty of delivering overly complex and attention-destroying products. Convincingly, he argues that innovators ought to do much more "to figure out how computers can be designed to help people think deeply and not be distracted." Pang also insists that, inevitably, clear creative thinking engages our bodies as well our minds. Riffing on the creative benefits of walking, Pang cites the seminal strolls taken by the biologist Charles Darwin and the philosophers Nietzsche and Kierkegaard. He deftly adds the Latin dictum *solvitur ambulando*—it is solved by walking—to his antidistraction toolkit (Pang, 2013, p. 185).

To be sure, much of Pang's advice is common sense, even pedestrian, reflecting the old folk wisdom *everything in moderation.* More troubling is his presumption that the entire burden of kicking your distraction addiction falls on the victim. Pang might have explored not only individual response to the problem, but societal or collective action. Just as laws exist to limit the sale of alcohol or drugs and to regulate pharmaceuticals—all done in the name of helping the individual—might government assist in reducing the incidence of "distraction addition" by imposing responsibilities on the providers of digital devices and Web services? If cigarettes carry warning labels, might the Web?

Of course, techno-libertarians routinely conceptualize most problems of consciousness and mortality in individual terms—and argue that new digital technologies "liberate" human energies in novel and compelling ways, presenting as plausible the utopian dreams of humans who are trending toward omniscience and immortality. These dreams of intellectual and physical powers clash with the sensibilities of the times. The historian Russell Jacoby has persuasively argued that humans live in "an anti-utopian age," where "blueprint utopians (who) map out the future in inches and minutes" face severe resistance amid what Jacoby describes as "the demise of utopian speculation" (Jacoby, 2005, p. viii).

Yet optimistic views of incremental enhancements persist. In a paean to the prospects for technology to improve the human experience, the popular writer Steven Johnson, an important American exponent of technolibertarianism, insists that digital advances automatically confer advantages to individuals and society. Modern technologies societies, Johnson argues, have a bias toward progress. "The trend is positive," he writes, and if we do fail to recognize steady incremental progress, we fail because of a "blind spot" caused by dramatic advances that render smaller gains unnoticeable (Johnson, 2012, xxvi—xxix).

The popular equation of technological change with progress is highly contested, drawing mocking critiques from skeptics. Evgeny Morozov, writing in *The New York Times*, squarely attacks the techno-utopian vision by highlighting what he terms "the perils of perfection." Morozov puts the blame for the popular mania for cognitive enhancements on the innovation enclave in northern California:

> Silicon Valley wants to give you an app to remember everything. If you occasionally prevaricate in order to meet your clashing obligations as a parent, friend or colleague, another app might spot inconsistencies in your behavior and inform your interlocutors if you are telling the truth. If you experience discomfort because you encounter people and things that you do not like, another app or gadget might spare you the pain by rendering them invisible. Sunny, smooth, clean: with Silicon Valley at the helm, our life will become one long California highway. (Morozov, 2013, p. SR1)

The eminent cultural historian Leo Marx once asked in the pages of a leading technology magazine,

> Does improved technology mean progress? Yes, it certainly *could* mean just that. But only if we are willing and able to answer the next question: progress towards what? What do we want our new technologies to accomplish? What do we want beyond such immediate, limited goals such as achieving efficiencies, decreasing financial costs and eliminating the troubling human element from our workplaces? In the absence of answers to these questions, technological improvements may very well turn out to be incompatible with genuine, that is to say, *social* progress. (Marx, 1987, p. 41)

Techno-optimists have generally swept aside Marx's concerns, with some even going so far as to challenge the presumption that humans can even question whether emerging technologies are subject to human choices. Kevin Kelly (2010), another prominent promoter of digital technologies, insists that human species are best when they enthusiastically submit to "what technology wants."

Extreme, uncritical enthusiasm for technological change remains the signal response to the prospect of cognitive enhancement by the technoscientific elite, at least in the United States and perhaps across the world. A more radical challenge to the claim that digital devices and networks will enhance human cognition is the counterclaim that, actually, these technologies will diminish humans, will make them dumber. Diminishment occurs, the technology critic Nicholas Carr has argued, through a process of atrophy; as machines do more of our thinking and judging, he argues in *The Glass Cage: Automation and Us* (2014), we become capable of less and less.

Carr applies his perspective to virtually every domain of life where computers and information networks are playing an ever-increasing role in work. From airplane pilots to doctors and architects, Carr finds that digital systems, while aiding productivity and enhancing human decisions, carry hidden costs, notably because they promote the decline of human skills and a growing sense of dependency on machines even for decisions and memory. Carr worries that digital dependence "erases the artistry from our lives." In short, our minds atrophy the more we depend on cloud-based assistance to remember, to choose, to act skillfully. Despite immediate benefits from automation, he continues, "making things easy for people can backfire." They became mentally lazy (Carr, 2014, 85–91).

Carr, while eloquent, offers a familiar complaint against the unfamiliar innovation. When radio began broadcasting recorded music in the 1930s, musicians worried that no one would want to listen to them perform in person. They were wrong. "Live" music isn't dead. In the 1960s, TV was called a "wasteland" by a high-ranking American government official; he worried the new technology would rot the minds of America's youth (Minow, 1961). TV survived, and so did American creativity. Similarly gloomy predictions about the pernicious effects of novel technologies on human minds have also proved exaggerated or wrong.

The lessons from the past do not prevent Carr from fretting that, in the future, robots will care for the elderly, driverless cars will rob humans of employment, and even education could go online, wiping out the teaching professions. Joblessness will spread far and wide, even as human cognition improves iteratively with the aid of machines.

A dystopian future of supersmart humans idly stagnating is hardly preordained. Even with digital devices gobbling up work, however, there remains plenty for humans to do. Clean the environment. Cure Ebola. Grow food. Educate one another. Invent new tools. Amuse ourselves until death (sustainably).

Carr worries about both the quality and quantity of jobs. He offers anecdotal and some research evidence to suggest that the quality of jobs is declining as digital systems—computers and networks—take over more duties, giving the worker—even the professional worker—a growing sense of being a passenger in her own life. He complains that navigational systems—and an emerging tool for "indoor mapping"—steal from people their laudable "way finding" capacities (Carr, 2014, p. 137). His *uber* case of computer degradation involves airline pilots whose jobs have greatly changed because of massive computerization of flight navigation, takeoff, and landing. Carr cites pilots complaining about their skills atrophying because the plane now does so much of the real work of flight. And he cites crashes where digital systems failed and pilots were forced to shift suddenly from managing instruments to flying the plane—and could not.

The trouble with Carr's concern is that, while flying may be less satisfying for pilots, the experience—heavily dependent on computer systems—is safer for paying customers. The same may be true about driverless cars. Driving, after all, is a leading killer, in the United States and globally. Just because driving tests the mind and body in ways Carr admires, is that any reason to sacrifice so many people to premature deaths?

Carr is on stronger ground when he argues that humans manage their digital lives best when striking a balance between self-reliance and digital assistance. He cites troubling evidence that electronic medical records interfere with the judgment of doctors. By presenting doctors with a menu of possible diagnoses, medical software may actually raise the cost of care by increasing the number and variety of unnecessary tests and medications. Rather than uncritically embrace "technology-centered automation," Carr favors what he calls "human-centered automation"—designs that put limits on digital devices in ways that promote human health and vigorous mind-body processes (Carr, 2014, pp. 160–164).

In his well-meaning exhortation against overreliance on computers, Carr declares that balancing the human and the machine "isn't hard." Yet he also concedes, "as society becomes ever more computerized, the technologist becomes its unacknowledged legislator" (2014, p. 161). Ultimately, Carr finds that the protean power of computers renders the idea of maintaining balance—the Goldilocks principle of neither too hot nor too cold—elusive, if not impossible. Unless we mount new forms of resistance against automating the mind, we are destined to lose what is most human about us.

Might there be parallels with the English Luddites, who battled textile machinery nearly 200 years ago? Historical accounts would compel a fair observer to say "no" (Hobsbawm, 1952). But the contest over who controls the production of consciousness is now so worrisome that even Jaron Lanier, once an uncritical troubadour of mind-machine merger through electronics, now questions the project because of the costs for individuals and society of the obsession with human enhancement through digital electronics. In his pithy manifesto *You Are Not a Gadget,* Lanier sides with digital heretics, declaring "something started to go wrong with the digital revolution around the turn of the 21st century" (2010, 3).

Or maybe striking the right balance is a matter of time and requires only further incremental "improvements" in digital performance brought about by ever-expanding processing power. Surveying the field of computer intelligence, journalist Jayson Greene, writing in the *New York Times*, warns that binary understandings of machines "versus" humans are losing their relevance: "Machines and humans work better with their heads together than apart. Jobs where workers and computers collaborate yield more effective results than when either goes it alone." The ideal of humans maintaining a boss–servant relationship with computer networks—so attractive, in principle—might be another illusion, however. Greene writes:

> Skeptics point to experiments like this as evidence that human interpretation still matters and that our greatest gift—creativity—remains behind a sealed-off door that machines can never penetrate. But the line between creativity and statistical analysis blurs the harder you look at it, and machines are looking hard: They will cross it eventually. Creativity is not some exalted milk that we alone drink; it is a chain of small, insignificant leaps leading to one small, significant one. Years of data processing in our memories smoothes into the murmur we call intuition, a pleasing hum that is easy to mistake for divine inspiration. (2014, MM56)

***

Whether the preferred mode is mind expansion through drugs or cognitive enhancement through computer networks, the expanders and the enhancers work within a capitalist framework. Enhancements and expansions are products and services to be bought and sold, advertised and promoted, priced, discounted, and auctioned. Human enhancements are pathways to profit. Mind and mental capacity, in short, are commodities, subject to the logic and "market failures" of the capitalist system.

In the terrain of political economy, perhaps no greater spectacle involves the race between two giant industries, two wealthy enclaves, two peddlers of tomorrow: these are the digital electronics industry and the biopharmaceutical industry. We have seen that the path toward superconsciousness is complex and contested, costly and competitive. What about the pharmaceutical path to (cognitive) enhancement?

Pharmaceuticals are much more tightly controlled and regulated than digital electronics—and, thus, slower to reach markets, legally, and much more likely to be traded in rogue, or outlaw, markets. Already cosmetic enhancements are proffered outside the law, and consumers routinely travel to nations where services and products unavailable in the United States are offered legally for sale. The experimental development and human adoption of biopharma enhancements also differ from digital enhancements because of definitional issues. Some pharmaceuticals can be considered foods and naturally occurring and, thus, not subject to the kinds of regulatory control that, for instance, blood-pressure drugs come under. Moreover, some specific foods—and diets, more broadly—influence the workings of the mind.

Perhaps the most famous literary example of the *natural* paths to enhanced consciousness is the arresting essay by Aldous Huxley *The Doors of Perception*. Huxley recounts his experience of mind-expanding hallucinations, brought on by eating a plant revered by Native Americans. In our own times, we routinely imbibe caffeinated drinks, which effect cognition. The most popular form of doing so is the practice of drinking coffee, which makes Starbucks a kind of unregulated cognitive-enhancement delivery system.

Ten years ago, a few months before my 50th birthday, I mounted my own Huxley-lite experiment, not with peyote or mescaline but with caffeine, delivered through a commonly available beverage that I had, nevertheless, shunned for decades. Yet now solidly in the middle years of my existence, I worried about declining cognitive quality (and quantity). Had I lost the zip in my thoughts the way an aging pitcher loses the top of his fastball? Had I lost some of the *élan vital* that the philosopher Henri Bergson had considered central to creative intuition and to the capacity of humans to transform mundane experience into insight? Was the sheer persistence of my existence rubbing the rough edges of my consciousness—the creative sources of discontent and discord that I credited with nourishing my individual cast of mind? Could I somehow reverse the softening of the outer shells of my consciousness through a seemingly trivial change in the beverage I drank in the morning? Could coffee, liberally imbibed, save my smarts? Might a daily dose of "Joe" make me more alert and cognitively muscular at least in the mornings, when I tended to write the most and (possibly) think the hardest and most creatively?

About this time, my inner debate over how safely to mimic Huxley on mind expansion was interrupted by a request from an editor I knew well. The editor approached me with a simple proposition: research the cognitive benefits of drinking caffeinated coffee, in large quantities—and then report my findings. Because the editor knew I was a faithful decaf drinker, the article might carry an added twist: the story of my personal conversion to daily infusions of caffeine.

I began skeptically (though I began earnestly because, frankly, the editor offered to pay me for the literary experiment). Indeed, I needed an incentive. For decades, caffeinated coffee had, after all, engendered in me a toxic combination of hyperactivity and mania that turned me into an obnoxious social butterfly. For whatever reason, I was highly anxious in my youth, so that what I considered to be a state of calm was viewed by others as pure panic. Experimenting with coffee in my early 20s, I had experienced numerous bad outcomes—not as bad as taking acid but distinctly suboptimal by any measure. Even when I managed to cope with the effects of coffee, say, by locking myself in my office, if I drank coffee in the afternoon, I would never fall asleep before midnight. If I took a brew in the late afternoon, I could reliably predict a sleepless night.

Given my history, the prospect of drinking regular coffee terrified me. Yet, in the months before turning 50, the prospect of turning back my cognitive clock, even for a few hours each day, compelled me to experiment with drinking caffeinated coffee as if my sanity depended on the practice. Before long, my assignment became more than a chance to earn another writing payday but a potential game changer. I could almost taste, when I sipped the morning brew, my thoughts improve. And along with drinking plentifully, as I worked through my coffee encounter, I also interviewed medical experts.

After I started drinking, I stopped at the medical school of the University of California in San Francisco. I visited the university's library and assembled a stack of research journals and books. Taken together, the literature strongly suggested that caffeinated coffee is benign and enhances certain valued forms of cognition (Nehlig, 2010). Coffee—the strong stuff—drunk in moderation—a few cups a day—sharpened the mind, improved memory, mood, and sent jolts of needed energy through the body. Or so I understood from my reading.

To reinforce my understanding, I visited Neal Benowitz, professor of medicine, psychiatry, and biopharmaceutical sciences who had studied the effects of caffeine beverages for years. "If people are drinking modestly, up to three cups a day, there's very little evidence of any adverse effects," he told me. Dr. Benowitz went further, dismissing most claims about the adverse effects coffee drinking as simply wrong. Perhaps only pregnant women ought to fret over excessive coffee drinking, and even they can drink in moderation, he said. The rest of us can enjoy coffee drinking as a "a good social ritual, a pleasurable experience." Dr. Benowitz also advocates that all caffeinated beverages disclose the exact amount of caffeine delivered by their beverage. "Give people the information," he said. "Let them choose just how much caffeine they want."

Well, I've decided. I want a lot of coffee, daily. In the years since I published my "conversion to caffeine" article in 2005, I have consumed an awful lot, though mostly before noon. I usually drink dark roast for the flavor and persist even though I know that lighter roasts contain higher levels of caffeine. I just drink more to compensate, sometimes even a quart a day. Once a month, at least, I skip coffee for a day; usually by the next morning I have a minor headache. A few years ago, I stopped drinking coffee for two full weeks during a bout of dysentery. When my stomach cleared, I resumed drinking with a vengeance, never stopping to evaluate my weeks off. I know I'm addicted, and I don't care, partly because "withdrawals" are uncomfortable but brief. I also accept that over time I need more coffee to achieve the same effect of sharper thoughts, brighter moods, more motivation to complete challenging cognitive tasks (Mayo Clinic, 2014, April 14). I accept what Roland Griffith, a professor at the Johns Hopkins University in Baltimore, told me a decade ago: a coffee "habit" can be shaken after a week of headaches and irritability, and the more coffee I drink, the harder my habit is to kick. Who cares? This is one addiction I'm grateful for (Zachary, 2005).

I have experienced a side effect that I cannot ignore, however. In recent years, my blood pressure has drifted higher, almost surely in part because of my high coffee consumption. I now take a daily dose of Lisinopril to counteract my elevated blood pressure. I consider the trade-off reasonable. I'm way gone on coffee these days. I have even made pilgrimages—or shall we call them *learning journeys*—to coffee farmers in East Africa, where I conduct research on other matters. On my most recent visit in the summer of 2014, I delighted in marching through the foothills of Mount Elgon, on the border of Kenya and Uganda, inspecting the year's crop. The berries were ripening, and my heart quickened at the thought that my favored cognitive enhancement came directly from Mother Earth.

I dream of growing coffee myself someday or at least visiting coffee farms again on my next trip to East Africa.

***

If cognitive enhancement can come through a drink, why not a pill? That premise animates the novel *The Dark Fields,* a novel by the Irish writer Alan Glynn. Transformed into the Hollywood movie *Limitless*, the story presents a struggling writer who receives a large supply of an underground drug that delivers wondrous benefits—vastly improved creativity, speed, and stamina. At $500 a pill, the drug is expensive, but the promise of no physical side effects prompts the writer to pop one and then another and then another. Before long, he is writing full-length manuscripts in a weekend and mastering demanding cognitive tasks—learning new languages and specialized subject knowledge—in a fraction of the time expected.

Unintended consequences arise for the writer. His dealer is murdered. He gets greedy. Literature isn't enough of a payoff. He tries stock trading and scores big with the help of the pills. A Russian mobster, who supplies a large loan to get the writer started in stock investing, realizes the cause of the writer's high trading returns and demands a share of his cache of pills. The situation spirals out of control when the writer experiences side effects; the worst is a loss of consciousness (spoiler alert: he *forgets* that he killed someone in a hotel room!). The whole thing makes you pine for a more effective and imaginative Food and Drug Administration (Glynn, 2002).

The fictional scenario provides a valuable heuristic. Already some students and mature adults take legal drugs, such as Adderall, "off label," meaning that they do not actually have the symptoms of "attention deficit disorder" (ADD) that are used to justify a legal prescription. Other stimulants provide similar benefits, such as improved working memory, attention tasks, motivation, and arousal. The temptation to sanction wider use grows larger by the year. Doctors and medical professionals increasingly define human health not in terms of treating or curing diseases but as a pathway toward enhancement. In short, *not* enhancing oneself may be an *unacceptable risk*. The concept of health as risk management has been well explored by scholars since the landmark book *Risk and Culture*, published in 1982. In this work, Mary Douglas and Aaron Wildavsky persuasively argue that the concept of risk is socially constructed, opening the door to the now-accepted belief that a person may incur adverse risks by not embracing well-established and safe enhancements: "Risk should be seen as a joint product of *knowledge* about the future and *consent* about the most desired prospects," they wrote (5).

Charting the shift in 2012, Joseph Dumit observed,

> The very concept of health has changed in ways that are both utterly familiar to us and unbelievably alien. In this new health [concept] we are dedicated to measuring our health by understanding our risks and taking practical steps to reduce them. This just makes sense to us. At the same time ... this approach to health has the actual consequence of causing us to spend ever-increasing amounts of attention, energy and money on health, vastly and continually increasing our pharmaceutical consumption. (2012, p. 197)

Humans will hardly stop at individual solutions to the challenge of preventing cognitive decline. Enhancement as an explicit goal will become a *normal* part of health care, if it isn't already, and that raises vexing concerns. "Be careful what you wish for" is a common reply to those who, like Icarus, wish to fly higher, run faster, and work smarter in defiance of human traditions. Younger people especially now face the specter of a kind of cognitive arms race, on an individual level, in which they may become positively disadvantaged if they do not participate in everyday consumption of potentially enhancing pharmaceuticals. Older folks, meanwhile, express naked yearning for pills that might improve their memory, even if their memory "loss" is not caused by any particular underlying pathology. If coffee works to improve cognitive performance, would not a memory-and-thought-enhancing pill—a kind of *Viagra* for the brain—be even more welcome?

None of these interventions may help us understand the nature of consciousness or how to *perfect* consciousness. Improving cognition, especially for aging humans who want to retain their cognitive edge, is an everyday concern. Humans seek an edge, whether for their own survival or for a temporary advantage that can deliver them status or material improvements or both. Philosophical doubts about the nature of consciousness, and about the presumed impossibility of reducing consciousness to material processes, can easily coexist with tactical movements of self-improvement that fall plainly within pragmatic traditions of making life better on the ground.

For striving nations, competing in a global knowledge community, the option of socially engineering enhanced consciousness is a tempting option. If cognitive enhancers work for individuals, why not mandate them across generations to sustain higher educational achievement societally? Might a government committed to top-down, technocratic management of their country embrace the cognitive-enhancement path as surely as it today embraces routine public-health measures such as vaccinations? Might these governments promote the use of "thinking pills" in the same way they take steps to enhance air quality, auto safety, and basic nutrition?

To be sure, the race between digital enhancement to the human mind and cognitive enhancements through biopharma interventions are not an either-or proposition. Both approaches will persist. Economically, culturally, and politically, there is so much momentum behind these sibling approaches to the transformation of the human that the question is not whether one or both of these approaches will grow in influence. The question is how to conceptualize and someday manage, or merely cope with, the twin paths to enhanced consciousness.

References

Brand, S. (1995, March 1). We owe it all to the hippies. *Time, 145* (12). Available at: http://content.time.com/time/magazine/article/0,9171,982602,00.html.

Bush, V. (1945, July). As we may think. *Atlantic Monthly*. Available at www.theatlantic.com.

Carr, N. (2014). *The glass cage: Automation and us*. New York, NY: Norton.

CNN. (2013, February 22). How to post to Facebook, Twitter after you die. Available at www.cnn.com.

Douglas, M., & Wildavsky, A. (1982). *Risk and culture*. Berkeley, CA: University of California Press.

Dumit, J. (2012). *Drugs for life: How pharmaceutical companies define our health*. Durham, NC: Duke University Press.

Glynn, A. (2002). *The dark fields*. New York, NY: Bloomsbury.

Greene, J. (2014, September 19). Why is our sci-fi so glum about A.I. *New York Times*, Sunday Magazine, p. MM56.

Hobsbawm, E. J. (1952). The machine breakers. *Past and Present, 1* (1), 57–70.

Huxley, A. (1954). *The doors of perception and heaven and hell*. New York, NY: Harper & Brothers.

Jacoby, R. (2000). *The last intellectuals: American culture in the age of academe*. (Rev. ed.). New York, NY: Basic Books.

James, W. (1890). *Principles of psychology*. (3rd ed.). London, England: Williams and Norgate.

Johnson, S. (2012). *Future perfect: The case for progress in a networked age.* New York, NY: Riverhead.
Isaacson, W. (2011). *Steve Jobs.* New York, NY: Simon & Schuster.
Kelly, K. (2010). *What technology wants.* New York, NY: Viking.
Lanier, J. (2010). *You are not a gadget: A manifesto.* New York, NY: Knopf.
Lee, M. A. & Schlain, B. (1992). *Acid dreams: The complete social history of LSD: The CIA, the sixties and beyond.* New York, NY: Grove Press.
Lovejoy, A. O. (1936). *The great chain of being: A study of the history of an idea.* Cambridge, MA: Harvard University Press.
Markoff, J. (2005). *What the dormouse said: How the sixties counter-culture shaped the personal computer industry.* New York, NY: Viking Penguin.
Marx, L. (1987). Does improved technology mean progress? *Technology Review,* 33–41.
Mayo Clinic. (2014, April 14). Caffeine: How much is too much? Healthy Lifestyle, Nutrition and Healthy Eating. Available at http://www. mayoclinic.org.
Minow, N. (1961). Speech at the National Association of Broadcasters Convention, transcript. Available at http://www.janda.org.
Morozov, E. (2013, March 2). The perils of perfection. *New York Times,* p. SR1.
Nehlig, A. (2010). Is caffeine and cognitive enhancer? *Journal of Alzheimer's Disease, 20,* 85–94.
Pang, A. S.-K. (2013). *The distraction addiction.* New York, NY: Little Brown.
Rheingold, H. (1985). *Tools for thought: The people and ideas behind the next computer revolution.* New York, NY: Simon & Schuster.
Roszak, T. (1969). *The making of a counter culture.* New York, NY: Doubleday.
van Dam, R. (2014). Ask the expert: Coffee and health. Harvard School of Public Health, The Nutrition Source. Available at http://www.hsph.harvard.edu/nutritionsource/coffee/.
Venter, C. (2013). *Life at the speed of light.* New York, NY: Penguin.
Wells, H. G. (1994[1937]). A note on world brain and world mind. In *World brain: H. G. Wells on the future of world education* (pp. 123–124). London, England: Adamantine Press.
Williams, R. (2014). Our technological age, from the inside out. Technology and Culture, *55* (2), 461–476.
Zachary, G. P. (1990, January 23). Artificial reality: Computer simulations one day may provide surreal experiences. *The Wall Street Journal,* pp. A1, A9.
Zachary, G. P. (1997). *Endless frontier: Vannevar Bush, engineer of the American century.* New York, NY: Free Press.
Zachary, G. P. (2005, May 15). In Starbucks country, caffeine hooks us and constantly lures us back to feed our addiction. *San Francisco Chronicle.* Available at www.sfgate.com.

Bibliography

Ackerman, S. (2012, October 3). Darpa offers a million dollars for crowdsourced amphibious tank.*Wired.co.UK*. Available from: http://www.wired.co.uk.

Agar, N. (2004). *Liberal eugenics: In defense of human enhancement*. Oxford, England: Blackwell.

Ahvenharju, S., Halonen, M., Uusitalo, S., Launis, V., & Hjelt, M. (2006). *Comparative analysis of opinions produced by National Ethics Councils*. Helsinki, Finland: Gaia Group Ltd.

American Foundation for the Blind. (2004). Ray Kurzweil: An oral history interview. Available from: http://www.afb.org.

American scientist who created artificial life denies "playing God." (2010, May 21). *The Telegraph* (UK), News.

Anders, G. (1956). *Die Antiquiertheit des Menschen*. München, Germany: Beck.

Anderson, B. (1983). *Imagined communities: Reflections on the origin and spread of nationalism*. London, England: Verso. Rev. ed. 1991.

Annas, J. (1993). *The morality of happiness*. Oxford, England: Oxford University Press.

Appadurai, A. (1990). Disjuncture and difference in the global cultural economy. *Theory, Culture & Society, 7* (2), 295–310.

Appadurai, A. (1996). *Modernity at large: Cultural dimensions of globalization*. Minneapolis, MN: University of Minnesota Press.

Asad, T. (1992). Religion and politics: An introduction. *Social Research, 59* (1), 3–16.

Asad, T. (1993) *Genealogies of religion: Discipline and reasons of power in Christianity and Islam*. Baltimore, MD: The Johns Hopkins University Press.

Asad, T. (2003). *Formations of the secular: Christianity, Islam, modernity*. Stanford, CA: Stanford University Press.

Armitage, J., & Roberts, J. (2002). Chronotopia. In J. Armitage and J. Roberts (Eds.), *Living with cyberspace: Technology and society in the 21st century* (pp. 43–56). New York, NY: Continuum.

Austen, I. (2013, July 7). Deadly derailment in Quebec underlines oil debate. *New York Times*, p. A4.

Baab, F. (2013). *Was ist Humanismus? Geschichte des Begriffes, Gegenkonzepte, säkulare Humanismen heute*. Regensburg, Germany: Pustet.

Badmington, N. (Ed.). (2000). *Posthumanism*. New York, NY: Routledge.

Badmington, N. (2004a). *Alien chic: Posthumanism and the other within*. New York, NY: Routledge.

Badmington, N.(2004b). Post, oblique, human. *Theology & Sexuality, 10* (2), 56–64.

Barben, D., Fisher, E., Selin, C., & Guston, D. (2008). Anticipatory governance of nanotechnology, foresight, engagement, and integration. In J. Hackett, O. Amsterdamska, M. E. Lynch, & J. Wajcman (Eds.), *The handbook of science and technology studies* (pp. 979–1000). Cambridge, MA: MIT Press.

Barrat, J. (2013). *Our final invention: Artificial intelligence and the end of the human era*. New York, NY: St. Martin's Press.

Barrow, J. D., & Tipler, F. J. (1998). *The anthropic cosmological principle*. New York, NY: Oxford University Press.

Barry, A. (2001). *Political machines: Governing a technological society*. London, England: Athlone Press.

Barry, A., & Walters, W. (2003). From EURATOM to "complex systems": Technology and European government. *Alternatives: Global, Local, Political, 28* (3), 305–329.

Barth, U. (2003). Was ist Religion? In U. Barth, *Religion in der Moderne* (pp. 3–27). Tübingen, Germany: Mohr Siebeck.

Barth, U. (2004). *Aufgeklärter Protestantismus*. Tübingen, Germany: Mohr Siebeck.

Barton, R. (1987). John Tyndall, pantheist: A rereading of the Belfast address. *Osiris*, 3, 2nd ser., 111–134.

Bauman, Z. (1993). *Modernity and ambivalence*. Oxford, England: Polity Press.

Beck, U. (1992). *Risk society: Towards a new modernity* (1st ed.). Newbury Park, CA: Sage Publications, Ltd.

Bechmann, G. (2007). Die Beschreibung der Zukunft als Chance oder Risiko? *Technikfolgenabschätzung—Theorie und Praxis, 16* (1), 24–31.

Beckford, J. A. (2012). Public religions and the postsecular: Critical reflections. *Journal for the Scientific Study of Religion, 51* (1), 1–19.

Behler, E. (1996). Nietzsche in the twentieth century. In Bernd Magnus & Kathleen M. Higgins (Eds.), *The Cambridge companion to Nietzsche* (pp. 281–322). Cambridge, England: Cambridge University Press.

Beiser, F. C. (2002). *German idealism: The struggle against subjectivism, 1781-1801*. Cambridge, MA: Harvard University Press.

Bell, D. (2000). Cybercultures rewriter. In D. Bell & B. M. Kennedy (Eds.), *The cybercultures reader* (2nd ed., pp. 1–9). London, England: Routledge.

Bell, D., & Kennedy, B. M. (Eds.). (2000). *The cybercultures reader* (2nd ed.). London, England: Routledge.

Bengtsson, B., & Klintmann, M. (2010). Stakeholder participation in the EU governance of GMO in the food chain. In K. Bäckstrand, J. Kahn, A, Kronsell, & E. Lövbrand (Eds.), *Environmental Politics and Deliberative Democracy* (pp. 105–122). Cheltenham, England: Edward Elgar Publishing. Available from Elgaronline.

Bennett, G. (2015). *Technicians of human dignity*. New York, NY: Fordham University Press.

Bennett, I., & Sarewitz, D. (2006). Too little, too late? Research policies on the societal implications of nanotechnology in the United States. *Science as Culture, 15* (4), 309–325.

Bensaude-Vincent, B. (2000). *L'opinion publique et la science: à chacun son ignorance*. Bruxelles, Belgium: Synthelabo.

Bensaude-Vincent, B. (2001). A genealogy of the increasing gap between science and the public. *Public Understanding of Science, 10* (1), 99–113.

Bensaude-Vincent, B. (2004). *Se libérer de la matière? Fantasmes autour des nouvelles technologies*. Paris, France: Institut national de la recherche agronomique.
Bensaude-Vincent, B., & Guichet, X. (2007): Nanomachine: One word for three different paradigms. *Techné, 11* (1), 71–89.
Berger, P. (1999). The desecularization of the world: A global overview. In P. Berger (Ed.), *The desecularization of the world: Resurgent religion and world politics* (pp. 1–18). Grand Rapids, MN: Wm. B. Eerdmans.
Berlin, I., & Hardy, H. (2013). *The roots of romanticism*. Princeton, NJ: Princeton University Press.
Bernard, A. (2014). *Kinder Machen: Neue Reproduktionstechnologien und die Ordnung der Familie. Samenspender, Leihmütter, Künstliche Befruchtung*. Frankfurt am Main, Germany: S. Fischer.
Biegelbauer, P., & Hansen, J. (2011). Democratic theory and citizen participation: Democracy models in the evaluation of public participation in science and technology. *Science and Public Policy, 38* (8), 589–597.
Bijker, W., Bal, R., & Hendriks, R. (2009). *The paradox of scientific authority: The role of scientific advice in democracies*. Cambridge, MA: MIT Press.
Bijker, W. E., Hughes, T. P., & Pinch, T. (1987) *The social construction of technological systems: New directions in the sociology and history of technology*. Cambridge, MA: The MIT Press.
Bimber, B. (1994). Three faces of technological determinism. In M. R. Smith & L. Marx (Eds.), *Does technology drive history? The dilemma of technological determinism* (pp. 79–100). Cambridge, MA: MIT Press.
Bimber, B. A. (1996). *The politics of expertise in Congress: The rise and fall of the Office of Technology Assessment*. Albany, NY: State University of New York Press.
Birnbacher, D. (2006). *Natürlichkeit*. Berlin, Germany: de Gruyter.
Blaffer Hrdy, S. (2000). *Mutter Natur: Die weibliche Seite der Evolution*. Berlin, Germany: Berlin Verlag.
Bloch, E. (1973 [1959]). *Das Prinzip Hoffnung*. Frankfurt, Germany: Suhrkamp. In English: *Principle of hope*. (N. Plaice, S Plaice, & P. Knight, Trans.). Cambridge, MA: MIT Press, 1986.
Bloch, E. (2000 [1923]). *The spirit of utopia*. Stanford, CA: Stanford University Press.
Blondiaux, L. (1998). *La fabrique de l'opinion: Une histoire sociale des sondages*. Paris, France: Seuil.
Blumenberg, H. (1988). *Die Legitimität der Neuzeit*. Frankfurt am Main, Germany, Suhrkamp.
Boltanski, L., & Thévenot, L. (2006). *On justification: Economies of worth*. Princeton, NJ: Princeton University Press.
Bonazzi, M. (2010). *Communicating nanotechnology: Why, to whom, saying what and how*. Brussels, Belgium: European Commission DG Research.
Borgmann, A. (1984). *Technology and the character of everyday life*. Chicago, IL: University of Chicago Press.
Borgmann, A. (2003). *Power failure: Christianity in the culture of technology*. Grand Rapids, MI: Brazos.
Bostrom, N. (2001). Transhumanist values. Version of April 18, 2001. Available from: http://www.nickbostrom.com/tra/values.html.
Bostrom, N. (2003a). *The transhumanist FAQ: A general introduction, version 2.1*, s.l. Available from: http://transhumanism.org/index.php/WTA/faq/.

Bostrom, N. (2003b). *The transhumanist FAQ: A general introduction, version 3.0.* Available from: http://humanityplus.org/philosophy/transhumanist-faq/.
Bostrom, N. (2005). A history of transhumanist thought. *Journal of Evolution and Technology, 14* (1), 1–25.
Bostrom, N. (2009). Why I want to be a posthuman when I grow up. In B. Gordijn and R. Chadwick (Eds.), *Medical enhancement and posthumanity* (pp. 107–136). New York, NY: Springer.
Bostrom, N., & Sandberg, A. (2009). Cognitive enhancement: Methods, ethics, regulatory challenges. *Science and Engineering Ethics, 15* (3), 311–341.
Bourdieu, P. (1990). *The logic of practice.* (R. Nice, Trans.). Palo Alto, CA: Stanford University Press.
Bowman, D. M., & Hodge, G. A. (2007). Nanotechnology and public interest dialogue: Some international observations. *Bulletin of Science, Technology & Society, 27* (April), 118–132.
Boy, D., Kamel, D. D., & Roqueplo, P. (2000). Un exemple de démocratie participative: La "conférence de citoyens" sur les organismes génétiquement modifies. *Revue française de science politique, 50* (4–5), 779–810.
Bracke, S. (2008). Conjugating the modern/religious, conceptualizing female religious agency: Contours of a "post-secular" conjuncture. *Theory, Culture and Society, 6* (2), 51–67.
Braidotti, R. (2008). In spite of the times: The postsecular turn in feminism. *Theory, Cuture and Society, 25* (6), 1–24.
Braidotti, R. (2013). *The posthuman.* Cambridge, England: Polity.
Brand, S. (1995, March 1). We owe it all to the hippies. *Time, 145* (12). Available from: http://content.time.com.
Braun, B. (2004). Querying posthumanisms. *Geoforum, 35* (3), 269–273.
Breul, W. (2012). „Hoffnung besserer Zeiten." Der Wandel der „Endzeit" im lutherischen Pietismus um 1700. In A. Landwehr (Ed.), *Frühe neue Zeiten. Zeitwissen zwischen Reformation und Revolution* (pp. 261–282).Bielefeld, Germany: Transcript.
Bringsjord, S., Bringsjord, A., & Bello, P. (2013). Belief in the singularity is fideistic. In A. H. Eden, J. H. Moor, J. H. Sorakher, & E. Steinhart (Eds.), *Singularity hypotheses: A scientific and philosophical assessment* (pp. 395–409). Heidelberg, Germany: Springer.
Brown, N., & Michael, M. (2003). A sociology of expectations: Retrospecting prospects and prospecting retrospects. *Technology Analysis and Strategic Managment, 15* (1), 3–18.
Brune, H., Ernst, H., Grunwald, A, Grünwald, W., Hofmann, H., Krug, H. , Janich, P., Mayor, M., Rathgeber, W., Schmid, G., Simon, U., Vogel, V., & Wrywa, D. (2006). *Nanotechnology — Perspectives and Assessment.* Berlin, Germany: Springer.
Buchanan, A. (2011). *Beyond humanity: The ethics of biomedical enhancement.* Oxford, England: Oxford University Press.
Buchanan, A., Brock, D. W., & Daniels, N. (2000). *From chance to choice: Genetics and justice.* Cambridge, England: Cambridge University Press.
Bukatman, S. (1993). *Terminal identity: The virtual subject in post-modern science fiction.* Durham, NC: Duke University Press.
Bülow, M. H. (2012). Das gute alte Hirn. Wie die Sorgen um eine alternde Gesellschaft und die Ideen zum kognitiven Enhancement in den Neurowissenschaften interagieren. In M. Eilers, K. Grüber, & C. Rehmann-Sutter (Eds.), *Verbesserte Körper—gutes Leben? Bioethik, Enhancement und die Disability Studies* (pp. 269–285). Frankfurt am Main, Germany: Peter Lang.

Burk, D. L., & Lemley, M. A. (2002). Is patent law technology-specific? *Berkeley Technology Law Journal, 17* (4), 1155–1206.

Bush, V. (1945a). *Science: The endless frontier.* Washington, DC: Government Printing Office.

Bush, V. (1945b, July). As we may think. *Atlantic Monthly.* Available from: www.theatlantic.com.

Butler, J. (2008). Sexual politics, torture, and secular time. *British Journal of Sociology, 59* (1), 1–23.

Cady, L. E. (2012). Religion and the technowonderland of transhumanism. In H. Tirosh-Samuelson & K. L. Mossman (Eds.), *Building better humans? Refocusing the debate on transhumanism* (pp. 83–104). Frankfurt am Main, Germany: Peter Lang.

Calhoun, C. (2010). Rethinking secularism. *Hedgehog Review, 12* (3), 34–48.

Calhoun, C. (2012). Time, world, and secularism. In P.S. Gorki, D.K. Kim, J. Torpey, & J. Van Antwerpen (Eds.), *The Post-secular in question: Religion in contemporary society* (pp. 335–364). New York, NY: New York University Press.

Callon, M., Lascoume, P., & Barthe, Y. (2009). *Acting in an uncertain world: An essay on technical democracy.* Cambridge, MA: MIT Press.

Caplan, R. (2012). *Selfish reasons to have more kids: Why being a great parent is less work and more fun than you think.* New York, NY: Basic Books.

Carlson, N. (2014, December 17). What happened when Marissa Mayer tried to be Steve Jobs. *New York Times*, Sunday Magazine, p. MM22.

Carr, N. (2014). *The glass cage: Automation and us.* New York, NY: Norton.

Casanova, J. (1994). *Religions in the modern world.* Chicago, IL: University of Chicago Press.

Cassirer, E. (1930). Form und Technik. In E. Cassirer, *Symbol, Technik, Sprache* (pp. 39–92). Hamburg, Germany: Meiner, 1985.

Cassirer, E. (1996). *Versuch über den Menschen. Einführung in eine Philosophie der Kultur.* Hamburg, Germany: Felix Meiner.

Cavalieri, P., & Singer, P. (1994). *Menschenrechte für die Großen Menschenaffen: Das Great Ape Projekt.* (H.-J. Koskull, Trans.). Munich, Germany: Wilhelm Goldmann. In English: *The great ape project: Equality beyond humanity.* London, England: Fourth Estate, 1993.

Chalmers, D. J. (1996). *The conscious mind: In search of a fundamental theory.* London, England: Oxford University Press.

Chalmers, D. J. (2010). *The character of consciousness.* Oxford, England: Oxford University Press.

Chasiotis, A. (1999). *Kindheit und Lebenslauf: Untersuchungen zur evolutionären Psychologie der Lebensspanne.* Bern, Switzerland: Hans Huber.

Cheng, Y. (2008). *Creating the "new man": From Enlightenment ideals to socialist realities.* Honolulu, HI: University of Hawaii Press.

Christensen, C. M. (2011). *The innovator's dilemma: The revolutionary book that will change the way you do business.* New York, NY: HarperBusiness.

Christidou, V., Dimopoulos, K., & Kouladis, V. (2004). Constructing social representations of science and technology: The role of metaphors in the press and popular scientific disciplines. *Public Understanding of Science, 13* (4), 347–362.

Chu, T. (2014). *Human purpose and transhuman potential: A cosmic vision for our future evolution.* San Rafael, CA: Origin Press.

Clark, A. (1997). *Being there: Putting brain, body, and the world together again.* Cambridge, MA: MIT Press

Clark, A. (2003). *Natural-born cyborgs: Minds, technologies, and the future of human intelligence.* Oxford: Oxford University Press.

Clark, A. (2013 [2007]). Reinventing ourselves: The plasticity of embodiment, sensing and mind. In M. More & N. Vita-More (Eds.),*The transhumanist reader: Classical and contemporary essays on the science, technology and philosophy of the human future* (pp. 113–127). Originally published in *The Journal of Medicine and Philosophy, 32* (3) (2007).

Clausen, J. (2006). Die „Natur des Menschen": Geworden *und* gemacht. Anthropologisch-ethische Überlegungen zum Enhancement. *Zeitschrift für medizinische Ethik. Wissenschaft, Kultur, Religion, 52,* 391–401.

CNN. (2013, February 22). How to post to Facebook, Twitter after you die. Available from: www.cnn.com.

Coenen, C. (2008a). *Konvergierende Technologien und Wissenschaften. Der Stand der Debatte und politischen Aktivitäten zu "Converging Technologies."* Büro fuer Technikfolgen- Abschätzung beim Deutschen Bundestag. Hintergrundpapier Nr.16, March 2008. Available from: http://www.itas.kit.edu.

Coenen, C. (2008b). Verbesserung des Menschen durch konvergierende Technologien? Christliche und posthumanistische Stimmen in einer aktuellen Technikdebatte. In H. Böhm and K. Ott (Eds.), *Bioethik:Menschliche Identität in Grenzbereichen* (pp. 41–123). Leipzig, Germany: Evangelische Verlagsanstalt.

Coenen, C. (2009). Transhumanismus und Utopie. Ein Abgrenzungsversuch aus aktuellem Anlass. In R. Steltemeier, S. Dickel, S. Gaycken, & T. Knobloch (Eds.), *Neue Utopien. Zum Wandel eines Genres* (pp. 135-168). Heidelberg, Germany: Manutius.

Coenen, C. (2010). Deliberating visions: The case of human enhancement in the discourse on nanotechnology and convergence. In M. Kaiser, M. Kurath, S. Maasen, & C. Rehmann-Sutter (Eds.), *Governing future technologies: Nanotechnology and the rise of an assessment* regime (pp. 73–88). Dordrecht, Netherlands: Springer.

Coenen, C. (2010b). Zum mythischen Kontext der Debatte über *Human Enhancement.* In C. Coenen, S. Gammel, R. Heil, & A. Woyke (Eds.), *Die Debatte über „Human Enhancement. Historische, philosophische und ethische Aspekte der technologischen Verbesserung des Menschen* (pp. 63–90). Bielefeld, Germany: Transcript.

Coenen, C. (2014). Transhumanism in emerging technoscience as a challenge for the humanities and technology assessment. *Teorija in praksa, 51* (5), 754–771.

Coenen, C., Gammel, S., Heil, R., & Woyke, A. (Eds.). (2010). *Die Debatte über „Human Enhancement". Historische, philosophische und ethische Aspekte der technologischen Verbesserung des Menschen.* Bielefeld, Germany: Transcript.

Coenen, C., Hennen, L., & Link, H. J. (2009). The ethics of synthetic biology: Contours of an emerging discourse. *Technikfolgenabschätzung—Theorie und Praxis, 18* (2), 82–86.

Coenen, C., Schuijff, M., Smits, M., Klaassen, P., Hennen, L., Rader, M., & Wolbring, G. (2009). *Human enhancement study.* Science and Technology Options Assessment (IP/A/STOA/FWC/2005-28/SC35, 41 & 45). Brussels: European Parliament. Available from https://www.itas.kit.edu/downloads/etag_coua09a.pdf.

Coenen, C., & Simakova, E. (2013). STS policy interactions, technology assessment and the governance of technovisionary sciences. *Science, Technology & Innovation Studies, 9* (2), 3–20.

Cole-Turner, R. (1993). *The new Genesis: Theology and the genetic revolution.* Louisville, KY: Westminster/John Knox Press.

Cole-Turner, R. (2012). The singularity and the rapture: Transhumanist and popular Christian views of the future. *Zygon, 47* (4), 777–796.
Collingridge, D. (1980). The social control of technology. New York, NY: St. Martin's Press.
Collins, H., & Evans, R. (2007). *Rethinking Expertise*. Chicago, IL: The University of Chicago Press.
Coolidge, G. (2013, April 23). Help me help 1 billion people." *IndieGoGo*. Available from: https://www.indiegogo.com.
Cooper, D. (1995). Technology: Liberation or enslavement? In R. Fellows (Ed.), *Philosophy and technology* (pp. 7–18). Cambridge, England: Cambridge University Press.
Cutler, K.-M. (2013, April 23). The first ever synthetic biology kickstarter is about growing "glowing plants". *TechCrunch*. Available from: http://techcrunch.com.
Dao, J. (2013, July 7). Taking on rules to ease Sikhs' path to the army. *New York Times*, p. A9.
DARPA. (2013). Darpa robotic challenge trials 2013. Available from: http://archive.darpa.mil.
Davie, G. (2002). *Europe: The exceptional case. Parameters of faith in the modern world.* London, England: Darton, Longman and Todd.
Davies, P. (2008). *The Goldilocks enigma: Why is the universe just right for life?* New York, NY: Mariner Books.
Dawkins, R. (2008). *The God delusion*. Boston, MA: Houghton Mifflin.
DEEPEN. (2009). *Reconfiguring responsibility: Deepening debate on nanotechnology.* Available from: http://dro.dur.ac.uk/6399/1/6399.pdf?DDD14+dgg1mbk.
De Grey, A. (2007). *Ending aging: The rejuvenation breakthroughs that could reverse human aging in our lifetime.* New York, NY: St. Martin's Press.
Delfgaauw, B., Holz, H. H., & Nauta, L. (Eds.) (1985). *Philosophische Rede vom Menschen: Studien zur Anthropologie Helmuth Plessners.* Frankfurt am Main, Germany: Peter Lang International Academic Publishers.
Delitz, H. (2011). *Arnold Gehlen*. Konstanz, Germany: UVK.
Del Val, J., & Sorgner, S. L. (2011). A metahumanist manifesto. *The Agonist, 4* (2), 1–4.
Dennett, D. (2007). *Breaking the spell: Religion as a natural phenomenon*. London, England: Penguin.
De Saille, S. (2013). Innovating innovation: Operationalising RRI in the ERA. Paper presented at the Devices of Responsibility Workshop, Paris, Mines ParisTech, September 12–13.
Diamandis, P. H., & Kotler, S. (2014). *Abundance: The future is better than you think*. New York, NY: Free Press.
Dickel, S. (2009). Utopische Positionierungen. Vorüberlegungen zu einer Wissensoziologie des Utopischen und deren Anwendungsmöglichkeiten am Beispiel populärwissenschaftlicher Enhancement-Visionen. In R. Steltemeier, S. Dickel, S. Gaycken, & T. Knobloch (Eds.), *Neue Utopien. Zum Wandel eines Genres* (pp. 169–200). Heidelberg, Germany: Manutius.
Dickel, S. (2011). *Enhancement-Utopien. Soziologische Analysen zur Konstruktion des Neuen Menschen*. Baden-Baden, Germany: Nomos.
Dierse, U. (2001). Utopie. In J. Ritter, K. Gründer, & G. Gabriel (Eds.), *Historisches Handbuch der Philosophie. Band 11: U–V* (pp. 510–526). Darmstadt, Germany: Wissenschaftliche Buchgesellschaft.
Dillon, M. (2010). Can post-secular society tolerate religious differences? *Sociology of Religion, 71* (2), 139–156.
Dinham, A. (2012). *Faith and social capital after the debt crisis*. London, England: Palgrave Macmillan.

Donaldson, S., & Kymlicka, W. (2013). *Zoopolis: Eine politische Ethik der Tierrechte*. (J. Schulte, Trans.). Berlin, Germany: Suhrkamp.

Dorf, M. C., & Sabel, C. (1998). A constitution of democratic experimentalism. *Columbia Law Review, 98*, 267–473.

Dorgelo, C. (2015, September 5). Challenge.gov: Two years and 200 prizes later. The White House, Office of Science and Technology Policy. Available from: http://www.whitehouse.gov.

Douglas, C., & Stemerding, D. (2013). Governing synthetic biology for global health through responsible research and innovation. *Systems and Synthetic Biology, 7* (3), 139–150.

Douglas, M., & Wildavsky, A. (1982). *Risk and culture*. Berkeley, CA: University of California Press.

Drengson, A. (2010). Four philosophies of technology. In C. Hanks (Ed.), *Technology and values* (pp. 26–38). Malden, MA: Wiley-Blackwell.

Drexler, E. (1986). *Engines of creation: The coming era of nanotechnology*. Oxford, England: Oxford University Press.

Drexler, E. (2003a). Drexler counters. *Chemical & Engineering News, 81*, 40<–41.

Drexler, E. (2003b). Open letter to Richard Smalley. *Chemical & Engineering News, 81*, 38–39.

Drexler, E. K. (2013). *Radical abundance: How a revolution in nanotechnology will change civilization*. New York, NY: PublicAffairs.

Dreyfus, H. (1988). Why Heideggerian AI failed and why fixing it would require making it more Heideggerian. *Philosophical Psychology, 20* (2), 247–268.

Dreyfus, H. (1992). *What computers still can't do*. New York, NY: MIT Press.

Drinkwater, K., Kuiken, T., Lightfoot, S., McNamara, J., & Oye, K. (2014). *Creating a research agenda for the ecological implications of synthetic biology*. Joint Workshops by the MIT Program on Emerging Technologies and the Wilson Center's Synthetic Biology Project. May 7. Available from: http://www.researchgate.net.

Dryzek, J. (2010). *Foundations and frontiers of deliberative governance*. Oxford, England: Oxford University Press.

Dumit, J. (2012). *Drugs for life: How pharmaceutical companies define our health*. Durham, NC: Duke University Press.

Dupuy, J.-P. (2000). *The mechanization of mind. On the origins of cognitive science* (translated by M. B. DeBevoise). New York, NY: MIT Press.

Dupuy, J.-P. (2007a). Complexity and uncertainty: A prudential approach to nanotechnology. In F. Allhoff, P. Lin, J. Moor, & J. Weckert. (Eds.). *Nanoethics: The ethical and social implications of nanotechnology* (pp. 119–132). Hoboken, NJ: Wiley-Interscience.

Dupuy, J.-P. (2007b). Some pitfalls in the philosophical foundations of nanoethics. *Journal of Medicine and Philosophy, 32* (3), 237–261.

Dupuy, J. -P. (2011, September 1). H-: Cybernetics is an antihumanism: Advanced technologies and the rebellion against the human condition. *Metanexus*. Available from: http://www.metanexus.net.

Dzieza, J. (2013, August 18). Plants that glow in the dark spark heated debate. *The Daily Beast*. Available from: http://www.thedailybeast.com.

Eagleton, T. (2014). *Culture and the death of God*. New Haven, CT: Yale University Press.
Eberlein, B. & Kerwer, D. (2004). New governance in the European Union: A theoretical perspective. *Journal of Common Market Studies, 42* (1), 121–142.

Eden, A., Moor, J. H., Soraker, J. H., & Steinhart, E. (Eds.). (2012). *Singularity hypotheses: A scientific and philosophical assessment*. Heidelberg, Germany: Springer.

Edwards, P. N. (1996). *The closed world: Computers and politics of discourse in cold war America*. Cambridge, MA: MIT Press.
Eilers, M., Grüber, K., & Rehmann-Sutter, C. (Eds.). (2012). *Verbesserte Körper—gutes Leben? Bioethik, Enhancement und die Disability Studies*. Praktische Philosophie kontrovers. Frankfurt am Main, Germany: Peter Lang.
Eisenstadt, S. (2000). The reconstruction of religious arenas in the framework of "multiple modernities." *Millennium: Journal of International Studies, 29* (3), 591–611.
Eliot, L. (2001). *Was geht da drinnen vor? Die Gehirnentwicklung in den ersten fünf Lebensjahren*. Berlin, Germany: Berlin Verlag.
Ellul, J. (1964 [1954]). *The technological society*. New York, NY: Vintage Books.
Erlemann, K. (1995). *Naherwartung und Parusieverzögerung im Neuen Testament. Ein Beitrag zur Frage religiöser Zeiterfahrung*. Tübingen, Germany: Francke.
ETC Group. (2013a, May 30). Kickstopper! Putting a stop to synthetic biology pollution. Available from: http://www.etcgroup.org/kickstopper.
ETC Group. (2013b, May 30). Kickstopper: Stopping synbio pollution. *IndieGoGo*. Available from: https://www.indiegogo.com.
EC (European Commission). (2000a). Toward a European research area: Communication from the Commission to the Council, the European Parliament, the Economic and Social Committee and the Committee of the Regions, Brussels, 18.1.2000 COM (2000) 6, final.
EC (European Commission). (2000b). *The Europeans and biotechnology—Eurobarometer 52.1*. Luxembourg: Office for Official Publications of the European Communities.
EC (European Commission). (2004). Towards a European strategy for nanotechnology. COM2004(338). Available from: ftp://ftp.cordis.europa.eu/pub/nanotechnology/docs/nano_com_en.pdf.
EC (European Commission). (2005a). Nanosciences and nanotechnologies: An action plan for Europe 2005–2009. Luxembourg: Office for Official Publications of the European Communities. Available from: ftp://ftp.cordis.europa.eu/pub/nanotechnology/docs/nano_com_en.pdf.
EC (European Commission). (2005b). Working together for growth and jobs: A new start for the Lisbon strategy, COM (2005) 24. Brussels, Belgium: Official Journal of the European Communities. Available from: http://cordis.europa.eu.
EC (European Commission). (2007, February). *Strategy for communication outreach in nanotechnology*. Brussels, Belgium.
EC (European Commission). (2008). Commission recommendation of 07/02/2008 on a code of conduct for responsible nanosciences and nanotechnologies research. Bruxelles, C (2008) 424, final.
EC (European Commission). (2010). *Europe 2020: A strategy for smart, sustainable and inclusive growth*. Communication from the Commission, Brussels, 3.3.2010 COM (2010).
EC (European Commission). (2011). *Innovation union competitiveness report*. Available from: ec.europa.eu.
EC (European Commission). (2012). Responsible research and innovation: Europe's ability to respond to societal challenges. KI-31-12-921-EN-C. Available from: ec.europa.eu.
EC (European Commission). (2013). Options for strengthening responsible research and innovation: Report of the expert group on the state of art in Europe on responsible research and innovation. KI-NA-25-766-EN-C. Available from: ec.europa.eu.
Evans, A. (2013, April 23). Glowing plants: Natural lighting with no electricity. *Kickstarter*. Available from: https://www.kickstarter.com.

Evans, G., & Durant, J. (1995). The relationship between knowledge and attitudes in the public understanding of science in Britain. *Public Understanding of Science, 4* (1), 57–74.
Evans, J. H. (2002). *Playing God? Human genetic engineering and the rationalization of public bioethical debate*. Chicago, IL: University of Chicago Press.
Evans, J. H. (2010). *Contested reproduction: Genetic technologies, religion, and public debate*. Chicago, IL: University of Chicago Press.
Evans, J. H. (2011). Epistemological and moral conflict between religion and science. *Journal for the Scientific Study of Religion, 50* (4), 707–727.
Evans, J. H. (2014). Faith in science in global perspective: Implications for transhumanism. *Public Understanding of Science, 23* (7), 813–832.
Ezrahi, Y. (1990). *The descent of Icarus: Science and the transformation of contemporary society*. Cambridge, MA: Harvard University Press.
Faber, B. (2006). Popularizing nanoscience: The public rhetoric of nanotechnology, 1986–1999. *Technical Communication Quarterly, 15* (2), 141–169.
FAQ 3.0 Transhumanist FAQ 3.0. Available from: http://humanityplus.org.
Felt, U. (2010). Vers la construction d'un public européen? Continuités et ruptures dans le discours politique sur les cultures scientifiques et techniques. *Questions de communication, 17*, 33–58.
Ferrari, A., Coenen, C., & Grunwald, A. (2012). Visions and ethics in current discourse on human enhancement. *Nanoethics, 6* (3), 215–229.
Ferrari, A., & Nordmann, A. (Eds.). (2009). *Reconfiguring responsibility: Lessons for nanoethics—Part 2 of the report on deepening debate on nanotechnology*. Durham, England: Durham University.
Fischer, S. (1988). Jewish salvational visions, utopias, and attitudes towards the halacha. *International Journal of Comparative Sociology, 39* (1–2), 62–75.
Fisher, E. (2005). Lessons learned from ELSI program: Planning societal implications research for the national nanotechnology program. *Technology in Society, 27* (3), 321–328.
Flanagan, O. (1993). *Consciousness reconsidered*. Cambridge, MA: Bradford Books/MIT Press.
Fleming, J. V. (2013). *The dark side of the Enlightenment: Wizards, alchemists, and spiritual seekers in the age of reason*. New York, NY: W.W. Norton and Company.
Foucault, M. (1970). *The order of things: An archaeology of the human sciences*. New York, NY: Pantheon Books.
Foucault, M. (1979). *Discipline and punish*. New York: Vintage.
Foucault, M. (1984). *L'usage des plaisirs*. Paris, France: Gallimard.
Franssen, T. (2012). Prometheus steigt herab: Beeinträchtigung oder Enhancement? In M. Eilers, K. Grüber, & C. Rehmann-Sutter (Eds.), *Verbesserte Körper—gutes Leben? Bioethik, Enhancement und die Disability Studies* (pp. 205–227). Frankfurt am Main, Germany: Peter Lang.
Fuchs, M. (2005), *Nationale Ethikräte. Hintergründe, Funktionen und Arbeitsweisen im Vergleich*. Berlin, Germany: Nationaler Ethikrat.
Fuchs, R.-P. (2012). Gegen die Apokalypse? Zukunftsdiskurse im Dreißigjährigen Krieg. In A. Landwehr (Ed.), *Frühe neue Zeiten. Zeitwissen zwischen Reformation und Revolution* (pp. 237–259). Bielefeld, Germany: Transcript.
Fukuyama, F. (Summer 1989). The end of history? *The National Interest*, pp. 3–18.
Fukuyama, F. (2002). *Our posthuman future: Consequences of the biotechnology revolution*. (1st Picador ed.). New York, NY: Picador.
Fukuyama, F. (2004). Transhumanism. *Foreign Policy, 144*, 42–43.

Fuller, S. (2012a). *Humanity 2.0: What it means to be human past, present and future*. New York, NY: Palgrave Macmillan.
Fuller, S. (2012b). Precautionary and proactionary as the new right and the new left of the twenty-first century ideological spectrum. *International Journal of Politics, Culture, and Society, 25* (4), 157–174.
Fuller, S. (2013a, October 24). Ninety-degree revolution. *Aeon*. Available from: http://aeon.co.
Fuller, S. (2013b). *Preparing for life in humanity 2.0*. New York, NY: Palgrave Macmillan.
Fuller, S., & Lipinska, V. (2014). The proactionary imperative. A foundation for transhumanism. New York, NY: Palgrave Macmillan.
Gadamer, H-G. (1972). *Wahrheit und Methode* (3rd ed.). Tübingen, Germany: Mohr.
Gane, N. (2006). When we have never been human, what is to be done? An interview with Donna Haraway. *Theory, Culture and Society, 23* (7–8), 135–158.
Garner, S. (2011). The hopeful cyborg. In R. Cole-Turner (Ed.), *Transhumanism and transcendence* (pp. 87–100). Washington, DC: Georgetown University Press.
Garreau, J. (2004). *Radical evolution: The promise and peril of enhancing our minds, our bodies*. New York, NY: Doubleday.
Gaskell, G., Stares, S., Allansdottir, A., Allum, N., Corchero, C., Fischler, C., Hampel, J., Jackson, J., Kronberger, N., Mejlgaard, N., Revuelta, G., & Schreiner, C. (1996). *Public perception of biotechnology in Europe: Eurobarometer 52.1*. Brussels, Belgium: European Commission. Available at http://ec.europa.eu/public_opinion/archives/ebs/ebs_244b_en.pdf.
Gaskell, G., Stares, S., Allansdottir, A., Allum, N., Castro, P., Esmer, Y., Fischler, C., Jackson, J., Kronberger, N., Hampel, J., Mejlgaard, N., Quintanilha, A., Rammer, A., Revuelta, G., Stoneman, P., Torgersen, H., & Wagner, W. (2010). *Europeans and biotechnology in 2010: Winds of change*. A report to the European Commission's Directorate-General for Research. Brussels, Belgium: European Commission. Available at http://ec.europa.eu/public_opinion/archives/ebs/ebs_341_winds_en.pdf..
Geere, D. (2013, August 2). Kickstarter bans project creators from giving away genetically-modified organisms. *The Verge*. Available from: http://www.theverge.com.
Gehlen, A. (1940). *Der Mensch: seine Natur und seine Stellung in der Welt*. Berlin, Germany: Junker and Dünnhaupt. In English: *Man, his nature and place in the world*. (C. McMillan & K. Pillemer, Trans.). New York, NY: Columbia University Press, 1988.
Geraci, R. M. (2008). Apocalyptic AI: Religion and the promise of artificial intelligence. *Journal of the American Academy of Religion*, 76 (1), 138–166.
Geraci, R. M. (2010). *Apocalyptic AI: Visions of heaven in robotics, artificial intelligence, and virtual reality*. Oxford, England: Oxford University Press.
Geraci, R. M. (2012). Video games and transhumanist inclination. *Zygon: Journal of Religion and Science, 47* (4), 735–756.
Giddens, A. (1998). Risk society: The context of British politics. In J. Franklin (Ed.), *The politics of risk society* (pp. 23–35). Cambridge, England: Polity Press.
Gies, E. (2013, July 7). In Europe, greener transit on existing infrastructure. *New York Times*, p. B4.
Gilbreth, F. B., Jr., & Carey, E. G. (1948). *Cheaper by the dozen*. Boston, MA: Thomas Y. Crowell.
Glynn, A. (2002). *The dark fields*. New York, NY: Bloomsbury.
Goel, V. (2013, July 7). A new tool aims to help Facebook users dig deep. *New York Times*, p. B1.
Goertzel, B. (2006). *The Path to posthumanity: 21st-century technology and its radical implications for mind, society and reality*. Bethesda, MD: Academica Press.

Goertzel, B. (2007). *Advances in artificial general intelligence*. Washington, DC: IOS Press.

Gomel, E. (2014). *Science fiction, alien encounters and the ethics of posthumanism: Beyond the golden rule*. New York, NY: Palgrave Macmillan.

Goodman, N. (1954). Fact, fiction, and forecast. London, England: Athelone Press.

Gordon, P. E. (2008). The place of the sacred in the absence of God: Charles Taylor's "a secular age." *Journal of the History of Ideas, 69* (4), 647–673.

Gorski, P., Kim, D. K., Torpey, J., & VanAntwerpe, J. (Eds). (2012). *The post-secular in question: Religion in contemporary society*. New York: Social Science Research Council and New York University Press.

Graham, E. (2002). *Representations of the post/human: Monsters, aliens and others in popular culture*. Manchester, England: Manchester University Press.

Graham, E. (2003). Frankensteins and cyborgs: Visions of the global future in an age of technology. *Studies in Christian Ethics, 16* (1), 29–43.

Graham, E. (2004). Bioethics after posthumanism: Natural law, communicative ethics and the problem of self-design. *Ecotheology, 9* (2), 178–198.

Graham, E. (2007). Embodying technology, becoming our tools: Discussing the post/human. In J. Baxter (Ed.), *Wounds that heal: Imagination, theology and health* (pp. 125–148). London, England: SPCK.

Graham, E. (2013). *Between a rock and a hard place: Public theology in a post-secular age*. London, England: SCM Press.

Gray, J. (2009). *Politik der Apokalypse. Wie Religion die Welt in die Krise stürzt*. Stuttgart, Germany: Klett-Cotta.

Greene, J. (2014, September 19). Why is our sci-fi so glum about A.I.? *New York Times*, Sunday Magazine, p. MM56.

Grossnick, R. A. (Ed.). (1987). Kite balloons to airships: The navy's lighter-than-air experience. Washington, DC: Deputy Chief of Naval Operations (Air Warfare) and the Commander, Naval Air Systems Command.

Grunwald, A. (2007a). Converging technologies: Visions, increased contingencies of the *conditio humana*, and search for orientation. *Futures, 39*, 380–392.

Grunwald, A. (2007b). Grenzmanagement. Kontingenzsteigerung und neuer Orientierungsbedarf am Beispiel der technischen Verbesserung des Menschen. *Lifis Online*. Available from: http://www.leibniz-institut.de/de/lifis-online.

Grunwald, A. (2009). Technology assessment. In A. Meijers (ed.), *Philosophy of technology and engineering sciences* (volume 9, pp. 1103–1146). Amsterdam, the Netherlands: Elsivier/North Holland.

Grunwald, A. (2010). From speculative nanoethics to explorative philosophy of nanotechnology. *NanoEthics*, 4 (2), 91–101.

Grunwald, A. (2012a). *Technikzukünfte als Medium von Zukunftsdebatten und Technikgestaltung*. Karlsruhe, Germany: KIT Scientific Publishing.

Grunwald, A. (2012b). *Responsible nanobiotechnology: Philosophy and ethics*. Singapore: Pan Stanford Publications Pte. Ltd.

Grunwald, A. (2013a). Techno-visionary sciences: Challenges to policy advice. *Science, Technology and Innovation Studies, 9* (2), 21–38.

Grunwald, A. (2013b). Modes of orientation provided by futures studies: Making sense of diversity and divergence. *European Journal of Futures Studies, 2* (1): 2-30.

Geuss, R. (2010). Realismus, Wunschdenken, Utopie. *Deutsche Zeitschrift für Philosophie, 58* (3), 419–429.

Guston, D. H., & Sarewitz, D. (2002). Real-time technology assessment. *Technology in Society, 24* (1–2), 93–109.
Habermas, J. (1971). *Toward a rational society: Student protest, science, and politics.* (J. J. Shapiro, Trans.). Boston, MA: Beacon Press.
Habermas, J. (2001). *Die Zukunft der menschlichen Natur. Auf dem Weg zu einer liberalen Eugenik?* Frankfurt am Main, Germany: Suhrkamp.
Habermas, J. (2003). *The future of human nature.* Cambridge, England: Polity.
Habermas, J. (2005). Religion in der Öffentlichkeit. In J. Habermas, *Zwischen Naturalismus und Religion. Philosophische Aufsätze* (pp. 119–154). Frankfurt am Main, Germany: Suhrkamp.
Habermas, J. (2008). Religion in the public sphere: Cognitive presuppositions for the "public use of reason" by religious and secular citizens. In J. Habermas, *Between naturalism and religion: Philosophical essays* (pp. 114–147). London, England: Routledge.
Habermas, J. (2010). An awareness of what is missing. In J. Habermas et al., *An awareness of what is missing: Faith and reason in a post-secular age* (pp. 15–23). Cambridge, England: Polity.
Habermas, J. (2012). Eine Hypothese zum gattungsgeschichtlichen Sinn des Ritus. In J. Habermas, *Nachmetaphysisches Denken* II (pp. 77–95). Berlin, Germany: Suhrkamp.
Hagendijk, R. (2004). The public understanding of science and public participation in regulated worlds. *Minerva, 42* (1), 41–59.
Haldane, J. B. S. (1940). Possible worlds. In J. B. S. Haldane, *Possible worlds* (pp. 241–263). London, England: Evergreen Books.
Hanks, C. (2010). General introduction. In C. Hanks (Ed.), *Technology and values* (pp. 1–6). Malden, MA : Wiley-Blackwell.
Hansell, G., & Grassy, W. (Eds.) (2011). *H+: Transhumanism and its critics.* Philadelphia, PA: Metanexus Institute.
Haraway, D. (1991). A cyborg manifesto: Science, technology, and socialist-feminism in the late twentieth century. In D. Haraway, *Simians, cyborgs and women: The reinvention of nature* (pp. 149–182). London, England: Free Association Books.
Haraway, D. (1992). The promises of monsters: A regenerative politics for inappropriate/d others. In L. Grossberg, C. Nelson, & P. Treichler (Eds.). *Cultural studies* (pp. 295–337). New York, NY: Routledge.
Haraway, D. (2008). *When species meet.* Minneapolis, MN: University of Minnesota Press.
Hardt, M., & Negri, A. (2000). *Empire.* Cambridge, MA: Harvard University Press.
Harris, J. (1998). *Clones, genes and immortality: Ethics and the genetic revolution.* New York, NY: Oxford University Press.
Harris, J. (2007). *Enhancing evolution: The ethical case for making better people.* Princeton, NJ: Princeton University Press.
Harris, J. (2010). *Enhancing evolution: The ethical case for making better people.* Princeton, NJ: Princeton University Press.
Harris, S. (2005). *The end of faith: Religion, terror, and the future of reason.* London, England: Free Press.
Harvard University, Office of Communications. (2013, May 30). Commencement 2013: Larry Brilliant's address. Available from: http://www.hsph.harvard.edu/news.
Hauskeller, M. (2012). Reinventing Cockaigne: Utopian themes in transhumanist thought. *The Hastings Center Report, 42* (2), 39–47.

Hawking, S., Tegmark, M., Russell, S., & Wilczek, F. (2014, April 19). Transcending complacency about superintelligent machines. *Huff Post. Science.* Available from: http://www.huffingtonpost.com.

Hayles, N. K. (1999). *How we became posthuman: Virtual bodies in cybernetics, literature and informatics.* Chicago, IL: Chicago University Press.

Hecht, G. (1998). *The radiance of France.* Cambridge, MA: MIT Press.

Hecht, G. (2012). *Being nuclear: Africans and the global uranium trade.* Cambridge, MA: MIT Press.

Hefner, P. (2003). *Technology and human becoming.* Minneapolis, MN: Fortress.

Hegel, G. W. F. (1911). *Grundlinien der Philosophie des Rechts.* Leipzig, Germany: F. Meiner.

Heidegger, M. (1953). *Die Technik und die Kehre.* Stuttgart, Germany: Klett Cotta.

Heidegger, M. (1954a). Die Frage nach der Technik. In M. Heidegger, *Vorträge und Aufsätze,* Teil I (pp. 5–36). Pfullingen, Germany: Günther Neske.

Heidegger, M. (1954b). Überwindung der Metaphysik. In M. Heidegger, *Vorträge und Aufsätze,* Teil I (pp. 63–91). Pfullingen, Germany: Günter Neske.

Heidegger, M. (1954c). Das Ding. In M. Heidegger, *Vorträge und Aufsätze,* Teil II (pp. 163–181). Pfullingen, Germany: Verlag Günther Neske.

Heidegger, M. (1962). *Being and time.* (J. Macquarrie & E. Robinson, Trans.). New York, NY: Harper & Row.

Heidegger, M. (1967a). Brief über den "Humanismus." In M. Heidegger, *Wegmarken* (pp. 145–194). Frankfurt am Main, Germany: Vittorio Klostermann.

Heidegger, M. (1967b). *Sein und Zeit.* Tübingen, Germany: Max Niemeyer Verlag.

Heidegger, M. (1971). The thing. In A. Hofstader (Trans.), *Poetry, language, thought* (pp. 178–179). New York, NY: Harper & Row.

Heidegger, M. (1972). Overcoming metaphysics. In J. Stambaugh (Trans.), *The end of philosophy* (pp. 84–110). New York: Harper & Row.

Heidegger, M. (1977). Letter on humanism. (D. F. Krell, Trans.) In D. F. Krell (Ed.), *Basic writings in Martin Heidegger* (pp. 213–265). New York, NY: Harper & Row.

Heidegger, M. (1983). *Die Grundbegriffe der Metaphysik. Welt—Endlichkeit—Einsamkeit.* In F.-W. von Herrmann (Ed.), *Gesamtausgabe,* vols. 29/30. Frankfurt am Main, Germany: Vittorio Klostermann.

Heidegger, M. (1987). *The will to power as knowledge.* In J. Stambaugh, D. F. Krell, & F. A. Capuzzi (Trans.), *Nietzsche* (pp. 100–158). San Francisco, CA: HarperSanFrancisco.

Heidegger, M. (1989a). *Beiträge zur Philosophie (Vom Ereignis).* (F.-W. von Hermann, Ed). Frankfurt am Main, Germany: Vittorio Klostermann.

Heidegger, M. (1989b). *Nietzsches Lehre vom Willen zur Macht als Erkenntnis* (Summer semester 1939). In E. Hanser (Ed.), *Gesamtausgabe,* vol. 47. Frankfurt am Main, Germany: Vittoria Klostermann.

Heidegger, M. (1995). *Fundamental problems of metaphysics: World, finitude, solitude* (W. McNeill & N. Walker, Trans). Bloomington, IN: Indiana University Press.

Heidegger, M. (2000). *Reden und andere Zeugnisse eines Lebensweges (1910–1976).* In H. Heidegger (Ed.), *Gesamtausgabe,* Vol. 16 (pp. 620–33). Frankfurt am Main, Germany: Vittorio Klostermann.

Heidegger, M. (2004). *Zu Ernst Jünger „Der Arbeiter."* In P. Trawny (Ed.), *Gesamtausgabe,* vol. 90. Frankfurt am Main, Germany: Vittorio Klostermann.

Heidegger, M. (2008 [1954]). The question concerning technology. In D. F. Krell (Ed.), *Basic Writings by Martin Heidegger* (pp. 307–341). New York: Harper Perennial Modern Thought.

Heidegger, M. (2010). On the question concerning the determination of the matter for thinking. (R. Capobianco & M. Göbel, Trans.). *Epoche, 14* (2), 213–233.
Heidegger, M. (2012). *Contributions to philosophy.* (R. Rojcewicz & D. Vallega-Neu, Trans.). Bloomington, IN: Indiana University Press.
Heinrichs, D., Krellenberg, K., Hansjürgens, B., & Martínez, F. (Eds.) (2012). *Risk habitat megacity.* Heidelberg, Germany: Springer-Verlag.
Heller, R. (2003). *GM nation? The findings of the public debate.* London, England: UK Department of Trade and Industry.
Herbrechter, S. (2013). *Posthumanism: A critical analysis.* London, England: Bloomsbury.
Hessel, A. (2013, June 6). Grow your own glowing plant. *Kurzweil Accelerating Intelligence.* Available from: http://www.kurzweilai.net.
Hilgartner, S. (1990). The dominant view of popularization: Conceptual problems, political uses. *Social Studies of Science, 20* (3), 519–539.
Hilgartner, S. (2015). Capturing the imaginary: Vanguards, visions, and the synthetic biology revolution. In S. Hilgartner, C. Miller, & R. Hagendijk (Eds.), *Science and democracy: Knowledge as wealth and power in the biosciences and beyond* (chap. 3). Abingdon, England: Routledge
Hitchens, C. (2007). *God is not great: How religion poisons everything.* New York, NY: Warner Twelve.
Hobsbawm, E. J. (1952). The machine breakers. *Past and Present, 1* (1), 57–70.
Hopkins, P. D. (2005). Transcending the animal: How transhumanism and religion are and are not alike. *Journal of Evolution and Technology, 14* (2),13–28.
Hughes, J. (2004). *Citizen cyborg: Why democratic societies must respond to the redesigned human of the future.* Boulder, CO: Westview Press.
Hughes, J. (2014). Politics. In R. Ranisch & S. L. Sorgner (Eds.), *Post- and transhumanism: An introduction* (pp. 133–148). Frankfurt am Main, Germany: Peter Lang.
Hügli, A. (1998). Tod. In J. Ritter & K. Gründer (Eds.), *Historisches Wörterbuch der Philosophie. Band 10: St–T* (pp. 1227–1242). Darmstadt, Germany: Wissenschaftliche Buchgesellschaft.
Hughes, T. P. (2005). *Human-built world: How to think about technology and culture.* Chicago, IL: University of Chicago Press.
Humanity+ (2002). *Transhumanist FAQ 3.0.* Available from: http://humanityplus.org.
Hurlbut, J. B. (2015a). Reimagining responsibility in synthetic biology. *Journal of Responsible Innovation, 2* (1), 113–116.
Hurlbut, J. B. (2015b). Religion and public reason in the politics of biotechnology. *Notre Dame Journal of Law, Ethics and Public Policy, 29* (2), 423–452.
Hurlbut, J. B. (2015c). Remembering the future: Science, law, and the legacy of Asilomar. In S. Jasanoff & S.-H. Kim (Eds.), *Dreamscapes of modernity: Sociotechical imaginaries and the fabrication of power* (pp. 126–151). Chicago, IL: University of Chicago Press.
Hurlbut, J. B. (2015d). Technologies of imagination: Secularism, transhumanism, and the idiom of progress. In D. Yerxa (Ed.), *Religion and Innovation: Antagonists or partners?* (chap. 14). London, England: Bloomsbury.
Huxley, A. (1954). *The doors of perception and heaven and hell.* New York, NY: Harper & Brothers.
Huxley, J. (1951). Knowledge, morality, and destiny—The William Alanson White Memorial Lectures, third series. *Psychiatry, 14* (2), 127–151.
Huxley, J. (1957). Transhumanism. In J. Huxley, *new bottles for new wine: Essays* (pp. 13–17). London, England: Chatto & Windus.

Huxley, J. (1963). The future of man: Evolutionary aspects. In G. Wolstenholme (Ed.), *Man and his future* (pp. 1–22). Boston, MA: Little Brown & Co.
Isaacson, W. (2011). *Steve Jobs*. New York, NY: Simon & Schuster.
Jacoby, R. (2000). *The last intellectuals: American culture in the age of academe*. (Rev. ed.). New York, NY: Basic Books.
James, W. (1890). *Principles of psychology*. (3rd ed.). London, England: Williams and Norgate.
Jantzen, G. M. (1998). *Becoming divine: Towards a feminist philosphy of religion*. Manchester, England: Manchester University Press.
Jasanoff, S. (2002): Citizens at risk: Cultures of modernity in the US and the EU. *Science as Culture, 11*, 363–380.
Jasanoff, S. (2003). Technologies of humility: Citizen participation in governing science. *Minerva, 41* (3), 223–244.
Jasanoff, S. (Ed.). (2004). *States of knowledge: The co-production of science and social order*. London, England: Routledge.
Jasanoff, S. (2005). *Designs on nature: Science and democracy in Europe and the United States*. Princeton, NJ: Princeton University Press.
Jasanoff, S. (2013). Epistemic subsidiarity: Coexistence, cosmopolitanism, constitutionalism. *European Journal of Risk Regulation, 4* (2), 133–141.
Jasanoff, S. (2014). Vulnerability and development: Bhopal's lasting legacy. In A. Hommels, J. Mesman, & W. E. Bijker (Eds.), *Vulnerability in technological cultures: New directions in research and governance* (pp. 89–108). Cambridge, MA: MIT Press.
Jasanoff, S. (2015). Future imperfect: Science, technology, and the imaginations of modernity. In S. Jasanoff & S.-H. Kim (Eds.), *Dreamscapes of modernity: Sociotechical imaginaries and the fabrication of power* (pp. 1–47). Chicago, IL: University Of Chicago Press.
Jasanoff, S., & Kim, S.-H. (2009). Containing the atom: Sociotechnical imaginaries and nuclear regulation in the U.S. and South Korea. *Minerva, 47* (2), 119–146.
Jasanoff, S., & Kim, S.-H. (2013). Sociotechnical imaginaries and national energy policies. *Science as Culture, 22* (2),189–196.
Jasanoff, S., & Kim, S.-H. (Eds.) (2015). *Dreamscapes of modernity: Sociotechnical imaginaries and the fabrication of power*. Chicago, IL: University of Chicago Press.
Johnson, B. (2009, September 2). A school for changing the world. *The Guardian*, Technology.
Johnson, S. (2012). *Future perfect: The case for progress in a networked age*. New York, NY: Riverhead.
Jonas, H. (1952). Gnosticism and modern nihilism. *Social Research, 19* (4), 430–452.
Jonas, H. (1958). "Gnosticism, nihilism, and existentialism." In H. Jonas, *The Gnostic religion* (pp. 320–340). Boston, MA: Beacon Press.
Jonas, H. (1979). *Das Prinzip Verantwortung. Versuch einer Ethik für die technologische Zivilisation*. Frankfurt, Germany: Suhrkamp Verlag.
Jonas, H. (1984). *The imperative of responsibility: In search of an ethics for the technological age*. Chicago, IL: University of Chicago Press.
Jones, C. F. (2014). *Routes of power: Energy and modern America*. Cambridge, MA: Harvard University Press.
Jones, R. (2004). *Soft machines: Nanotechnology and life*. Oxford, England: Oxford University Press.
Jones, R. (2011). What has nanotechnology taught us about contemporary technoscience? In T. B. Zülsdorf, C. Coenen, A. Ferrari, U. Fiedeler, C. Milburn, & M. Wienroth (Eds.).

Quantum engagements: Social reflections of nanoscience and emerging technologies (pp. 13–26). Heidelberg, Germany: AKA.
Joss, S., & Durant, J. (Eds.). 1995. *Public participation in science: The role of consensus conferences in Europe.* London, England: Science Museum with the support of the European Commission Directorate General XII.
Joy, B. (2000, April). "Why the future doesn't need us." *Wired, 8* (4). Available from: http://archive.wired.com.
Jüngel, E. (1971). *Tod.* Stuttgart, Germany: Kreuz.
Jüngel, E. (2005). Tod VII. Dogmengeschichtlich und dogmatisch. In H. D. Betz, D. S. Browning, B. Janowski, & E. Jüngel (Eds.), *Religion in Geschichte und Gegenwart. Handwörterbuch für Theologie und Religionswissenschaft. Band 8 T–Z* (pp. 439–441). Tübingen, Germany: Mohr Siebeck.
Jünger, E. (2013 [1932]). *Der Arbeiter: Herrschaft und Gestalt.* Stuttgart, Germany: Klett-Cotta-Verlag.
Kamlah, W. (1973). *Philosophische Anthropologie: Sprachkritische Grundlegung und Ethik.* Mannheim, Germany: Bibliographisches Institut.
Kass, L. (2001, May). L'Chaim and its limits: Why not immortality? *First Things, 113,* 17–24.
Kass, L. (2003). Ageless bodies, happy souls: Biotechnology and the pursuit of happiness. *The New Atlantis: A Journal of Technology & Society, 1* (1), 9–28.
Kearnes, M., Macnaghten, P., & Wilsdon, J. (2006). *Governing at the nanoscale: People, policies and emerging technologies.* London, England: Demos.
Kelly, K. (2010). *What technology wants.* New York, NY: Viking.
Kelsey, D. (1998). Human being. In P. Hodgson and R. H. King (Eds.), *Christian theology: An introduction to its traditions and tasks* (pp. 167–193). London, England: SPCK.
Kennedy, D. (2009). *The rights of spring: A memoir of innocence abroad.* Princeton, NJ: Princeton University Press.
Kipke, R. (2011). *Besser werden. Eine ethische Untersuchung zu Selbstformung und Neuro-Enhancement.* Paderborn, Germany: Mentis.
Koch, T. (2005). Tod VIII. Ethisch. In H. D. Betz, D. S. Browning, B. Janowski, & E. Jüngel (Eds.), *Religion in Geschichte und Gegenwart. Handwörterbuch für Theologie und Religionswissenschaft. Band 8 T–Z* (pp. 441–443). Tübingen, Germany: Mohr Siebeck.
Körtner, U. H. J. (2005). *Lasset uns Menschen machen. Christliche Anthropologie im biotechnologischen Zeitalter.* München, Germany: Beck.
Kristeller, P. O. (1979). The dignity of man. In M. Mooney (Ed.), *Renaissance thought and its sources* (pp. 168–181). New York, NY: Columbia University Press.
Kroker, A. (2014). *Exit to the posthuman future.* Cambridge, England: Polity Press.
Kumar, K. (1991). *Utopianism.* Minneapolis, MN: University of Minnesota Press.
Kurath, M., & Gisler, P. (2009). Informing, involving or engaging? Science communication, in the ages of atom-, bio- and nanotechnology. *Public Understanding of Science,18* (5), 559–573.
Kurzweil, R. (1999). *The age of spiritual machines: When computers exceed human intelligence.* New York, NY: Penguin Books.
Kurzweil, R. (2001, March 7). The law of accelerating returns. *Kurzweil Accelerating Intelligence.* Available from: http://www.kurzweilai.net.
Kurzweil, R. (2005). *The singularity is near: When humans transcend biology.* New York, NY: Viking.

Kurzweil, R. (2012). *How to create a mind: The secret of human thought revealed.* New York, NY: Viking.

Landwehr, A. (Ed.). (2012). *Frühe neue Zeiten. Zeitwissen zwischen Reformation und Revolution.* Mainzer historische Kulturwissenschaften. Bielefeld, Germany: Transcript.

Lanier, J. (2010a, August 9). The first church of robotics. *The New York Times*, p. A19.

Lanier, J. (2010b). *You are not a gadget: A manifesto.* New York, NY: Vintage Books.

Latour, B. (1988). *The pasteurization of France.* (A. Sheridan & J. Law, Trans.) Cambridge, MA: Harvard University Press.

Latour, B. (1992). Where are the missing masses? In W. E. Bijker & J. Law (Eds.), *Shaping technology/building society: Studies in sociotechnical change* (pp. 225–258). Inside Technology. Cambridge, MA: MIT Press.

Latour, B. (1993). *We have never been modern.* (C. Porter, Trans.). Cambridge, MA: Harvard University Press.

Latour, Bruno. (2012, Winter). Love your monsters: Why we must care for our technologies as we do our children. *The Breakthrough*. Available from: http://thebreakthrough.

Law, J. (2009). Seeing like a survey. *Cultural Sociology, 3* (2), 239–256.

Lazier, B. (2012). *God interrupted: Heresy and the European imagination between the world wars.* Princeton, NJ: Princeton University Press.

Leach, A. (2013, May 7). Glow-in-the-dark plants make activists see red. *The Wall Street Journal*, TechEurope.

Lederberg, J. (1963). Biological future of man. In G. Wolstenholme (Ed.), *Man and his future: A Ciba Foundation volume* (pp. 263–273). Boston, MA: Little, Brown, and Co.

Lee, M. A., & Schlain, B. (1992). *Acid dreams: The complete social history of LSD: The CIA, the sixties and beyond.* New York, NY: Grove Press.

Levidow, L., & Carr, S. (2009). *GM food on trial: Testing European democracy.* London, England: Routledge.

Levin, Y. (2008). *Imagining the future: Science and American democracy.* New York, NY: Encounter Books.

Levi-Strauss, C. (1970 [1955]). *Traurige Tropen.* Frankfurt am Main, Germany: Suhrkamp.

Levitas, R. (2011). *The concept of Utopia.* Oxford, England: Peter Lang Oxford.

Lewis, C. S. (1994). *The discarded image: An introduction to medieval and renaissance literature.* Cambridge, England: Cambridge University Press.

Lightman, A. P., Sarewitz, D., & Desser, C. (Eds.) (2004). *Living with the genie: Essays on technology and the quest for human mastery.* Washington, DC: Island Press.

Lösch, A. (2007). Nanoroboter und Mini-U-Boote—Mediale Vermittlung nanomedizinischer Visionen. In D. Korczak & A. Lerf (Eds.), *Zukunftspotentiale der Nanotechnologien: Erwartungen, Anwendungen, Auswirkungen* (pp. 127–145). Kröning, Germany: Asanger.

Lösch, A., Gammel, S., & Nordmann, A. (2009). *Jenseits von Regulierung: Zum politischen Umgang mit der Nanotechnologie.* Heidelberg, Germany: AKA.

Lovejoy, A. O. (1936). *The great chain of being: A study of the history of an idea.* Cambridge, MA: Harvard University Press.

Loveridge, D., Dewick P., & Randles, S. (2008). Converging technologies at the nanoscale: The making of a new world? *Technology Analysis & Strategic Management, 20* (1), 29–43.

Luhmann, N. (1997). *Die Gesellschaft der Gesellschaft.* Frankfurt, Germany: Surkamp Verlag.

Lundvall, B.-A., & Lorenz, E. (2011). From the Lisbon strategy to Europe 2020. In N. Morel, B. Palier, & J. Palme (Eds.), *Toward a social investment welfare state? Ideas, policies and challenges* (pp. 333–351). Bristol, England: Policy Press.

Maasen, S. (2010). Converging technologies—diverging reflexivities? Intellectual work in knowledge-risk-media-audit societies. In M. Kaiser, M. Kurath, S. Maasen & C. Rehmann-Sutter (Eds.). *Governing future technologies: Nanotechnology and the rise of an assessment regime* (pp. 303–319). (Sociology of the Sciences Yearbook 27). Dordrecht, the Netherlands: Springer.

MacIntyre, A. (2001). *Die Anerkennung der Abhängigkeit: Über menschliche Tugenden*. (C. Goldmann, Trans.). Hamburg, Germany: Rotbuch.

Macnaghten, P., & Kearnes, M. (2007). DEEPEN deliverable 4: Working paper—Scenario planning and draft design of focus groups. Unpublished. Durham, England.

Macnaghten, P., Kearnes, M., & Wynne, B. (2005). Nanotechnology, governance, and public deliberation: What role for the social sciences? *Science Communication, 27* (2), 268–291.

Magee, G. (2008). *Hegel and the hermetic tradition*. Ithaca, NY: Cornell University Press.

Mali, F. (2009). Bringing converging technologies closer to civil society: The role of the precautionary principle. *Innovation: The European Journal of Social Science Research, 22* (1), 53–75.

Mali, F., & Pustovrh, T. (2014). Anticipatory governance of synthetic biology: Some challenges for Slovenia as a country with weak stakeholder networks. Paper at STS Conference Graz, Critical Issues in Science and Technology Studies, May 5–6, 2014, Graz, Austria. Available from: http://www.ifz.aau.at/ias/IAS-STS/.

Mali, F., Pustovrh, T., Groboljsek, B., & Coenen, C. (2012). National ethics advisory bodies in the emerging landscape of responsible research and innovation. *Nanoethics, 6* (2), 167–184.

Malik, S. (2000). *Nihilism now! Monsters of energy*. (K. Ansell-Pearson & D. Morgan, Eds.). New York, NY: St. Martin's Press.

Malykhina, E. (2014, April 2). New DARPA office merges biology and technology. *Information Week, Government*. Available from: http://www.informationweek.com.

Mannheim, K. (1986a). Das utopische Bewußtsein. In A. Neusüß (Ed.), *Utopie. Begriff und Phänomen des Utopischen* (pp. 265–285). Frankfurt/Main, Germany: Campus-Verlag.

Mannheim, K. (1986b). Utopie. In A. Neusüß (Ed.), *Utopie. Begriff und Phänomen des Utopischen* (pp. 113–120). Frankfurt/Main, Germany: Campus-Verlag.

Marcuse, H. (1967). *Der eindimensionale Mensch. Studien zur Ideologie der fortgeschrittenen Industriegesellschaft*. Neuwied, Germany: Hermann Luchterhand Verlag.

Marcuse, H., & Kellner, D. 1991. *One-dimensional man: Studies in the ideology of advanced industrial society*. Boston, MA: Beacon Press.

Markoff, J. (2005). *What the dormouse said: How the sixties counter-culture shaped the personal computer industry*. New York, NY: Viking Penguin.

Marris, C. (2001). La perception des OGM par le public: Remise en cause de quelques idées recues. *Economie Rurale, 266*, 58–79.

Max More: Transhuman and the singularity. (2012). https://www.youtube.com.

Maynard, A. (2014). Responsible innovation. Manuscript of a talk at Arizona State University, April 1.

Mayo Clinic. (2014, April 14). Caffeine: How much is too much? Healthy Lifestyle, Nutrition and Healthy Eating. Available from: http://www. mayoclinic.org.

Mayr, O. (1986). *Authority, liberty and automatic machinery in early modern Europe*. Baltimore, MD: Johns Hopkins University Press.

Mazlish, B. (1993). *The fourth discontinuity: The co-evolution of humans and machines*. New Haven, CT: Yale University Press.

McGrath, S. J. (2005). The facticity of being godforsaken: The young Heidegger's accommodation of Luther's theology of the cross. *American Catholic Philosophical Quarterly, 79* (2), 273–290.

McIntosh, S. (2012). *Evolution's purpose*. New York, NY: Select Books.

McKenny, G. (2011). Transcendence, technological enhancement, and Christian theology. In Ronald Cole-Turner (Ed.), *Transhumanism and transcendence: Christian hope in an age of technological enhancement* (pp. 177–192). Washington, DC: Georgetown University Press.

Mercer, C., & Trothen, T. J. (Eds.) (2014). *Religion and transhumanism: The unknown future of human enhancement*. New York, NY: Praeger.

Meumann, M. (2012). Der Zeitgeist vor dem Zeitgeist. *Genius saeculi* als historiographisches, mnemonisches und gegenwartsdiagnostisches Konzept im 17. und 18. Jahrhundert. In A. Landwehr (Ed.), *Frühe neue Zeiten. Zeitwissen zwischen Reformation und Revolution* (pp. 283–317). Bielefeld, Germany: Transcript.

Minow, N. (1961). Speech at the National Association of Broadcasters Convention, transcript. Available from: http://www.janda.org.

Minsky, M. (1986). *The society of mind*. New York, NY: Simon & Schuster.

Minsky, M. (2006). *The emotion machine: Commonsense thinking, artificial intelligence, and the future of the human mind*. New York, NY: Simon and Schuster.

Mitcham, C. (1994). *Thinking through technology: The path between engineering and philosophy*. Chicago, IL: University of Chicago Press.

Monod, P. K. (2013). *Solomon's secret arts: The occult in the age of Enlightenment*. New Haven, CT: Yale University Press.

Moos, T. (2005). *Staatszweck und Staatsaufgaben in den protestantischen Ethiken des 19. Jahrhunderts*. Münster, Germany: Lit.

Moos, T. (2012). Die kosmologische Verlegenheit moderner Religion. Zur praktischen Notwendigkeit einer theologischen Kosmologie. In F. Vogelsang, H. Meisinger, & T. Moos (Eds.), *Gibt es eine Ordnung des Universums? Der Kosmos zwischen Messung, Anschauung und religiöser Deutung. Festschrift für Professor Dr. Jürgen Hübner* (pp. 201–229). Bonn, Germany: Evangelische Akademie im Rheinland.

Moravec, H. (1988). *Mind children: The future of robot and human intelligence*. Cambridge, MA: Harvard University Press.

Moravec, H. (1999). *Robot: More machine to transcend mind*. Oxford, England: Oxford University Press.

More, M. (1996 [1990]). Transhumanism: Towards a futurist philosophy. Available from : http://www.scribd.com.

More, M. (2013). The philosophy of transhumanism. In M. More and N. Vita-More (Eds.), *The transhumanist reader: Classical and contemporary essays on the science, thechnology, and philosophy of the human future* (pp. 3–17). West Susssex, England: John Wiley & Sons.

More, M., & Vita-More, N. (2003). Principles of extropy, version 3.11.2003. Original version. The extroprian principles. *Extropy, 5* (5) (May 1990). Available from: www.extropy.org/principles.htm.

More, M., & Vita-More, N. (Eds.) (2013). *The transhumanist reader: Classical and contemporary essays on the science, technology and philosophy of the human future*. Malden, MA: Wiley Blackwell.

Morozov, E. (2013, March 2). The perils of perfection." *New York Times*, p. SR1.

Mumford, L. (1967). *The myth of the machine: Technics and human development (volume 1)*. New York: Harcourt Brace Jovanovich.

Mumford, L. (1970). *The myth of the machine: The pentagon of power (volume 2)*. London, England: Secker and Warburg.
Münch, N. (2012). Körperverachtung oder Phänomenologie der Leiblichkeit? Eine Kritik am Transhumanismus. In M. Eilers, K. Grüber & C. Rehmann-Sutter (Eds.), *Verbesserte Körper—gutes Leben? Bioethik, Enhancement und die Disability Studies* (pp. 287–310). Frankfurt am Main, Germany: Peter Lang.
Musil, R. (1970). *Der Mann ohne Eigenschaften* (special ed.). Reinbek bei Hamburg, Germany: Rowohlt.
Myers, N. (2008): Conjuring machinic life. In *Spontaneous Generations, 2* (1), 112–121.
Naam, R. (2005). *More than human: Embracing the promise of biological enhancement*. New York, NY: Broadway Books.
Nagel, T. (1974). What is it like to be a bat? *Philosophical Review, 83* (4), 435–450.
National Academies. (2013). Forum on synthetic biology. Washington, DC, October 21. Available from: http://sites.nationalacademies.org.
Nayar, P. K. (2014). *Posthumanism*. Cambridge, England: Polity Press.
Nehlig, A. (2010). Is caffeine and cognitive enhancer. *Journal of Alzheimer's Disease, 20*, 85–94.
Nelkin, D. (1987). *Selling science: How the press covers science and technology*. New York, NY: W. H. Freeman.
Nelkin, D., & Pollak, M. (1981). *The atom besieged*. Cambridge, MA: MIT Press.
Nerlich, B. (2005, December 22). From nautilus to nanobo(a)ts: The visual construction of nanoscience. *AZoNano—Online Journal of Nanotechnology*. Available from: www.azonano.com.
Neusüß, A. (Ed.). (1986). *Utopie: Begriff und Phänomen des Utopischen*. Frankfurt/Main, Germany: Campus-Verlag.
Nida-Rümelin, J., & Kufeld, K. (Eds.). (2011). *Die Gegenwart der Utopie. Zeitkritik und Denkwende*. Freiburg, Germany: Alber.
Niemeyer, C. (2011). *Nietzsche verstehen: Eine Gebrauchsanweisung*. Darmstadt, Germany: Lambert Schneider.
Nietzsche, F. (1968). *The will to power*. (W. Kaufmann & R. J. Hollingdale, Trans.). London, England: Weidenfeld and Nicholson.
Nietzsche, F. (1974). *The gay science*. (W. Kaufmann, Trans.). New York, NY: Vintage Press.
Nietzsche, F. (1988). *Also sprach Zarathustra*. Munich, Germany: dtv.
Noble, D. (1999). *The religion of technology: The divinity of man and the spirit of invention* (2nd ed.). New York, NY: Penguin.
Nordmann, A. (2004). *Converging technologies: Shaping the future of European societies*. European Commission Research Report. Available from: http://ec.europa.eu.
Nordmann, A. (2007). If and then: A critique of speculative nanoethics. *NanoEthics, 1*, 31–46
Nordmann, A. (2008). *Technikphilosophie zur Einführung*. Hamburg, Germany: Junius.
Nordmann, A. (2009). European experiments. *Osiris, 24* (1), 278–302.
Nordmann, A. (2010a). Enhancing material nature. In K. L. Kjølberg & F. Wickson (Eds.), *Nano meets macro: Social perspectives on nanoscale sciences and technologies* (pp. 283–306). Singapore: Pan Stanford.
Nordmann, A. (2010b). A forensics of wishing: Technology assessment in the age of technoscience. *Poiesis & Praxis: International Journal of Technology Assessment and Ethics of Science, 7* (1–2), 5–15.
Nordmann, A. (2014a). Responsible innovation: The art and craft of anticipation. *Journal of Responsible Innovation, 1* (1), 87–98.

Nordmann, A. (2014b): Sanfte Technik: Vom Mythos der Maschine zum Mythos nicht-maschineller Maschinen. In A. Gelhard & A. Kaminski (Eds.). *Zur Philosophie informeller Technisierung* (pp. 21–40). Darmstadt, Germany: Wissenschaftliche Buchgesellschaft.

Nordmann, A. (Forthcoming). Werkwissen oder: How to express things in works. In *Jahrbuch Technikphilosophie* (vol. 1). Zürich, Switzerland: Diaphanes.

Nordmann, A., & Rip, A. (2009). Mind the gap revisited. *Nature Nanotechnology, 4* (5), 273–274.

Nordmann, A., & Schwarz, A. (2010). Lure of the "yes": The seductive power of technoscience. In M. Kaiser, M. Kurath, S. Maasen, & C. Rehmann-Sutter (Eds.), *Governing future technologies: Nanotechnology and the rise of an assessment regime* (pp. 255–277) (Sociology of the Sciences Yearbook 27). Dordrecht, the Netherlands: Springer.

Nowotny, H., Scott, P., & Gibbons, M. (2004). *Wissenschaft neu denken: Wissenschaft und Öffentlichkeit in einem Zeitalter der Ungewißheit.* Weilerswist, Germany: Velbrück.

Nussbaum, Martha C. (1999). *Gerechtigkeit oder das gute Leben.* (I. Utz, Trans.). Frankfurt am Main, Germany: Suhrkamp Verlag.

Nye, D. (1994). *American technological sublime.* Cambridge, MA: MIT Press.

Nynäs, P., Lassander, M., & Utriainen, T. (Eds). (2012). *Post-secular society.* New Brunswick, NJ: Transactions Publishers.

Nye, D. (2006). *Technology matters: Questions to live with.* Cambridge, MA: MIT Press.

Owen, R., Macnaghten, P., & Stilgoe, J. (2012). Responsible research and innovation: From science in society to science for society, with society. *Science and Public Policy, 39* (6), 751–760.

Pang, A. S.-K. (2013). *The distraction addiction.* New York, NY: Little Brown.

Paschen, H., & Petermann, Th. (1992). Technikfolgenabschätzung—ein strategisches Rahmenkonzept für die Analyse und Bewertung von Technikfolgen. In T. Petermann (Ed.), *Technikfolgen-Abschätzung als Technikforschung und Politikberatung* (pp. 19–42). Frankfurt, Germany: Campus.

PCSBI (U.S. Presidential Commission for the Study of Bioethical Issues). (2010). *New directions: The ethics of synthetic biology and emerging technologies.* Washington, DC: PCSBI. Available from: www.bioethics.gov.

Pence, G. E. (1998). *Who's afraid of human cloning?* Lanham, MD: Rowman and Littlefield.

Pew Forum on Religion & Public Life. (2012). *The global religious landscape: A report on the size and distribution of the world's major religious groups as of 2010.* Washington, DC: Pew Research Center.

Picht, G. (1971). *Prognose Utopie Planung.* Stuttgart, Germany: Klett.

Pico della Mirandola, G. (1988). *Über die Würde des Menschen.* (A. Buck, Ed.). Zürich, Switzerland: Manesse. English: (1948). Oration on the dignity of man. In E. Cassirer, P. O. Kristeller, & J. H. Randall Jr. (Eds.), *The Renaissance philosophy of man* (pp. 223–255). Chicago, IL: University of Chicago Press.

Pinkard, T. (2002). *German philosophy, 1760-1860: The legacy of idealism.* Cambridge, England: Cambridge University Press.

Pitt, J. (2011). *Doing philosophy of technology: Essays in a pragmatist spirit.* Dordrecht, the Netherlands: Springer.

Plessner, H. (1975). *Die Stufen des Organischen und der Mensch.* Berlin, Germany: de Gruyter.

Pollack, A. (2013, May 7). A dream of trees aglow at night. *New York Times*, Energy & Environment, B-1.

Pollack, A. (2014, March 5). A genetic entrepreneur sets his sights on aging and death. *New York Times*, B1.

Pollack, A. (2015, January 1). By "editing" plant genes, companies avoid regulation. *The New York Times*, Energy & Environment, B-1.
Polt, R. (2104). A Heideggerian critique of cyberbeing. In H. Pedersen & M. Altman (Eds.) *The horizons of authenticity* (pp. 179–199). Dordrecht, the Netherlands: Springer.
Popper, R. K. (1957). *The poverty of historicism*. London, England: Routledge.
Popper, R. K. (1962). *The open society and its enemies*. (Vol. 1). London, England: Routledge.
Popper, R. K. (1963). *Conjectures and refutations: The growth of scientific knowledge*. London, England: Routledge.
Portola Institute. (1969). *Whole earth catalog*. Menlo Park, CA: Portola Institute, Inc.
President's Commission for the Study of Ethical Problems in Medicine and Biomedical and Behavioral Research. (1982). *Splicing life: A report on the social and ethical issues of genetic engineering with human beings*. Washington, DC.
Presidential Commission on Bioethical Issues in Biomedical Research. (2010). "New directions: The ethics of synthetic biology and emerging technologies." Washington, DC. Available from: http://bioethics.gov/sites/default/files/PCSBI-Synthetic-Biology-Report-12.16.10_0.pdf.
Prisco, G. (2013). Transcendent engineering. In M. More & N. Vita-More (Eds.), *The transhumanist reader: Classical and contemporary essays on the science, technology and philosophy of the human future* (pp. 234-242). Malden, MA: Wiley-Blackwell.
Ptolemy, R. B. (Director). (2009). *Transcendent man: The life and times of Ray Kurzweil* [Documentary]. United States: Ptolemaic Films.
Pustovrh, T., & Mali, F. (2013). Exploring some challenges of the pharmaceutical cognitive enhancement discourse: Users and policy recommendations. *Neuroethics*, *7* (2): 137–158.
Rabinbach, A. (1992). *The human motor: Energy, fatigue and the origins of modernity*. Los Angeles, CA: University of California Press.
Rabinow, P., and Bennett, G. (2009). Synthetic biology: Ethical ramifications. *Systems and Synthetic Biology, 3* (1–4), 99–108.
Rajan, K. S. R. 2006. *Biocapital: The constitution of postgenomic life*. Durham, NC: Duke University Press.
Rampell, C. (2013, July 8). Coveting not a corner office, but time at home. *New York Times*, p. A1.
Ranisch, R., & Sorgner, S. L. (Eds.). (2014). *Post- and transhumanism: An introduction*. Frankfurt am Main, Germany: Peter Lang.
Ray Kurzweil: *A University for the Coming Singularity*. (2009, February). *TED*. Available from: http://www.ted.com/talks.
Reider, R. (2009). *Dreaming the biosphere: The theater of all possibilities*. Albuquerque, NM: University of New Mexico Press.
Rein, L. (2009, June 29). Singularity University—day one. *Humanity+*. Available from: http://hplusmagazine.com.
Reuleaux, F. (1875). *Theoretische Kinematik. Grundzüge einer Theorie des Maschinenwesens*. Braunschweig, German: Vieweg. In English: *Kinematics of machinery: Outlines of a theory of machines* (A. W. B. Kennedy, Trans. & Ed.). London: Macmillan and Co., 1876.
Rheingold, H. (1985). *Tools for thought: The people and ideas behind the next computer revolution*. New York, NY: Simon & Schuster.
Rip, A., Misa, T., & Schot, J. (Eds.) (1995). *Managing technology in society*. London, England: Pinter Publishers.
Risen, J. (2013, July 7). Privacy group to ask Supreme Court to stop N.S.A.'s phone spying program. *New York Times*.

Riskins, J. (2003). Eighteenth-century wetware. *Representations, 83*, 97–125.
Ritter, J. (1972). Fortschritt. In J. Ritter (Ed.), *Historisches Wörterbuch der Philosophie. Band 2: D–F* (pp. 1032–1059). Darmstadt, Germany: Wissenschaftliche Buchgesellschaft.
Roco, M.C., & Bainbridge, W. S. (Eds.) (2002). *Converging technologies for improving human performance*. Arlington, VA: National Science Foundation, Department of Commerce.
Rohan, T. (2013, July 7). Beyond the finish line. *New York Times*, p. D1.
Rose, N. (2007). *The politics of life itself: Biomedicine, power, and subjectivity in the twenty-first century*. Princeton, NJ: Princeton University Press.
Rosenblum, B., & Kuttner, F. (2011). *Quantum enigma: Physics encounters consciousness*. New York, NY: Oxford University Press, 2011.
Rosenthal, B. G. (2002). *New myth, new worlds: From Nietzsche to Stalinism*. University Park, PA: Pennsylvania State University Press.
Roszak, T. (1969). *The making of a counter culture*. New York, NY: Doubleday.
Rowan, D. (2013, May 6). On the exponential curve: Inside Singularity University. *Wired UK*. Available from: http://www.wired.co.uk.
Saage, R. (2009). Zur Differenz zwischen dem konvergenztechnologischen Futurismus und der klassischen Utopietradition. In R. Steltemeier, S. Dickel, S. Gaycken, & T. Knobloch (Eds.), *Neue Utopien. Zum Wandel eines Genres* (pp. 116–134). Heidelberg, Germany: Manutius.
Saenz, A. (2012, June 16). Singularity University 2012: The graduate studies program is near and we have insider coverage. *SingularityHUB*. Available from: http://singularityhub.com.
Sandal, N. A. (2012). The clash of public theologies? Rethinking the concept of religion in global politics. *Alternatives: Global, Local, Political, 37* (1), 66–83.
Sandberg, A. (2013). An overview of models of technological singularity. In M. More & N. Vita-More (Eds.), *The transhumanist reader: Classical and contemporary essays on the science, technology and philosophy of the human future* (pp. 376-394). Malden, MA: Wiley-Blackwell.
Sandberg, S., with N. Scovell. (2013). *Lean in: Women, work, and the will to lead*. New York, NY: Knopf.
Sandel, M. (2007). *The case against perfection: Ethics in the age of genetic engineering*. Cambridge, MA: Belknap Press of Harvard University Press.
Sandercock, L. (2006). Spirituality and the urban professions: The paradox at the heart of planning. *Planning Theory and Practice, 7* (1), 65–97.
Sargent, L. T. (2010). *Utopianism: A very short introduction*. Oxford, England: Oxford University Press, 2010.
Savulescu, J. (2001). Procreative beneficence: Why we should select the best children. *Bioethics, 15* (5–6), 413–426.
Savulecu, J., & Bostrom, N. (Eds.). (2009). *Human enhancement*. Oxford, England: Oxford University Press.
Savulescu, J., & Kahane, G. (2009). The moral obligation to create children with the best chance of the best life. *Bioethics, 23* (5), 274–290.
Schaper-Rinkel, P. (2013). Politiken des Human Enhancement. Transhumanistische Versprechen und die Analyse von technowissenschatlichen Zukünften. *Technikfolgenabschätzung, Theorie und Praxis, 22* (1), 16–31.
Scharff, R. C., & Dusek, V. (2005). *Philosophy of technology: The technological condition, an anthology*. Oxford, England: Blackwell Publishing.
Scheler, M. (2005). *Die Stellung des Menschen im Kosmos*. Bonn, Germany: Bouvier.

Scheunpflug, A. (2001). *Biologische Grundlagen des Lernens*. Berlin, Germany: Cornelsen Scriptor.
Schmidt, M., Kelle, A., Ganguli-Mitra, A., & Vriend, H. (Eds.). (2010). *Synthetic biology: The technoscience and its societal consequences*. Dordrecht, the Netherlands: Springer.
Schölderle, T. (2012). *Geschichte der Utopie: eine Einführung*. Stuttgart,Germany: Böhlau.
Schumacher, E. F. (1973). *Small is beautiful: Economics as if people mattered*. New York, NY: Harper and Row.
Schummer, J. (2009). *Nanotechnologie. Spiele mit Grenzen*. Frankfurt am Main, Germany: Suhrkamp.
Schumpeter. (2014, June 24). The not so golden state. *The Economist*. Available from: http:// www.economist.com.
Scott, J. C. (1998). *Seeing like a state: How certain schemes to improve the human condition have failed*. New Haven, CT: Yale University Press.
Scott, N. (2009). Research ethics: European and Asian perspective, global challenges. In M. Ladikas (Ed.), *Embedding society in science and technology policy: European and Chinese perspectives* (pp. 21–39). Luxembourg: Office for Official Publications of the European Communities.
Scully, J. L. (2012). Auf moralisch unsicherem Terrain: Über Embodiment, Enhancement, und Normativität. In M. Eilers, K. Grüber, & C. Rehmann-Sutter (Eds.), *Verbesserte Körper— gutes Leben? Bioethik, Enhancement und die Disability Studies* (pp. 147–163). Frankfurt am Main, Germany: Peter Lang.
Segal, H. P. (2005). *Technological utopianism in American culture*. Syracuse, NY: Syracuse University Press.
Seligman, A. (1988). Christian utopias and Christian salvation: A general introduction. *International Journal of Comparative Sociology, 39* (1–2), 13–29.
Sen, S. (1974). *A richer harvest: New horizons for developing countries*. New Delhi, India: Tata McGraw-Hill.
Sheehan, T. (1998). Nihilism: Heidegger/Jünger/Aristotle.I In B. C. Hopkins (Ed.), *Phenomenology: Japanese and American perspectives* (pp. 273–316). Dordrecht, the Netherlands: Kluwer Academic Publishers.
Sheehan, T. (2002). Nihilism and its discontents. In F. Raffoul & D. Pettigrew (Eds.), *Heidegger and practical philosophy* (pp. 275–300). Albany, NY: State University of New York Press.
Sheehan, T. (2014). *Making sense of Heidegger*. Lanham, MD: Rowman & Littlefield.
Sidiqqi, A. (2010). Competing technologies, national(ist) narratives, and universal claims: Toward a global history of space exploration. *Technology and Culture, 51* (2), 425–443.
Silver, L. M. (2000). Reprogenetics: How reproductive and genetic technologies will be combined to provide new opportunities for people to reach their reproductive goals. In G. Stock & J. Campbell (Eds.), *Engineering the human germline* (pp. 57–71). New York, NY: Oxford University Press.
Simmel, G. (1998). *Philosophische Kultur: Über das Abenteuer, die Geschlechter und die Krise der Moderne*. Berlin, Germany: Wagenbach.
Simon, B. (2003). Toward a critique of posthuman futures. *Cultural Critique, 53* (1), 1–9.
Simondon, G. (1958). *On the mode of existence of technical objects*. London, England: University of Western Ontario.
Singer, P. (2002). *Animal liberation*. New York, NY: Harper.
Singer, P. (2011a). *Practical ethics* (3rd ed.). Cambridge, England: Cambridge University Press.

Singer, P. (2011b). Vorwort [Foreword]. (S. L. Sorgner, Trans.). In T.-L. Eissa & S. L. Sorgner (Eds.), *Geschichte der Bioethik: Eine Einführung* (pp. 13–15). Paderborn, Germany: Mentis.

Singularity University. (2009, July 30). SU names GSP-09 students. *Singularity University Blog*. Available from: http://www.kurzweilai.net.

Singularity University. (2012, December 6). Singularity University acquires Singularity Summit. *Singularity University Blog*. Available from: http://singularityu.org.

Singularity University. (2015a). Core curriculum. Available from: http://singularityu.org.

Singularity University. (2015b). Corporate founders. Available from: http://singularityu.org.

Singularity University. (2015c). Graduate studies program. Available from: http://singularityu.org.

Sloterdijk, P. (2008). *Regeln für den Menschenpark: Ein Antwortschreiben zu Heideggers Brief über den Humanismus*. Frankfurt am Main, Germany: Suhrkamp.

Sloterdijk, P. (2009). *Du mußt dein Leben ändern. Über Anthropotechnik*. Frankfurt am Main, Germany: Suhrkamp.

Smalley, R. (2003a). Smalley responds. *Chemical & Engineering News, 81*, 39–40.

Smalley, R. (2003b). Smalley concludes. *Chemical & Engineering News, 81*, 41–42.

Smith, C. 2010. *What is a person? Rethinking humanity, social life, and the moral good from the person up*. Chicago, IL: University Of Chicago Press.

Smith, J. Z. (1971). Native cults in the Hellenistic period. *History of Religions, 11* (2), 236–249.

Smith, J. Z. (1982) *Imagining religion: From Babylon to Jonestown*. Chicago, IL: The University of Chicago Press.

Smith, J. Z. (1990). *Drudgery divine: On the comparison of early Christianities and the religions of late antiquity*. Chicago, IL: University of Chicago Press.

Smith, M. R., & Marx, L. 1994. *Does technology drive history? The dilemma of technological determinism*. Cambridge, MA: MIT Press.

Socrates. (2011, September 11). Salim Ismail on Singularity 1 on 1: We are already gods, we might as well start acting as such. Available from: http://www.singularityweblog.com.

Solomon, M., & Bostrom, N. (2005, January–February). Transhumanist dream. *Foreign Policy, 146*, 4.

Sorgner, S. L. (2007). *Metaphysics without truth: On the importance of consistency within Nietzsche's philosophy* (2nd rev. ed.). Milwaukee, WI: University of Marquette Press.

Sorgner, S. L. (2009). Nietzsche, the overhuman, and transhumanism. *Journal of Evolution and Technology, 20* (1), 29–42.

Sorgner, S. L. (2010). *Menschenwürde nach Nietzsche. Die Geschichte eines Begriffs*. Darmstadt, Germany: WBG.

Sorgner, S. L. (2013a): Evolution, education, and genetic enhancement. In S. L. Sorgner & B.-R. Jovanovic (Eds.), *Evolution and the future: Anthropology, ethics, religion* (pp. 85–100). Frankfurt am Main, Germany: Peter Lang.

Sorgner, S. L. (2013b). Human dignity 2.0: Beyond a rigid version of anthropocentrism. *Trans-Humanities, 6* (1), 135–159.

Sorgner, S. L. (2013c): Paternalistic cultures versus nihilistic cultures. *European Journal of Science and Theology, 9* (1), 55–60.

Sorgner, S. L. (2014). Is there a "moral obligation to create children with the best chance of the best life"? *Humana Mente. Journal of Philosophical Studies, 26*, 199–212.

Spaemann, R. (1984). Tierschutz und Menschenwürde." In Ursula M. Händel (Ed.). *Tierschutz—Testfall unserer Menschlichkeit* (pp. 71–81). Frankfurt am Main, Germany: Fischer Taschenbuchverlag.

Spencer, H. (1857). Progress: Its law and cause. *Westminster Review, 67* (132), 445–485.
STAC (EC President's Science and Technology Advisory Council). (2013): Science for an informed, sustainable and inclusive knowledge society. Policy paper by President Barroso's Science and Technology Advisory Council, Brussels, August 29, 2013. Available from: http://ec.europa.eu.
Stiegler, B. (1998). *Technics and time, 1: The fault of Epimetheus.* (R. Beardsworth & G. Collins, Trans.). Stanford, CA: Stanford University Press.
Stiegler, B. (2009). *Technics and time, 2: Disorientation.* (S. Barker, Trans.). Stanford, CA: Stanford University Press.
Stiegler, B. (2010). *Technics and time, 3: Cinematic time and the question of malaise.* (S. Barker, Trans.). Stanford, CA: Stanford University Press.
Stock, G. (1993) *Metaman: The merging of humans and machines into a global superorganism.* New York, NY: Simon & Schuster.
Stock, G. (2002). *Redesigning humans: Our inevitable genetic future.* New York, NY: Houghton Mifflin.
Storey, D. (2015). *Naturalizing Heidegger: His confrontation with Nietzsche, his contributions to environmental philosophy.* Albany, NY: State University of New York Press.
Strauss, L. L. (1954, September 17). Speech to the National Association of Science Writers, New York City, September 16. *New York Times.*
Swimme, B. T., & Tucker, M. E. (2011). *Journey of the universe.* New Haven, CT: Yale University Press.
Szyszczak, E. (2006). Experimental governance: The open method of coordination. *European Law Journal, 12* (4), 486–502.
Taubes, J. (1991). *Abendländische Eschatologie. Mit einem Anhang.* München, Germany: Matthes und Seitz.
Taubes, S. A. (1954). The Gnostic foundation of Heidegger's nihilism. *The Journal of Religion, 24* (3), 155–172.
Taylor, C. (2003). *Modern social imaginaries.* Durham, NC: Duke University Press.
Taylor, C. (2007). *A secular age.* Cambridge, MA: Belknapp Press of the Harvard University Press.
Teilhard de Chardin, P. (1965). *The phenomenon of man.* (B. Wall, Trans.). New York, NY: Harper & Row.
Thompson, D. (1997). *Das Ende der Zeiten. Apokalyptik und Jahrtausendwende.* Hildesheim, Germany: Claassen.
Thweatt-Bates, J. (2012). *Cyborg selves: A theological anthropology of the posthuman.* Burlington, VT: Ashgate.
Tillich, P. (1971). Critique and justification of Utopia. In F. E. Manuel (Ed.), *Utopias and utopian thought* (pp. 296–309). London, England: Souvenir Press.
Tirosh-Samuelson, H. (2003). *Happiness in premodern Judaism: Virtue, knowledge and well-being.* Cincinnati, OH: Hebrew Union College Press.
Tirosh-Samuelson, H. (2011). Engaging transhumanism. In G. R. Hansell & W. Grassie (Eds.), *H+/: Transhumanism and its critics* (pp. 19–54). Philadelphia, PA: Metanexus.
Tirosh-Samuelson, H. (2012a). Science and the betterment of humanity: Three British prophets of transhumanism. In H. Tirosh-Samuelson and K. L. Mossman (Eds.), *Building Better Humans? Refocusing the Debate on Transhumanism* (pp. 55–82). Frankfurt am Main, Germany: Peter Lang.
Tirosh-Samuelson, H. (2012b). Transhumanism as a secularist faith. *Zygon, 47* (4), 710–734.

Tirosh-Samuelson, H. (2014a). Religion. In R. Ranisch & S. L. Sorgner (Eds.), *Post- and transhumanism: An introduction* (vol. 1, pp. 49-71). Frankfurt, Germany: Peter Lang International Academic Publishers.

Tirosh-Samuelson, H. (2014b). Utopianism and eschatology: Judaism engages transhumanism." In C. Mercer & T. J. Trothen (Eds.), *Religion and transhumanism: The unknown future of human enhancement* (pp. 161–180). New York, NY: Praeger.

Tirosh-Samuelson, H., & Mossman, K. (Eds.) (2012). *Building better humans? Refocusing the debate on transhumanism*. Frankfurt, Germany: Peter Lang.

Transhumanism. (2012). Special issue of *Zygon: Journal of Religion and Science, 47* (5), 659–795. This issue contains four essays of H. Tirosh-Samuelson, Robert Geraci, James Hughes, and Ron Cole-Turner.

Turner, F. M. (1974). *Between science and religion: The reaction to scientific naturalism in late Victorian England*. New Haven, CT: Yale University Press.

Turner, F. M. (1981). John Tyndall and Victorian scientific naturalism. In W. H. Brock, N. D. McMillan, & R. C. Mollan (Eds.), *John Tyndall: Essays on a natural philosopher* (pp. 169–180). Dublin, Ireland: The Royal Dublin Society.

van Dam, R. (2014). Ask the expert: Coffee and health. Harvard School of Public Health, The Nutrition Source. Available from: http://www.hsph.harvard.edu.

Vattimo, G (1997).*Glauben–Philosophieren*. (C. Schultz, Trans.). Stuttgart: Reclam. Original Italian: *Credere di Credere* (Garzanti Editore, 1996). English: *Belief* (L. D'Isanto and D. Webb, Trans.) (Stanford University Press, 1999).

Veblen, T. 1906. The place of science in modern civilization. *American Journal of Sociology, 11* (5), 585–609.

Venter, C. (2013). *Life at the speed of light*. New York, NY: Penguin.

Veronico, N. A. (2006). *Moffett Field*. Charleston, S.C: Arcadia Publishing.

Vinge, V. (1993). The coming technological singularity. *Whole Earth Review*, (Winter). Reprinted in 2013 as "Technological singularity" in M. More & N. Vita-More (Eds.), *The transhumanist reader: Classical and contemporary essays on the science, technology and philosophy of the human future* (pp. 365–375). Malden, MA: Wiley-Blackwell.

von Schomberg, R. (2009). Organizing collective responsibility: On precaution, codes of conduct and understanding public debate. Keynote lecture at the meeting of the Society for the Study of Nanoscience and Emerging Technologies, Seattle, WA, September 11. Available from: nanotechproject.org.

von Schomberg, R. (2012). Prospects for technology assessment in a framework of responsible research and innovation. In M. Dusseldorp & R. Beecroft (Eds.), *Technikfolgen abschätzen lehren: Bildungspotenziale transdisziplinärer Methoden* (pp. 39–62). Wiesbaden, Germany: Vs Verlag.

von Schomberg, R. (2013). A vision of responsible research and innovation. In R. Owen, J. Bessant, & M. Heintz (Eds.), *Responsible innovation: Managing the responsible emergence of science and innovation in society* (pp. 51–74). London, England: John Wiley.

von Schomberg, R., & Davies, S. (Eds.) (2010). *Understanding public debate on nanotechnologies*. EUR 24169. Brussells, Belgium: European Commission.

Ward, G. (2000). *Cities of God*. London, England: Routledge.

Waters, B. (2011). Whose salvation? Which eschatology? Transhumanism and Christianity as contending salvific religions. In R. Cole-Turner (Ed.), *Transhumanism and transcendence: Christian hope in an age of technological enhancement* (pp. 163–175). Washington, DC: Georgetown University Press.

Watson, J. D. E. (2009). The harm of premature death: Immortality—the transhumanist challenge. *Ethical Perspectives, 16* (4), 435–458.
Weber, M. (2004 [1918]). The vocation of science. In S. Whimster (Ed.), *The essential Weber: A reader* (pp. 270–287). London, England: Routledge.
Weingart, P. (2001). *Die Stunde der Wahrheit? Zum Verhältnis der Wissenschaft zu Politik, Wirtschaft und Medien in der Wissensgesellschaft.* Goettingen, Germany: Velbrueck Verlag.
Wells, H. G. (1994 [1937]). A note on world brain and world mind. In *World brain: H. G. Wells on the future of world education* (pp. 123–124). London, England: Adamantine Press.
Wickler, W., & Seibt, U. (1998). *Männlich—Weiblich: Ein Naturgesetz und seine Folgen.* Heidelberg, Germany: Spektrum.
Wickson, F., & Wynne, B. (2012). The anglerfish deception: The light of proposed reform in the regulation of GM crops hides underlying problems in EU science and governance. *EMBO reports, 13* (2), 100–105.
Wietteriede, H. (2009). *Eine Einführung in die Philosophische Anthropologie: Max Scheler, Helmuth Plessner, Arnold Gehlen.* Frankfurt am Main, Germany: Peter Lang International Acaemic Publishers.
Wilber, K. (1995). *A brief history of everything.* Boston, MA: Shambhala.
Williams, R. (2014). Our technological age, from the inside out. *Technology and Culture, 55* (2), 461–476.
Wilson, C. (2013). *Defcad.com.* Available https://www.youtube.com.
Winner, L. (1986). *The whale and the reactor: A search for limits in an age of high technology.* Chicago, IL: University of Chicago Press.
Winner, L. (2005). Resistance is futile: The posthuman condition and its advocates. In H. W. Baillie & T. Casey (Eds.), *Is human nature obsolete? Genetics, bioengineering, and the future of the human condition* (pp. 385–411). Cambridge, MA: MIT Press.
Winterhager-Schmid, L. (2000). "'Groß' und 'klein'—Bedeutung der Erfahrung mit Generationendifferenz im Prozeß des Heranwachsens." In L. Winterhager-Schmid (Ed.), *Erfahrung mit Generationendifferenz* (pp. 15–37). Weinheim, Germany: Deutscher Studien-Verlag.
Wolbring, G. (2007). New and emerging sciences and technologies, ableism, transhumanism and religion, faith, theology, and churches. *Madang—International Journal of Contextual Theology, 7,* 79–112.
Wolfe, A. J. (2002). Germs in space: Joshua Lederberg, exobiology, and the public imagination, 1958–1964. *Isis, 93* (2), 183–205.
Wolfe, C. (2010). *What is posthumanism?* Minneapolis, MN: University of Minnesota.
World Commission on Environment and Development. (1987). *Our common future.* Oxford, England: Oxford University Press.
Wynne, B. (2006). Public engagement as a means of restoring public trust in science: Hitting the notes, but missing the music? *Public Health Genomics, 9* (3), 211–220.
Young, S. (2006). *Designer evolution: A transhumanist manifesto.* Amherst, NY: Prometheus Books.
Zachary, G. P. (1990, January 23). Artificial reality: Computer simulations one day may provide surreal experiences. *The Wall Street Journal,* pp. A1, A9.
Zachary, G. P. (1997). *Endless frontier: Vannevar Bush, engineer of the American century.* New York, NY: Free Press.
Zachary, G. P. (2005, May 15). In Starbucks country, caffeine hooks us and constantly lures us back to feed our addiction. *San Francisco Chronicle.* Available from: www.sfgate.com.

Zhang, J. Y. (2012). The art of trans-boundary governance: The case of synthetic biology. *Systems and Synthetic Biology, 7* (3), 107–114.
Zimmerman, M. E. (1990). *Heidegger's confrontation with modernity: Technology, politics, art*. Bloomington, IN: Indiana University Press.
Zimmerman, M. E. (2008). The singularity: A crucial phase in divine self-actualization?" *Cosmos and History: The Journal of Natural and Social Philosophy, 4* (1–2), 347–370.
Žižek, S. (2003). *The puppet and the dwarf: The perverse core of Christianity*. Cambridge, MA: MIT Press.
Zülsdorf, T. B., Coenen, C., Ferrari, A., Fiedeler, U., Milburn, C., & Wienroth, M. (Eds.) (2011). *Quantum engagements: Social reflections of nanoscience and emerging technologies*. Heidelberg, Germany: AKA.

Contributors

Margarita Boenig-Liptsin is a Ph.D. candidate at Harvard University in the History of Science Department with a secondary field in science and technology studies and in philosophy at the Sorbonne. She studies the role of technology in the human condition by investigating how material technologies and peoples' imaginations of them are coproduced with the sense of self and of the world. Her doctoral dissertation focuses on the global rise of computer literacy as ideal and practice deployed at the end of the 20th century to form a particular kind of citizen for the "information age." Her research is distinctly interdisciplinary, informed by the methods and ideas of science and technology studies, intellectual and cultural history, anthropology, and philosophy.

Micha Brumlik (Ph.D., University of Heidelberg in social work, 1973) is Professor Emeritus of education at the Johann Wolfgang Goethe University of Frankfurt am Main, Germany. From 1981 to 2000, he taught education at the University of Heidelberg and, from 2000 to 2005, served as director of the Fritz Bauer Institute for the Study and Documentation of the History of the Holocaust. Writing on education, human rights, Jewish history, Jewish identity in contemporary Germany, and science and culture, Brumlik is an influential public intellectual. Among his many publications are *Infantile sexuality* (2012); *Sigmund Freud: Der Denker des 20. Jahrhunderts* (2006); *Aus Katastrophen lernen: Grundlagen zeitgeschichtlicher Bildung in menschenrechtlicher Absicht* (2004); *Bildung und Glück: Versuch einer Theorie der Tugenden* (2002); and *Deutscher Geist und Judenhass: Das Verhältnis des philosophischen Idealismus zum Judentum* (2000).

John H. Evans (Ph.D., Princeton University, 1998) is professor of sociology at the University of California-San Diego. His research focuses on religion, culture, politics, and science, especially on public bioethical debate. His first book, *Play-*

ing God? Human genetic engineering and the rationalization of public bioethical debate (2002), won the Distinguished Book Award from the Religion Section of the American Sociological Association. This book explains the emergence of the profession of bioethics, as well as why public debates are structured as they are. His second book, *Contested reproduction: Genetic technologies, religion and public debate* (2010), examines how ordinary religious people in the United States talk about reproductive genetic technologies. His third book, *The history and future of bioethics: A sociological view* (2012), uses sociological reasoning to examine the history of bioethics in detail, offering a solution to the crisis in bioethics debate. Dr. Evans is also the author of numerous essays and book reviews.

Elaine Graham (Ph.D., University of Manchester, 1993) is the Grosvenor Research Professor of Practical Theology at the University of Chester in Chester, United Kingdom. Before joining the University of Chester, she was (until October 2009) the Samuel Ferguson Professor of Social and Pastoral Theology at the University of Manchester. In March 2014, she was installed as Canon Theologian of Chester Cathedral. She is a practical theologian who writes on media, culture, and religion; theology and practice; religion and public life, especially "postsecular" public theology; religion, culture, and gender; and technology and the posthuman condition. Professor Graham was the president of the International Academic of Practical Theology (2005–2007) and a member of the Archbishops' Commission on Urban Life and Faith. Her books include *Between a rock and a hard place: Public theology in a post-secular age* (2013); *What makes a good city? Public theology and the urban church* (with Stephen R. Lowe) (2009); *Representations of the post/human: monsters, aliens and others in popular culture* (2002); and *Transforming practice: Pastoral theology in an age of uncertainty* (1996).

Armin Grunwald (Ph.D., University of Cologne in solid state physics, 1987; habilitation, University of Marburg in philosophy, 1998) is the director of the Institute for Technology Assessment and Systems Analysis (ITAS) at the Karlsruhe Institute of Technology (KIT), Germany. He also holds the chair of philosophy of technology and technology ethics at KIT (since 2007) and directs the Office of Technology Assessment at the German Bundestag in Berlin (since 2002). Professor Grunwald has written widely on technology assessment, science policy, and the interplay of technology and culture. Among his most recent publications are *The hermeneutic side of responsible research and innovation* (2014); *Handbuch Technikethik* (2013); *Modes of orientation provided by futures studies: Making sense of diversity and divergence* (2013); and *Responsible nanobiotechnology: Philosophy and ethics* (2012).

J. Benjamin Hurlbut (Ph.D. 2010 Harvard University) is assistant professor of bioscience ethics in the School of Life Sciences at Arizona State University. He is trained in the history of the modern biomedical and life sciences and in science and technology studies (STS). His research lies at the intersection of STS, bioethics, and political theory. Hurlbut studies the changing relationships between science, politics, and law in the governance of biomedical research and innovation in the 20th and 21st centuries. Focusing on controversy around morally and technically complex problems in areas like human embryonic stem cell research and genomics, he examines the interplay of science and technology with shifting notions of democracy, of religious and moral pluralism, and of public reason. He was a postdoctoral fellow in the Program on Science, Technology and Society at the John F. Kennedy School of Government at Harvard University.

Sheila Jasanoff (Ph.D., Harvard University in linguistics, 1973; J.D., Harvard University, 1976) is Pforzheimer Professor of Science and Technology Studies at the John F. Kennedy School of Government at Harvard University and director of the Program on Science, Technology and Society at Harvard. Her research on science and the state in contemporary democratic societies has influenced thinking in disparate fields, including science and technology studies, comparative politics, law and society, political and legal anthropology, and policy analysis. Noted for its comparative approach, her work spans the United States, the United Kingdom, Germany, the European Union, and India, as well as emerging global regimes in areas such as climate and biotechnology. She has contributed greatly to building science and technologies studies as an academic field and is the founder of the Science and Democracy Network, a group of scholars committed to deepening our understanding of the relations between science, technology, and politics. Her research has been recognized with many awards, including the Bernal Prize from the Society for Social Studies of Science and a Guggenheim Fellowship. Her books include *Designs on nature: Science and democracy in Europe and the United States* (2005); *Science at the bar: Law, science and technology in America* (1995); *The fifth branch: Science advisers as policymakers* (1990); and *Controlling chemicals: The Politics of regulation in Europe and the United States* (1985).

Brice Laurent (Ph.D., Centre de Sociologie de l'Innovation, Paris, 2011) is a sociologist and a researcher at Mines ParisTech. He teaches at Mines ParisTech and Sciences Po Paris. His work focuses on the relationship between the making of science and the construction of democratic order. Using the methodology of science and technology studies, he analyzes the devices that connect research programs, the making of technical objects, and the production of various types of publics. The overall object of his research is to study the formation of political spaces through the analysis of sites where scientific objectivity and democratic legitimacy are questioned. His publications include *Les politiques des nanotechnologies: Pour un traitment democratique d'une science émergent* (2010).

Franc Mali (Ph.D., University of Ljubljana, 1993) teaches sociology of science, science and society, and epistemology of the social sciences at the Faculty of Social Sciences at the University of Ljubljana. As a visiting researcher, he worked at the University of Graz (2004, 2009), University of Bielefeld (1998), University of Vienna (1989), University of Klagenfurt (1994), and Johannes Kepler University in Linz (1992). From 1999 to 2008, he was a member of the executive committee of the Section for Sociology of Science and Technology in the European Sociological Association (ESA). He is the author of *Science as a systemic part of society* (1994); *Development of modern science? Social mechanisms* (2002), and *Epistemology of social science* (2006), in addition to more than 100 articles published in Slovenian and international journals and in conference proceedings.

Thorsten Moos (Th.D., Humboldt University of Berlin in Protestant theology, 2002; diploma in Theoretical Physics, Freie Universität Berlin, 1995) is a minister and chair of the Department of Religion, Law and Culture in the Protestant Institute of Interdisciplinary Research (FEST) in Heidelberg. His research areas include cultural theory, bioethics, medical ethics, and theological ethics. He has published several edited volumes, including *Genetisches Wissen: Formation und Übersetzungen zwischen Wissenschaft und Gesellschaft* (2011); *Schöpfung bewahren: Theologische Ethik der Ökologie im interdisziplinären Gesprach* (2013); and *Theologie und Naturwissenschaften: Eine interdisziplinäre Werkstatt* (2006). His doctoral thesis was *Staatszweck und Staatsaufgaben in den protestantischen Ethiken des 19. Jahrhunderts* (2005).

Alfred Nordmann (Ph.D., Hamburg University in philosophy, 1986) is professor of philosophy at Darmstadt University of Technology and an adjunct professor in the Department of Philosophy at the University of South Carolina. He has written extensively on analytic philosophy, history and philosophy of science, nanotechnology, and science policy. Among his numerous publications are the co-edited volumes *Research objects in their technological setting* (2015); *Science transformed? Debating claims of an epochal break* (2011; German ed., 2014); and *Science in the context of application* (2010).

Stefan L. Sorgner (Ph.D., Friedrich-Schiller-Universität Jena in philosophy) is the director and co-founder of the Beyond Humanism Network and a fellow at the Institute for Ethics and Emerging Technologies (IEET). Dr. Sorgner is associate professor of philosophy at John Cabot University/Rome and previously taught philosophy at the University of Erfurt. He is author and editor of more than 10 books, among them *Menschenwürde nach Nietzsche* (WBG, 2010). Furthermore, he is in great demand as a speaker in all parts of the world (for example, Phil.cologne; *ICISTS-KAIST*; TEDx; World Humanities Forum) and a regular contact person of national and international journalists and media representatives (for example, *Die Zeit*). His main fields of research are Nietzsche, the philosophy of music, bioethics and meta-, post- and transhumanism.

Hava Tirosh-Samuelson (Ph.D., Hebrew University, 1978) is Irving and Miriam Lowe Professor of Modern Judaism, the director of Jewish Studies, and professor of history at Arizona State University. Her research focuses on Jewish intellectual history, Judaism and ecology, science and religion, and feminist theory. In addition to numerous articles in academic journals and book chapters in edited volumes, she is the author of the award-winning *Between worlds: The life and work of Rabbi David ben Judah Messer Leon* (1991) and *Happiness in premodern Judaism: Virtue, knowledge and well-being in premodern Judaism* (2003). She is also the editor of several volumes, including *Judaism and ecology: Created world and revealed world* (2002); *Judaism and the phenomenon of life: The legacy of Hans Jonas* (2008); *Building better humans? Refocusing the debate on transhumanism* (with Kenneth L. Mossman; 2011), and *Jewish philosophy for the twenty-first century: Personal reflections* (with Aaron Hughes; 2014). Professor Tirosh-Samuelson is the editor-in-chief of the Library of Contemporary Jewish Philosophers (2013–), a set of 20 volumes, featuring outstanding Jewish thinkers.

G. Pascal Zachary is a professor of practice at the School for the Future of Innovation in Society at Arizona State University. He studies technology and culture, science and politics, across three domains: war, security, and science in 20th-century U.S. history; technology and sociotechnical systems in contemporary sub-Saharan Africa; global media, computers and the future of consciousness. Zachary is the author of *Hotel Africa: The politics of escape* (2012) and, under a grant from the National Science Foundation, is documenting the emergence of computer science as a research discipline in East Africa. He is also the author of *Endless frontier: Vannevar Bush, engineer of the American century* (1997) and *Showstopper* (1994), about the creation of Microsoft's Windows operating system. From 1986 to 1995, Zachary reported on Silicon Valley for *The Wall Street Journal* and *The San Jose Mercury News*. For *The*

New York Times, Zachary wrote the "Ping" column on innovation (2007–2008). He currently writes on technology and society for *Spectrum*, the monthly magazine of the Institute of Electrical and Electronics Engineers (IEEE).

Nasser Zakariya (Ph.D., Harvard University in history of science, 2010) is an assistant professor in the Department of Rhetoric at the University of California, Berkeley. He has previously held postdoctoral fellowships at the Chemical Heritage Foundation, at the Max Planck Institute for the History of Science, and at New York University; and was a lecturer at Michigan State University. His research interests concern science, narrative, and documentary, as well as various topics in the history and philosophy of mathematics and physics and the interplay of science, law, and race. His publications include "Is history still a fraud?" and "Making knowledge whole: Genres of synthesis and grammars of ignorance," both in the journal *Historical Studies in the Natural Sciences*; and "Poetics of brotherhood: Organic and mechanistic narrative in late Tolstoi" (*Slavic Review*) and "Organic and mechanistic time and the limits of narrative" (*Configurations*), each coauthored with Ilya Kliger.

Michael E. Zimmerman (Ph.D., Tulane University in philosophy, 1974) is professor of philosophy and the former director of the Center for Humanities and the Arts at Colorado University-Boulder. He has written extensively on environmental problems, examining the metaphysical, cultural, ethical, cognitive, political, and religious dimensions of such problems. His approach is multidisciplinary, encompassing the natural sciences, the social sciences, the humanities, and the arts. He has criticized the command-and-control attitudes toward nature that have characterized modernity, while also warning of the danger posed by antimodernist attitudes that characterize some versions of environmentalism. His books include *Integral ecology: Uniting multiple perspectives on the natural world* (with Sean Esbjörn Hargens; 2009); *Environmental philosophy: From animal rights to radical ecology* (4th ed., 2004); *Contesting earth's future: Radical ecology and postmodernity* (1994); and *Heidegger's confrontation with modernity: Technology, politics and art* (1990).

Index

P

MIX
Papier aus verantwortungsvollen Quellen
Paper from responsible sources
FSC® C105338

FSC
www.fsc.org

If you have any concerns about our products,
you can contact us on
ProductSafety@springernature.com

In case Publisher is established outside the EU,
the EU authorized representative is:
Springer Nature Customer Service Center GmbH
Europaplatz 3, 69115 Heidelberg, Germany

Printed by Libri Plureos GmbH
in Hamburg, Germany